Springer Series on Environmental Management

David E. Alexander
Series Editor

Springer
*New York
Berlin
Heidelberg
Barcelona
Budapest
Hong Kong
London
Milan
Paris
Santa Clara
Singapore
Tokyo*

Springer Series on Environmental Management
David E. Alexander, Series Editor

Gradient Modeling: Resources and Fire Management (1979) S.R. Kessell

Disaster Planning: The Preservation of Life and Property (1980) H.D. Foster

Air Pollution and Forests: Interactions between Air Contaminants and Forest Ecosystems (1981) W.H. Smith

Natural Hazard Risk Assessment and Public Policy: Anticipating the Unexpected (1982) W.J. Petak and A.A. Atkisson

Environmental Effects of Off-Road Vehicles: Impacts and Management in Arid Regions (1983) R.H. Webb and H.G. Wilshire (eds.)

Global Fisheries: Perspectives for the '80s (1983) B.J. Rosthschild (ed.)

Heavy Metals in Natural Waters: Applied Monitoring and Impact Assessment (1984) J.W. Moore and S. Ramamoorthy

Landscape Ecology: Theory and Applications (1984) Z. Naveh and A.S. Lieberman

Organic Chemicals in Natural Waters: Applied Monitoring and Impact Assessment (1984) J.W. Moore and S. Ramamoorthy

The Hudson River Ecosystem (1986) K.E. Limburg, M.A. Moran, and W.H. McDowell

Human System Responses to Disaster: An Inventory of Sociological Findings (1986) T.E. Drabek

The Changing Environment (1986) J.W. Moore

Balancing the Needs of Water Use (1988) J.W. Moore

The Professional Practice of Environmental Management (1989) R.S. Dorney and L. Dorney (eds.)

Chemicals in the Aquatic Environment: Advanced Hazard Assessment (1989) L. Landner (ed.)

Inorganic Contaminants of Surface Water: Research and Monitoring Priorities (1991) J.W. Moore

Chernobyl: A Policy Response Study (1991) B. Segerståhl (ed.)

Long-Term Consequences of Disasters: The Reconstruction of Friuli, Italy, in its International Context, 1976-1988 (1991) R. Geipel

Food Web Management: A Case Study of Lake Mendota (1992) J.F. Kitchell (ed.)

Restoration and Recovery of an Industrial Region: Progress in Restoring the Smelter-Damaged Landscape near Sudbury, Canada (1995) J.M. Gunn (ed.)

Limnological and Engineering Analysis of a Polluted Urban Lake: Prelude to Environmental Management of Onondaga Lake, New York (1996) S.W. Effler (ed.)

Assessment and Management of Plant Invasions (1997) J.O. Luken and J.W. Thieret (eds.)

James O. Luken, Editor
Department of Biological Sciences
College of Arts and Sciences
Northern Kentucky University
Highland Heights, Kentucky

John W. Thieret, Editor
Department of Biological Sciences
College of Arts and Sciences
Northern Kentucky University
Highland Heights, Kentucky

Assessment and Management of Plant Invasions

With 60 Illustrations

James O. Luken
John W. Thieret
Department of Biological Sciences
Northern Kentucky University
Highland Heights, KY 41009
USA

Series Editor:
David E. Alexander
Department of Geology and Geography
University of Massachusetts
Amherst, MA 01003
USA

Cover photo: Kudzo. Photo courtesy of James O. Luken.

Sources of line drawings used on chapter opening pages: Chapters 1, 2, 3, 4, 5, 7, 8, 9, 12, and 13 from *Selected Weeds of the United States*, United States Department of Agriculture, 1970. Chapter 6 from Parker KF, *An Illustrated Guide to Arizona Weeds*, The University of Arizona Press, Tucson, AZ, 1972. Illustration by Lucretial B. Hamilton. Used by permission of University of Arizona Press. Chapter 10 from Mohlenbrock RH, *The Illustrated Flora of Illinois, Flowering Plants, Willows to Mustards*, Southern Illinois University Press, Carbondale, IL, 1980. Used by permission of the author. Chapter 11 from Hao K-s, Caprifoliaceae. In Liou T-n, Flore Illustree du nord de la Chine 3:1–94, 1934. Chapter 14 from United States Department of Agriculture herbarium sheet. Chapter 15 from Cooperrider TS, *The Dicotyledoneae of Ohio*, Ohio State University Press, Columbus, OH. Used by permission of Ohio State University Press. Chapter 16 from Cronk QCB and Fuller JL, *Plant Invaders*, Chapman and Hall, London, 1995. Used by permission of International Thomson Publishing Services. Chapter 17 from Cope TA, *Flora of Pakistan* No. 143, 1982. Chapter 18 from Chittenden FJ (ed.), *The Royal Horticultural Society Dictionary of Gardening*, Clarendon Press, Oxford, 1956. Appendix 1 from Thieret JW, personal drawing.

Library of Congress Cataloging-in-Publication Data
Assessment and management of plant invasions/[edited by] James O.
 Luken, John W. Thieret.
   p. cm. — (Springer series in environmental management)
  Includes bibliographical references (pp. 268–316) and index.
  ISBN 0-387-94809-0 (hbk.: alk. paper)
  1. Invasive plants—Control. 2. Invasive plants—Ecology.
3. Plant invasions. 4. Plant introduction. I. Luken, James O.,
1955– . II. Thieret, John W. III. Series.
SB613.5.A77 1996
639.9'9—dc20   96-19131

Printed on acid-free paper.

© 1997 Springer-Verlag New York, Inc.
All rights reserved. This work may not be translated or copied in whole or in part without the written permission of the publisher (Springer-Verlag New York, Inc., 175 Fifth Avenue, New York, NY 10010, USA), except for brief excerpts in connection with reviews or scholarly analysis. Use in connection with any form of information storage and retrieval, electronic adaptation, computer software, or by similar or dissimilar methodology now known or hereafter developed is forbidden. The use of general descriptive names, trade names, trademarks, etc., in this publication, even if the former are not especially identified, is not to be taken as a sign that such names, as understood by the Trade Marks and Merchandise Marks Act, may accordingly be used freely by anyone.

Acquiring Editor: Robert C. Garber.
Production coordinated by Chernow Editorial Services, Inc., and managed by Francine McNeill; manufacturing supervised by Jeffrey Taub.
Typeset by Best-set Typesetter Ltd., Hong Kong.
Printed and bound by Maple-Vail Book Manufacturing Group, York, PA.
Printed in the United States of America.

9 8 7 6 5 4 3 2 1

ISBN 0-387-94809-0 Springer-Verlag New York Berlin Heidelberg   SPIN 10540418

Series Preface

This series is concerned with humanity's stewardship of the environment, our use of natural resources, and the ways in which we can mitigate environmental hazards and reduce risks. Thus it is concerned with applied ecology in the widest sense of the term, in theory and in practice, and above all in the marriage of sound principles with pragmatic innovation. It focuses on the definition and monitoring of environmental problems and the search for solutions to them at scales that vary from the global to the local according to the scope of analysis. No particular academic discipline dominates the series, for environmental problems are interdisciplinary almost by definition. Hence a wide variety of specialties are represented, from oceanography to economics, sociology to silviculture, toxicology to policy studies.

In the modern world, increasing rates of resource use, population growth, and armed conflict have tended to magnify and complicate environmental problems that were already difficult to solve a century ago. Moreover, attempts to modify nature for the benefit of humankind have often had unintended consequences, especially in the disruption of natural equilibria. Yet, at the same time, human ingenuity has been brought to bear in developing a new range of sophisticated and powerful techniques for solving environmental problems, for example, pollution monitoring, restoration ecology, landscape planning, risk management, and impact assessment. Books in this series will shed light on the problems of the modern environment and contribute to the further development of the solutions. They will contribute to the immense effort by ecologists of all persuasions to nurture an environment that is both stable and productive.

David E. Alexander
Amherst, Massachusetts

Preface

The impetus for this book was provided when one of us (JOL) attended an annual meeting of the Natural Areas Association held in southern Florida. Not surprisingly, the subject of many presentations was the problem of nonindigenous plants in national forests, parks, and nature reserves. Clearly, a majority of resource managers attending this meeting perceived plant invasion as a direct threat to the integrity of preserved nature. Furthermore, the threat was severe enough to warrant direct management (i.e., biological, chemical, or physical methods) in numerous situations. Management of plant communities in an effort to eradicate or control nonindigenous plants appears on the surface as a relatively simple process of matching method and target: apply the proper control method at a point in time or space when the target plant is most vulnerable. However, this approach, firmly rooted in classic weed science, ignores many important interactions that emerge prior to plant invasion as well as interactions that emerge after management occurs.

This book attempts to cast the issue of nonindigenous plant invasion in a broader ecological context that includes humans acting as managers of natural resources, designers of regulations, and dispersers of organisms. We have chosen authors in an attempt to address the following questions: When is a plant invasion formally and scientifically deemed an ecological problem? What methods are available to prioritize the myriad problems that may exist due to plant invasion? What ecological interactions must be considered when assessing effects of plant invasion or the long-term effects of management? What types of research are required to assess ecological effects of invasion and management? And last, what regulations and human activities are critical to the modification of plant invasions?

We hope that the information and questions posed here will inspire greater interaction among horticulturists in the plant industry (who may introduce new invaders), ecologists (who may elucidate the ecological effects of plant invasion), and resource managers (who may use designed disturbances to achieve goals for natural resources).

As we were developing the Appendix and checking various sources of information, several individuals provided us with important data: Barbara

Ertter, Barney L. Lipscomb, Jerry Lorenz, Julia F. Morton, and Richard P. Wunderlin. The extensive reference list includes citations as we received them from the authors except for correction of obvious errors and adjustments to achieve uniform format. We owe much to the staff and collections of the Lloyd Library, Cincinnati. Our work with plant invasions and thus the production of this book has been supported in part by Northern Kentucky University and the National Science Foundation. Finally, we thank our wives for enduring the somewhat difficult times that were associated with editing this book.

James O. Luken
John W. Thieret
Highland Heights, Kentucky

Contents

Series Preface	v
Preface	vii
Contributors	xiii
Introduction	1
Carla M. D'Antonio	

SECTION I. Human Perceptions

1. Defining Indigenous Species: An Introduction 7
   Mark W. Schwartz
   - Defining Native Species 8
   - Mechanisms for Change in Species Distributions 12
   - Humans as a Component of the Natural 13
   - Consequences of Choosing a Definition 14
   - Conclusions 16

2. Defining Weeds of Natural Areas 18
   John M. Randall
   - Definitions of Weed 18
   - Definition and Examples of Natural-Area Weeds 21
   - Adaptive Management of Natural-Area Weeds 24
   - Conclusions 25

3. Potential Valuable Ecological Functions of
       Nonindigenous Plants 26
   Charles E. Williams
   - Assessing Potential Ecological Values of Nonindigenous Plants 28
   - Case Histories and Examples: Anecdotal, Hypothetical,
       and Otherwise 31
   - Future Ecological Values of Nonindigenous Plants in a Changing
       Global Environment 33
   - Conclusions 34

SECTION II. Assessment of Ecological Interactions

4. Documenting Natural and Human-Caused Plant Invasions Using
   Paleoecological Methods 37
   Stephen T. Jackson
      Nature of the Paleoecological Record 38
      Paleoecological Assessments of Human-Related Plant Invasions 42
      Long-Term Records of Plant Invasions and Environmental Change 46
      Interactions Between Environmental Change and Human Disturbance ... 48
      Conclusions 54

5. Community Response to Plant Invasion 56
   Kerry D. Woods
      Potential Mechanisms of Community Effects 57
      Possible Instances of Effects on Community Composition and Structure .. 58
      Invasion by Indigenous Species and the Paleoecological Record 64
      Generalizations 65
      Conclusions 67

6. Impacts of Invasive Plants on Community and Ecosystem Properties 69
   Lawrence R. Walker and Stanley D. Smith
      Primary Productivity 71
      Soil Nutrients 73
      Soil Water and Salinity 74
      Disturbance Regimes 79
      Community Dynamics 80
      Case Studies 81
      Conclusions 85

7. Animal-Mediated Dispersal and Disturbance: Driving Forces Behind
   Alien Plant Naturalization 87
   Paula M. Schiffman
      Naturalization 89
      Conclusions 93

8. Outlook for Plant Invasions: Interactions with Other Agents of
   Global Change 95
   Laura Foster Huenneke
      Primary Agents of Global Change 96
      Interactions Among Factors 101
      Conclusion 102

9. Experimental Design for Plant Removal and Restoration 104
   Michael L. Morrison
      Scientific Methods 105
      Principles of Design 106
      Experimental Design 112
      Applications: Managing Study Plots and Data 115
      Conclusions 116

10. Response of a Forest Understory Community to Experimental Removal
    of an Invasive Nonindigenous Plant (*Alliaria petiolata*, Brassicaceae) 117
    Brian C. McCarthy
       Biology of the Study Species 118

Contents

    Methods 118
    Results 121
    Discussion 126
    Conclusions 127
    Appendix 10.1. Species list 128

Section III. Direct Management

11. Management of Plant Invasions: Implicating Ecological Succession 133
    James O. Luken
    Succession: The Rise and Fall of Populations 134
    Plant Invasion: A Successional Interpretation 136
    The Decision to Manage 139
    The Response to Management 141
    A Paradigm for Management 143
    Conclusions 144

12. Methods for Management of Nonindigenous Aquatic Plants 145
    John D. Madsen
    Why Manage Nonindigenous Aquatic Plants? 146
    Management Techniques 151
    Developing an Integrated Management Plan 164
    Conclusions 170

13. Biological Control of Weeds in the United States and Canada 172
    C. Jack DeLoach
    Comparison of Control Methods 175
    Community-Level Effects of Invasion and Response to
      Biological Control 177
    Research Protocol 179
    Regulations and Safeguards 182
    Control of Major Weeds in the United States and Canada 182
    Future Directions of Biological Control 190
    Conclusions 193

14. Prioritizing Invasive Plants and Planning for Management 195
    Ronald D. Hiebert
    Why Prioritize Invasive Plants? 196
    How to Prioritize 197
    Steps in the Decision-Making Process 197
    Decision-Making Tools 200
    A Generalized Nonindigenous Plant Ranking System 203
    Conclusions 208
    Appendix 14.1. A system for ranking nonindigenous plants ... 210

Section IV. Regulation and Advocacy

15. Prevention of Invasive Plant Introductions on National and Local Levels ... 215
    Sarah E. Reichard
    Modes of Species Entry 216
    Current Laws 217
    Developing Predictive Methods 218

      Monitoring for New Invasions 225
      Preventing Invasions on the Local Scale 226
      Conclusions 227

16. Exotic Pest Plant Councils: Cooperating to Assess and Control Invasive
    Nonindigenous Plant Species 228
    Faith Thompson Campbell
      Measuring the Impact of Invasions 229
      Overall Significance of Invasions 237
      The Need for Coordination 238
      Role of Exotic Pest Plant Councils 239
      Conclusions 242

17. Team Arundo: Interagency Cooperation to Control Giant Cane
    (*Arundo donax*) 244
    Paul R. Frandsen
      Historical Setting 244
      Plant Biology 245
      The Problem 245
      The Solution? Team Arundo 246
      The Future 247
      Conclusions 247

18. A Multiagency Containment Program for Miconia (*Miconia calvescens*),
    an Invasive Tree in Hawaiian Rain Forests 249
    Patrick Conant, Arthur C. Medeiros, and Lloyd L. Loope
      Distribution and Ecology of Miconia 249
      Agencies and Citizen Groups Involved in Miconia Control in Hawaii ... 250
      Strategies and Tactics 252
      Prospects for Success 254

Appendix: Selected Plant Species Interfering with Resource Management
Goals in North American Natural Areas 255

References 268

Index 317

Contributors

Faith Thompson Campbell, National Coalition of Exotic Pest Plant Councils, Springfield, VA 22152, USA

Patrick Conant, Hawaii Department of Agriculture, Honolulu, HI 96814, USA

Carla M. D'Antonio, Department of Integrative Biology, University of California at Berkeley, Berkeley, CA, 94720-3140, USA

C. Jack DeLoach, United States Department of Agriculture, Agricultural Research Service, Soil and Water Research Laboratory, Temple, TX 76502-9601, USA

Paul R. Frandsen, Riverside County Regional Park and Open-Space District, Riverside, CA 92519, USA

Ronald D. Hiebert, National Park Service, Omaha, NE 68102, USA

Laura Foster Huenneke, Department of Biology, New Mexico State University, Las Cruces, NM 88002, USA

Stephen T. Jackson, Department of Botany, University of Wyoming, Laramie, WY 82071-3265, USA

Lloyd L. Loope, National Biological Service, Haleakala National Park Field Station, Makawao, HI 96768, USA

James O. Luken, Department of Biological Sciences, Northern Kentucky University, Highland Heights, KY 41099, USA

John D. Madsen, United States Army Engineer Waterways Experiment Station, Vicksburg, MS 39180-6199, USA

Brian C. McCarthy, Department of Environmental and Plant Biology, Ohio University, Athens, OH 45701-2979, USA

Arthur C. Medeiros, National Biological Service, Haleakala National Park Field Station, Makawao, HI 96768, USA

Michael L. Morrison, Department of Biological Sciences, California State University, Sacramento, CA 95819, USA

John M. Randall, The Nature Conservancy, Wildland Weeds Management and Research, Section of Plant Biology, University of California at Davis, Davis, CA 95616, USA

Sarah E. Reichard, Department of Zoology, University of Washington, Seattle, WA 98195, USA

Paula M. Schiffman, Department of Biology, California State University at Northridge, Northridge, CA 91330-8303, USA

Mark W. Schwartz, Center for Population Biology, University of California at Davis, Davis, CA 95616, USA

Stanley D. Smith, Department of Biological Sciences, University of Nevada, Las Vegas, NV 89154-4004, USA

John W. Thieret, Department of Biological Sciences, Northern Kentucky University, Highland Heights, KY 41099, USA

Lawrence R. Walker, Department of Biological Sciences, University of Nevada, Las Vegas, NV 89154-4004, USA

Charles E. Williams, Department of Biology, Clarion University of Pennsylvania, Clarion, PA 16214-1232, USA

Kerry D. Woods, Division of Natural Sciences, Bennington College, Bennington, VT 05201, USA

Introduction

Carla M. D'Antonio

Although 20 years ago this book may have been as necessary as it is today, nothing like it was written. It has been only in the last 10 years that basic ecologists and land managers have begun to work together to understand the ecology of nonindigenous (NI) plant species and to think about the control of such species from community and ecosystem perspectives. During this time we have witnessed a proliferation of books on NI species (e.g., Groves and Burdon 1986; MacDonald et al. 1986; Mooney and Drake 1986; Drake et al. 1989; di Castri et al. 1990; McKnight 1993; Pyšek et al. 1995), including an impressive 400-page, U.S. government–sponsored report (OTA 1993) documenting the threats posed by numerous NI plants, animals, and microbes. Why have we seen such a dramatic change in attitude and such an outpouring of literature and concern over NI species during the last two decades?

During the 1960s and into the 1980s, many academic ecologists were concerned with developing niche theory, exploring the Lotka–Volterra equations, determining the relative importance of competition, predation, and physical factors in determining species composition, and, overall, understanding the coexistence of species that were assumed to have evolved together. Charles Elton's *Ecology of Invasions by Animals and Plants* (1958) went relatively unnoticed. Instead, value was placed on working in "natural communities," i.e., communities relatively uninfluenced by humans. However, by the late 1980s most ecologists recognized that a large percentage of the world's ecosystems were tremendously influenced by humans. In addition to the effects of chemical pollution, overharvesting of plant and animal species, and habitat destruction, NI species were beginning to be recognized as common and often dominant modifiers of ecological systems. Indeed it has become difficult to find communities that do not suffer from either direct human alteration of the disturbance regimen or ones in which NI species are absent. A 1988 survey of nature reserves throughout the world (see *Biological Conservation* 44[1]) revealed that even places set aside for their natural beauty and biological diversity were being invaded and sometimes threatened by NI species.

While invasion and range expansion were indeed a part of the biological history of the earth before humans (see Jackson, Chapter 4, this volume), the rate and scale at which invasions occurred was not similar to that resulting from human activity. Prior to the last decade, invasions were primarily considered to be an island phenomenon. But the mere existence of the numerous books on invasions into countries throughout the world demonstrates that invasions are an important phenomenon on continents as well as on islands.

The International Union for the Conservation of Nature has now formed an international Exotic Species Task Force; reports of harmful NI species have emerged from all around the world. Data on numbers of NI species on a country-by-country basis are confined primarily to North America, Europe, Australia, and Oceania and are variable in quality. For plants, extensive data are available for the United States and European countries all of which have at least some NI species. All of the islands of Oceania have large numbers of NI species (Given 1992), and data are beginning to emerge from Africa that show that the more temperate countries such as South Africa and Namibia harbor large numbers of NI species (Rejmánek 1995). As a result of emerging data, Vitousek and I suggested (D'Antonio and Vitousek 1992) that because the human-caused breakdown of barriers to species dispersal was a global phenomenon and because this had ramifications at all levels of biological organization from the populations and genes to ecosystems and landscapes, invasions should be recognized as one of the major global environmental changes of our time. After land-use change and habitat loss, biological invasion is one of the major contributors to the local and global loss of indigenous biological diversity.

As reserve managers and ecologists know only too well, numbers of NI species do not tell the whole story. Many successfully established introduced species spread throughout their new ranges without any apparent effect on the communities they enter. Indeed, Simberloff (1981) found that, of the 854 species he surveyed, fewer than 20% had a measurable population- or system-level impact. Because of the limited resources available to governmental agencies and local managers, it is important that control and eradication efforts be aimed at those species or types of organisms posing the greatest threat to the persistence of either agriculture or wildland resources (see, e.g., Hiebert, Chapter 14, this volume). In one of the few attempts to provide a conceptual framework for predicting species impact, Peter Vitousek suggested that species having the potential to alter ecosystem processes include those that (1) alter the disturbance regimen, (2) alter the trophic structure of an area, and (3) alter the rate of accumulation or supply of resources. Examples of such species are discussed in chapters 5 and 6 of this volume. Other important species are those altering the physical structure of the habitat (e.g., sand-binding dune grasses) and those causing direct mortality through disease or predation and lacking natural controls in their new areas. The latter are well known in the agricultural world but are also important in impacting wildland resources.

While many governments have lists of noxious species whose importation should be controlled or whose eradication is a top priority, these lists are far from complete and rarely include species that are threats to wildland ecosystems (see chapters 15 and 16, this volume). Thus, there is a strong need for information exchange among scientists, government officials, the nursery industry, reserve managers, and the public to facilitate more effective control, set realistic enforceable targets, and curb further potentially destructive invasions before they happen. Analyses of problem invaders and their prominent attributes, such as that described by Reichard (Chapter 15, this volume), are an important step in identifying species whose import and export should be banned and which should be placed at the top of management priority lists.

In addition to such lists, there are many unanswered questions that hinder our ability to manage our ecosystems. For example, we still do not know what governs community susceptibility to invasion or whether we can manage ecosystems in a way that reduces the risk of NI species. We need synthetic, process-based approaches to understanding temporal and spatial patterns of abundance of NI species (see Luken, Chapter 11, this volume). We also do not understand how and why established NI species have altered community and ecosystem processes and whether their effects are reversible after removal (see, e.g., McCarthy, Chapter 10, this volume).

This book brings together a diversity of individuals to achieve a more complete understanding of the ecology and management of biological invasions and to demonstrate how interagency management and concerned citizens can effectively help to reverse the tide of species invasions and change that we are witnessing. Thus, it is part of an essential information exchange that must take place if we are to preserve the regional distinctiveness of our flora and fauna.

Section I
Human Perceptions

1
Defining Indigenous Species: An Introduction

Mark W. Schwartz

Marsh-elder
(*Iva xanthifolia*)

It may seem curious to begin a book on management of plant invasions with a chapter on defining indigenous (IN) species. The first objective, however, in making a decision to manage a potentially problematic species invasion is to first determine whether a species is a natural component of, or is indigenous to, the community in question. It is important that specific criteria for determining species membership (specific status) within a plant community are defined. Using specific status as one among many tools, scientists and land managers can begin to assess invading species and develop management strategies. There is an abundance of terms (often differently defined and applied by various researchers, land managers, and regulatory agencies) describing specific status (native or indigenous versus exotic, alien, or nonindigenous [NI]), mode of entry (locally persisting, waif species, escaped, adventive, naturalized, or introduced), and population trend (noninvasive versus weedy or invasive) in local, natural vegetation. As a rule, ecological terminology is messy; discrete ecological categories rarely exist. Ecologists continue to debate the precise meanings of fundamental terms such as *ecosystem*, *community*, and even *species*. The need for unambiguous definitions and categorizations may seem obvious, but the actual categorization of species is often quite difficult. Terms such as *indigenous* and *nonindigenous* that appear to be clear and concise are often not.

The problem of a plant's status is of foremost concern to systematists. Many floras (40 out of 100 randomly chosen North American floras as surveyed by Palmer et al. [1995]) include plant status descriptors such as *native* and *introduced*. Land managers typically rely on these descriptions to guide management decisions. Accordingly, Palmer et al. (1995) recommended that floristic studies contain information on plant status in order to meet minimum standards for acceptable floras. Following determination of plant status, it is the business of land managers to determine if plant species pose a threat to the maintenance of broad-scale biological diversity (Luken 1994). While this volume is devoted to assessing problematic NI species and managing them, discerning plant status is commonly a prerequisite. The focus of this chapter is to discuss criteria that have been and should be used to make decisions on whether species are considered indigenous or not (see Table 1.1). This chapter draws significance from recognizing that (1) definitive categorization of the species is often difficult, (2) plant status information is often not presented in local floras (Palmer et al. 1995), and (3) plant status information found in floras is occasionally ambiguous, conflicting, or wrong.

Defining Native Species

While this volume adheres to a uniform practice of using the terms *indigenous* and *nonindigenous* to refer to a plant's status, the term *native* is synonymous with *indigenous* and more common in colloquial usage. The definition of native in standard dictionaries, such as the *Random House Webster's College Dictionary* (1991), often makes explicit reference to undefined criteria regarding temporal and spatial extent of species distributions: "originating naturally in a particular country or region." Under this definition, when was "originally," and is "country" meant to refer to national boundaries? In the *Concise Oxford Dictionary of Ecology*, a source that ought to be more explicit

TABLE 1.1. Nine criteria for presuming indigenous (IN) and nonindigenous (NI) status of species. Criteria are listed in order of importance for assessing indigenous status.

Criteria*	Evidence for indigenous status
1. Fossil evidence	Continuous presence of fossil evidence from some post-Pleistocene period. Absence of a fossil record accompanied by present occurrence is suggestive, but not conclusive, of NI status.
2. Historical evidence	Historical documentation of introduction can verify NI status. Historical documentation of early presence cannot verify IN status.
3. Habitat	Species restricted to artificial environments are more likely to be nonindigenous. Caution is advised, however, because anthropogenic environments tend frequently to be disturbed, and one can confuse IN ruderals capitalizing on this disturbance with NI weeds.
4. Geographic distribution	Although geographic disjunctions are not uncommon in plants, the occurrence of a species as a disjunct population may be an indicator of NI status.
5. Frequency of known naturalization	Species naturalized in many other locations are more likely to be nonindigenous rather than indigenous for any particular site.
6. Genetic diversity	Genetic differentiation among isolated populations suggests support for a claim of IN status; many NI weeds are genetically depauperate and uniform among sites.
7. Reproductive pattern	Few IN plants reproduce entirely vegetatively. Lack of seed production supports a claim of NI status.
8. Possible means of introduction	NI species require a dispersal mechanism for invasion. A plausible hypothesis for species introduction supports a contention of NI status.
9. Relationship to oligophagous insects	NI plants tend to have few herbivores relative to closely related IN species.

*Criteria 1–8 are modified from Webb (1985); criterion 9 is from Preston (1986).

1. Defining Indigenous Species: An Introduction

than a standard dictionary, Allaby (1994) defined the term *native* as "applied to a species that occurs naturally in an area, and therefore one that has not been introduced either accidentally or intentionally." Other important terms such as *exotic*, *naturalized*, and *adventive* are either ignored or defined in terms of *native*. These definitions of *native* are insufficient for this discussion and serve to highlight problems with assigning specific status. In particular, the definitions presented above assume both temporal and spatial boundaries of natural distributions of species but do not suggest any guidelines for defining the boundaries.

Webb (1985), dealing with the British Isles, defined IN species as those that "evolved in these islands, or which arrived there . . . before the beginning of the Neolithic period, or which arrived there since that [period] by a method entirely independent of human activity." This definition neatly exemplifies some of the problems in defining IN species. First, spatial boundaries are somewhat arbitrary and problematic except when one considers the special case of islands. Second, species vary in the ease with which they can be assigned membership status. Those with long historical records suggesting pre-Neolithic distributions are readily categorized as indigenous. Likewise, those with recent distributions, and with origins from other continents, are readily categorized as nonindigenous. Species with no historical record are problematic. Even more problematic are those species with recent local distributions and long-term nearby distributions, because the means of immigration (with or without human assistance) must be determined (Table 1.1).

Webb (1985) suggested that the appropriate temporal standard for determining status is evidence of presence prior to the rise of global travel in the sixteenth century. North Americans purport to use this definition by using terms such as *pre-Columbian* or *presettlement*. That is, if a species is found in a region after, and not before, European settlement it is an NI species. There are two problems with this guideline. First, some human-assisted changes in distribution predate the temporal cutoff on all continents. Second, we frequently have insufficient data from the European settlement period to make a determination regarding status and, therefore, we often use a much later date (e.g., mid-1800s) when the flora for a region was first cataloged. The U.S. National Park Service, in contrast, relegates species introduced by humans in pre-Columbian times to NI status. For example, the coconut palm brought by Polynesians to Hawaii is considered an NI species (U.S. Department of the Interior, National Park Service 1991).

Similarly, problems arise from the use of political boundaries as proxies for biological boundaries. Because of the lack of clear standards for temporal or spatial scales and a lack of thorough information, some (Luken 1994; Salisbury 1961) have suggested that attempts to define species status be dropped entirely. In my mind, this only makes the situation worse, because for most species in most locations plant status is clear (Schwartz and Randall 1995). If a system works for most species (probably in excess of 90%), then why discard it? We simply need to be apprised of concerns regarding how we classify problematic species.

Original Vegetation

Species are constantly shifting in both their distribution and abundance (see Jackson, Chapter 4, this volume). An often used and logical criterion for a temporal time frame in delineating IN species is presence prior to European settlement. In North America, this temporal definition for these species is not completely satisfactory because it would include corn and squash along with agricultural weeds associated with native American agriculture (Bender 1975). Numerous species were introduced as crops by indigenous peoples prior to European colonization (Williams 1989), muddying our practical definition of species status. To exclude species brought in through agricultural development by indigenous peoples we could choose a temporal reference point after the most recent glaciation but prior to extensive agriculture in

North America. This would establish a cutoff date sometime between 3000 and 10,000 years ago. With this standard, black spruce (*Picea mariana*) becomes an IN species, but one long extirpated, in Kentucky (Davis 1981b). Further, beech (*Fagus grandifolia*) becomes an NI species in Wisconsin, Michigan, and Ontario because of the recency of invasion (e.g., about 3000 to 5000 years ago [Davis 1981b]). Insuring the exclusion of species associated with indigenous agricultural development results in a very unsatisfying definition of IN species owing to the long time frame for many postglaciation changes in distribution driven by natural causes (see Jackson, Chapter 4, this volume).

Species boundaries are in continual flux. Even when they are more or less static there is a constant process of local extinctions balancing colonization events (Carter and Prince 1988). Past climatic changes have resulted in dramatic shifts in plant distributions as demonstrated through the pollen record for trees (see Jackson, Chapter 4, this volume). Even during the past 100 years, changes in distribution limits of about a dozen conifers have been attributed to climatic fluctuation (Graumlich and Brubaker 1995). Given that distribution changes are a constant and natural process for species, it is impossible to pinpoint any static benchmark for definition of IN species. All definitions result in some nonsensical determinations. Nonetheless, the pre-European settlement period is the most frequently used time horizon in North America and elsewhere. This time horizon is used because it is biologically important; it represents the beginning of a time period marked by a dramatic increase in the rate at which species were transported by humans across formerly insurmountable distribution barriers (Webb 1985).

The Boundaries of a Location

All definitions of specific status make reference to a location. In the abstract we can think of a location as one of an infinite number of points covering the surface of the earth. At our reference time frame, species were either present (IN) or absent (NI) at each point. Obviously we do not have this level of detailed information. In addition, this spatial scale is not particularly useful for the practical purpose of defining boundaries of species distributions. To accommodate for this lack of reality we aggregate points into larger units in which species are either present or absent. Yet we must recognize that any scale of spatial boundaries (e.g., nations, states, or counties) that we use to define citizenship, or status, is artificial. We are, by necessity, aggregating information and thereby making generalizations that include exceptions to the rules. This problem is well illustrated by the following example.

Florida torreya (*Torreya taxifolia*), a federally listed endangered gymnospermous tree, occurs in ravine habitats on the eastern side of the Apalachicola River with one small outlying population west of the Apalachicola (Godfrey 1988). This area of distribution is less than $400\,km^2$ and is predominantly in Florida (Schwartz and Hermann 1993). The ravine habitats and the distribution of the tree penetrate into southern Georgia approximately $1\,km^2$ (Fig. 1.1). The current Georgia population consists of 25 trees (Schwartz et al. 1995). By virtue of these few trees, all within sight of the Florida state line, torreya is an IN species in Georgia. Because of its declining and endangered population (Schwartz and Hermann 1993), informal proposals have been made to introduce the tree into seemingly appropriate habitats in northern Georgia. There are no historical observations of torreya in Georgia other than in the previously described location. Using biological criteria, moving the tree to habitats in northern Georgia would be considered an introduction. To the federal and state governments, where species distributions are defined by state boundaries, however, this is not a species introduction. In the eyes of governmental jurisdiction, torreya is an IN species in all of Georgia. In contrast, there are appropriate habitats within $50\,km$ of the historic distribution of torreya that happen to fall in Alabama. Without historical documentation of the species presence in Alabama, creating populations in these nearby ravines

FIGURE 1.1. The natural distribution of *Torreya taxifolia* (northern Florida, southern Georgia).

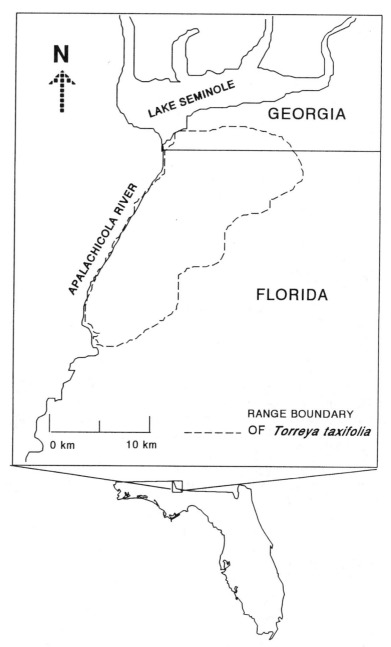

would, both legally and biologically, be considered a species introduction. Yet, biologically, this 50-km range expansion seems rather inconsequential and certainly more acceptable than creating populations more than 500 km distant but within Georgia. These extreme views of spatial scales highlight the observation that specific boundaries of location often provide arbitrary and misleading definitions for IN species.

Many botanists are using county records to define distributions (e.g., in the Carolinas [Radford et al. 1968]). Yet, different counties have received different amounts of attention from botanists. For example, Champaign County, home of the campus of the University of Illinois and the Illinois Natural History Survey, boasts 1190 plant species. No surrounding county contains more than 1000 species (Iverson 1992). The sizes and habitats of these counties are all quite similar. One presumes this pattern to be a function of the density of botanists in Champaign County. Thus, county-level information can, and must, be smoothed to incorporate such anomalies and to estimate species richness (Palmer 1995). All smoothing requires redefining specific status within counties. For individual species, however, the specific distribution as outlined on any distribution map is best viewed as a working hypothesis (Palmer 1995).

Historical data provide the backdrop for our current understanding of the distribution boundaries for many tree species. Land Office Survey records, along with vast personal experience, allowed Little (1971) to define range boundaries for most trees in the United States. Davis et al. (1991a) described a discrepancy of about 10 km for eastern hemlock (*Tsuga canadensis*) between detailed notes in Land Office Survey data (collected in the mid-1800s) and actual current distributions of trees along a road in a randomly selected township in Wisconsin. While a resolution of within 10 km seems quite good, it exceeds our resolution for most species in most locations. Thus, determining whether presence of a species is more accurately viewed, historically, as indigenous or nonindigenous becomes increasingly difficult as one approaches the distribution limit of a species.

Mechanisms for Change in Species Distributions

In general, we can define four means by which the distributions of species change. Species can change distribution (1) as a natural process (i.e., independent of humans) of dispersal and colonization of new habitats; (2) as a result of habitat changes caused by human land use; (3) through accidental or incidental human dispersal (e.g., most agricultural weeds); or (4) through intentional dispersal by humans (e.g., purple loosestrife [*Lythrum salicaria*]). Ecologists readily recognize the first case, natural range expansion, as an attribute of species; they define new locations occupied through this process as the natural distribution of a species. On the other extreme, accidental and intentional dispersal by humans are clearly considered as NI distributions.

Most problems associated with determining specific status are related either to insufficient historical information regarding the aforementioned factors explaining the presence of a species in a locale or the fact that human-caused habitat change includes a very complex and nearly endless suite of possibilities. A series of examples serves to highlight this contention.

The sea lamprey (*Petromyzon marinus*) is indigenous to the St. Lawrence Seaway and Lake Ontario with a distribution that was formerly limited, as with many other aquatic species, by Niagara Falls. Once the Welland Canal was constructed, providing a channel around the falls, the lamprey moved into Lake Ontario, Lake Erie, and Lake Huron and now extends into Lake Michigan and Lake Superior (Hubbs and Pope 1937). Biologists treat the lamprey, along with six other species of fishes that moved into Lake Michigan as a result of the Welland Canal, as an NI species (Laird and Page n.d.) despite the fact that the dispersal was self-motivated (a mechanism most ecologists term "naturalized").

Similarly, the Old World cattle egret (*Bubulcus ibis*) established a breeding population in South America during the 1880s, presumably by scattered wayward migrating birds accidentally crossing the Atlantic Ocean (Ehrlich et al. 1988). Cattle egrets have flourished and are currently among the most common egrets in the New World with a distribution extending north into the United States. The egret expanded its range on its own, taking advantage of human-induced changes in habitat. As with the sea lamprey, the cattle egret is considered an NI species in the New World (Ehrlich et al. 1988).

Several dry-prairie plants (e.g., the sunflowers *Helianthus annuus* and *Helianthus salicifolius*; the poppy mallow *Callirhoe involucrata* [Mohlenbrock 1986]) are listed as naturalized in Illinois, taking advantage of human-induced landscape change. These species are all indigenous to western United States short-grass prairie. In Illinois, most of them are found along railroad rights-of-way. While perhaps taking advantage of the wind currents of passing trains and the right-of-way habitat, the seeds of these plants, for the most part, dispersed on their own. The aforementioned species are treated as NI species in Illinois.

There is considerable confusion, however, that may arise from species expanding into new areas as a result of human-altered habitats. One example is that of cottonweed (*Froelichia gracilis*). Mohlenbrock (1986) in Illinois, Swink and Wilhelm (1994) in the Chicago region, and Deam (1940) in Indiana all considered cottonweed non-indigenous in Illinois and Indiana. In contrast, range descriptions of cottonweed by Steyermark (1963) in Missouri and by the Great Plains Flora Association (1986) in the Great Plains described cottonweed indigenous in Illinois and Indiana. Similarly, Mohlenbrock (1986) treated sumpweed (*Iva ciliata*) as indigenous to all of Illinois, while Swink and Wilhelm (1994) considered it nonindigenous in the Chicago region. The problem is that we do not have historical information to determine precisely where these distributions ended prior to human-induced habitat alteration.

The recent movement of species leads to confusion in defining not just common species but also rare species within states. Most states keep a list of plant species of special status because of rarity within the state. Species on these lists are often rare because they are on the edges of their distributions. Scientists must then decide whether particular range-edge species are indigenous to the state, for example, water pennywort (*Hydrocotyle ranunculoides*). This species is listed as very rare in Illinois by Mohlenbrock (1986) and is included in Herkert's (1991) list of Illinois endangered and threatened plants. Water pennywort is known from just two Illinois sites, both artificial ponds. In one, the species is considered nonindigenous. In the other, where it is considered indigenous, the first observation was in 1925 (Nyboer et al. 1976). It is thought to have been dispersed to Illinois by waterfowl. Water pennywort is treated as a threatened IN plant within Illinois even though it inhabits only artificial impoundments.

Humans as a Component of the Natural

One focal point in distinguishing between IN and NI species rests on how we define human intervention in changing the distribution of species. In North America, we feel that the definition of specific status is linked to a relatively recent colonization by European settlers accompanied by a rapid acceleration of land-use change and species movement. This historical backdrop compels us to use pre-European vegetation as the benchmark for natural community composition and the definition of IN species. This may provide a false sense of security regarding determination of specific status. There is ample evidence that species such as giant ragweed (*Ambrosia trifida*), sunflower (*Helianthus annuus*), marsh-elder (*Iva xanthifolia*), amaranth (*Amaranthus*), and lamb's-quarters (*Chenopodium*) were cultivated for food, dye, medicine, or fiber (Bender 1975).

Broad-scale movement of these species by native Americans places the indigenous distribution of these plant species, all currently very widespread, into question. Consider the more widespread situation in the Mediterranean area, southwestern Asia, and northern Africa, where cultivation of cereal crops and the domestication of grazing animals extends back at least 6000 years and probably closer to 9000 years (Bender 1975; Clark and Brandt 1984). Extensive alteration of natural habitats and complete disruption of natural ecosystems in many areas dates back at least 3000 years (Perlin 1991).

At what point do we cease to define human dispersal of species as being external to natural processes, therefore creating NI species? Clearly our knowledge of the distributions of species is far more recent than the actions of human industry that alter natural distributions. Likewise, substantial climatic change has occurred during this long period of agrarian use of plants. It may, therefore, be impossible to make a distinction between natural range shifts linked to climatic forcing and anthropogenic dispersal (in itself often tracking climate [Bender 1975]).

To summarize, we can describe three discrete phenomena: First, species express some prehuman distribution. Second, climate change during the past several thousand years would likely have resulted in this prehuman distribution changing irrespective of human impact. Third, at some time prior to recorded history, we recategorized humans from acting as natural dispersal agents to becoming the driving force creating NI species. Clearly we have little hope of determining the natural distribution of many of these long-cultivated plants.

Webb (1985) argued that one can identify three periods of human impact on plant distributions: prehuman (no impact); paleosynanthropic (before about 1550, when global travel increased); and neosynanthropic (since 1550, when human alteration of species distributions became common). Webb recommended that plants with paleosynanthropic distributions be considered indigenous. This definition, however, obscures the anthropogenic movement of species between North and South America as well as across broad stretches of Eurasia resulting from commerce and trade among primarily agrarian cultures. One consequence of Webb's definition is that it lumps together species that are relatively recent additions to their environments with those that evolved in the context of their environments (Webb 1985). While Webb's criterion for a temporal cutoff may incorrectly categorize a few species, it is often used because it is functional and it results in fewer misclassifications than any other time frame that we could adopt.

Consequences of Choosing a Definition

Any choice of a definition for what constitutes IN species will leave some ambiguity. Much of this ambiguity may arise from how we define human-induced changes in species distributions. Suppose for a moment that global warming proceeds as predicted by global circulation models (Mitchell et al. 1990) and that over the next century the mean annual temperature in eastern United States increases 3°C. This global warming is a change in the environment that clearly is driven by human activity. In response to this warming, a rare and restricted plant shifts its distribution northward 300 km from northern Georgia, where it is an IN species, into Kentucky, where it is not. As a result of warmer temperatures, this species no longer survives in Georgia. Because the plant shifted its range northward as a result of human-induced climatic change, the species is considered nonindigenous in Kentucky; it no longer has an IN range. Extending this scenario, let us assume that our hypothetical species fails to disperse on its own accord, is going extinct in Georgia, and is intentionally introduced into Kentucky to prevent its extinction. Once in Kentucky, the species finds many more habitats available, with few predators, and becomes a pest species: a classic example of the behavior of an NI species. While this extreme example of creating a pest species may not be a likely scenario, managing biological diversity under a changing global climate is likely to present managers with conflicting conservation objectives regarding shifting distributions of species.

In North America, the problem of ill-defined citizenship is probably best exemplified in determining which circumboreal species are indigenous (as opposed to those that are early arrivals through European settlement and thus are nonindigenous). Consider the case of the buttercup *Ranunculus pensylvanicus*. This perennial is indigenous from Newfoundland to Delaware west to British Columbia and Alberta. It is listed as adventive in Alaska by Moss (1983) and indigenous to that state by Porsild and Cody (1980) (Fig. 1.2). Climatic amelioration could trigger a relatively slight range expansion across the Bering Strait and allow the species to become circumboreal. There are hundreds of species of boreal plants that are circumboreal or nearly so (Porsild and Cody 1980).

If climatic warming represents a human alteration of habitats, then responses of plants to climatic change may result in a multitude of species invasions in arctic regions. Most of these introductions, like many others, may have relatively little ecological significance in that they will not be characterized by uncontrolled population growth in their new habitats. In defining specific status, however, it is difficult to determine which attributes of specific climate changes are driven by humans versus natural phenomena. Defining the consequences of climate change on species distributions presents a formidable problem. In the past, and likely in the future, we will treat range changes as a result of climate change as natural shifts, despite the obvious inaccuracy of this decision.

Similarly, we must make explicit decisions regarding how to classify distribution shifts resulting from human-induced alterations in frequency and intensity of natural disturbances. Hobbs and Huenneke (1992) described how management practice, e.g., fire and grazing, or how human-generated disturbance, e.g., nutrient inputs, trampling, and fragmentation, might facilitate plant invasion. Thus, land managers may be faced with tough decisions when land-use and conservation goals are in conflict (Hobbs and Huenneke 1992; Schwartz 1994). Less obvious anthropogenic changes in habitats (e.g., atmospheric deposition of nitrogen or increasing atmospheric CO_2 [Vitousek 1994]) may have nearly undetectable effects on abilities of plants to expand into new locations. Putting aside difficulties in assessing past dispersal events for which we have little historical evidence, we are now faced with a broad array of human disturbances not easily categorized with respect to how they may influence species distributions or interact with natural factors facilitating plant distribution changes. Thus, virtually all future shifts in plant distributions must be viewed as somewhat suspect no matter how "natural" the expansion may appear.

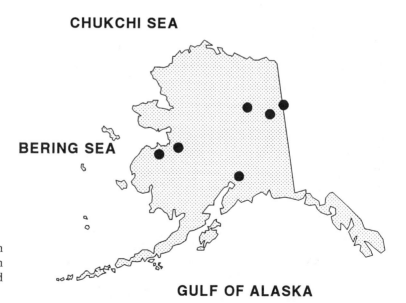

FIGURE 1.2. The distribution of *Ranunculus pensylvanicus* in Alaska. (Redrawn from Porsild and Cody 1980)

Conclusions

During the past millennium there has been a continual increase in the rate of species invasions attributable to human intervention. Further, the breadth of mechanisms by which humans facilitate this homogenization of the earth's biota continues to increase. As a consequence, the rate at which we are introducing species to new environments is increasing (Havera and Suloway 1994; Hickman 1993; Raven 1988). It is a popular notion that humans facilitated the bulk of species introductions by the early twentieth century, prior to our knowledge of the consequences of these actions. If one examines the size of floras, however, one finds that the rate of increase appears exponential with no leveling off in the twentieth century (Fig. 1.3). Fortunately, most of these species introductions appear inconsequential. One must caution, however, that there are many examples of formerly benign NI species mysteriously becoming weedy and of weeds appearing to lose their weediness (Salisbury 1961).

How should conservation biologists approach management problems regarding NI species in light of the aforementioned problems? The answer lies in developing explicit conservation goals (Schwartz 1994). There are two basic types of conservation targets that would entail different approaches to addressing specific status. First, if the goal of a conservation program is to maintain genetic or species diversity, then concerns regarding whether a globally endangered species is indigenous become secondary. Similarly, if broadscale biodiversity is the primary target of management, then peripheral populations of species may represent genetic novelty of conservation value. Again, we should be only secondarily concerned about status of IN species. Second, if the goal of a conservation program

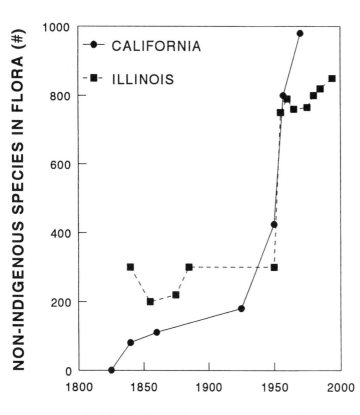

FIGURE 1.3. The number of non-indigenous species in California (solid line) and Illinois (dashed line) as a function of time. (Data for California are from Hickman [1993] and Raven [1988]. Data for Illinois are from Havera and Suloway [1994])

is to create historical representations of what is best estimated as pre-European habitat (a common operational goal in land management), then a strict definition of IN species is required. The problem for conservation managers is that there are no clear criteria to suggest which of these occasionally conflicting conservation goals ought to take precedence (Schwartz 1994).

The pre-European standard, though somewhat gratifying, makes relatively little biological sense. One may argue that historical community composition affords species the best opportunity to evolve and adapt within their evolutionary context. The magnitude of human-induced environmental change, from the local to global scale, however, is enormous. I recommend that the conservation target of historical representation, while often a practical goal, is a lesser goal than maintenance of biological diversity within complex ecosystems. Most natural habitats now lack many of their top predators (e.g., wolves or panthers) or dominant herbivores (e.g., bison), thus full historical accuracy is impossible. When questions about specific status arise, I recommend conferring putative residency and managing to maximize IN species diversity, using historical accuracy to guide, but not direct, management. The more important question resides in whether the species is causing management problems as a result of exponential population growth, a question further discussed by Randall (Chapter 2, this volume).

Acknowledgments. I thank John Randall and Jim Luken for valuable discussions on the topic. Carol Lippincott invited me to present a talk on this subject at the 1994 Natural Areas Conference, which gave me the opportunity to think critically about these issues.

2
Defining Weeds of Natural Areas

John M. Randall

Leafy spurge
(*Euphorbia esula*)

Weeds are usually defined in the negative: they are plants that are not wanted. Of course, which plants are wanted and which are unwanted depends on the setting and sometimes on individual prejudices and tastes. In this sense, the term *weed* is subjective and personal. Where there is a clear, agreed-upon vision of which species are wanted and, better yet, the level of productivity desired—as in a corn field—much of this subjectivity disappears. Here, any plants that might suppress the corn or otherwise limit its productivity are unwanted and therefore weeds. In this chapter I will argue that managers of parks, preserves, and other natural areas should likewise clearly articulate what they are managing *for* and only then attempt to determine which species interfere with management goals or have the potential to do so in the future (Schwartz and Randall 1995). These species can appropriately be designated as *weeds* or *pests* or some other term with negative connotations. The label chosen is less important than the process used to identify and manage the problem-causing plants.

Definitions of Weed

Weed may be used as a verb or as a noun. Here we are interested in its definition as a noun. According to *Webster's Collegiate Dictionary* a *weed* is "1. a plant of no value and usually of rank growth, especially one that tends to overgrow or choke out more desirable plants" and "2. an obnoxious growth, thing or person . . . like a weed in detrimental quality; especially an animal unfit to breed from." Many other authors have provided their own definitions, most of which include the idea of an unwanted plant: "any plant, either native or introduced, interfering with the objectives or requirements of people" (Binggeli 1993); "a plant out of place" (Blatchley 1912); or more poetically "a plant whose virtues have not yet been appreciated" (Emerson 1878). Other attributes included in many definitions of *weed* are competitive and aggressive behavior, persistence and resistance to control, and unsightliness (Harlan and deWet 1965; Radosevich and Holt 1984). Table 2.1 contains a sample of various other definitions of *weed*.

Some definitions restrict *weeds* to disturbed areas and/or agricultural lands: a plant that "in any specified geographical area, its populations grow entirely or predominantly in situations disturbed by man" (Baker 1974); "pest plants of agricultural and other managed ecosystems, regardless of their geographical origin. They reduce the value of these systems" (Bazzaz 1986). For many speakers of English the term *weed* brings to mind images of rapidly growing pests choking out valuable species in agricultural lands, gardens, or lawns. Because management of these areas usually involves regular disturbance of the vegetation cover and soil, most unwanted species on them are characterized by their ability to colonize disturbed areas and to reproduce quickly. Plants of this type often build up large populations that disperse propagules far and wide, increasing chances that some will drop in a freshly disturbed area before the population itself is overtaken by late-successional species (Frenkel 1970; Harper 1977). Species adapted to colonize and exploit ephemeral habitats, including recently disturbed sites, are often referred to as "r-selected" (MacArthur and Wilson 1967). Other authors prefer the label "ruderals" for plants like lamb's quarters (*Chenopodium album*) or bull thistle (*Cirsium vulgare*) that are adapted to colonize disturbed sites (Grime 1977; Grime et al. 1988). Disturbance, as used here, is defined as physical damage to the vegetation or soil. Ruderals are sometimes equated with weeds or described as "weedy."

Some authors have recommended that use of the word *weed* be avoided in ecological literature. They regard the term as imprecise and subjective, based on human perceptions of what is desirable or undesirable under given circumstances rather than on inherent characteristics of the species (Bazzaz 1986). I disagree with this recommendation, although I recognize that *weed* carries slightly different meanings for different people and is most often used to designate pests of farms, lawns, and gardens (Perrins et al. 1992a). *Weed* is a familiar term to people from all walks of life and is therefore useful for communicating the image of a troublesome, unwanted plant to a wide audience more readily than any other word or phrase. The type of habitat where the species is troublesome can be specified by placing a modifier before the term. Agricultural weeds interfere with growth and productivity of crops; range *weeds* reduce forage value, poison livestock, or make products derived from the animals inedible or less valuable. Natural-area *weeds* threaten valued indigenous (IN) species or otherwise disrupt natural biological communities and systems. Australian writers commonly refer to the last category as environmental *weeds* (Humphries et al. 1991 (see Table 2.1b)), but this pairing has rarely been used by North American authors (for an exception see James 1994). In all of these situations, plants are labeled *weeds* because of their behavior in a particular environment or habitat.

Use of a commonly understood word such as *weed* or *pest* to refer to plants troublesome in parks, preserves, and other natural areas may have the added benefit of helping to dispel the all-too-common belief that these areas are unmanaged. Much of the public is unaware that natural-area management requires more than posting fences, keeping out poachers, and leading tours. Displays,

TABLE 2.1. Definitions of *weed* and related terms.

a. Definitions of *weed*

A plant not wanted and therefore to be destroyed (Bailey and Bailey 1941).

A plant is a weed if, in any specified geographical area, its populations grow entirely or predominantly in situations markedly disturbed by man (without, of course, being deliberately cultivated plants) (Baker 1965).

Plant invaders are weeds even when they dominate regional vegetation as do the Mediterranean annual plants in the grasslands of California (Baker 1986).

A plant which contests with man for possession of the soil (Blatchley 1912).

A weed can be defined as a plant out of place. In the native bush, exotics like the small wandering jew, right up to the large camphor laurel, are all growing out of place and are all classed as weeds (Bradley 1988).

Any plant other than the crop sown. A plant that grows so luxuriantly or plentifully that it chokes out all other plants that possess more valuable properties (Brenchley 1920).

Weeds are pioneers of secondary succession, of which the weedy arable field is a special case (Bunting 1960).

Introduced plant species which take possession of cultivated or fallow fields and pastures (Dayton 1950).

Any plant which grows where it is not wanted (Fogg 1945)

1. (1): an introduced plant growing in ground that is or has been in cultivation usu. to the detriment of the crop or to the disfigurement of the place: an economically useless plant: a plant of unsightly appearance; esp. one of wild or rank growth (2): a tree or shrub of low economic value that tends to grow freely and by its presence exclude or retard more valuable plants <gray birch is a common ——— species in much of New England> (3): a form of vegetable life or exuberant growth and injurious effect (as various molds or bacteria frequently contaminating cultures) (4): a forb in rangeland b. wild growth usu. in the nature of rank grass or undergrowth <the land must be cleared of ——— Emil Levgyel>
2. a marine or freshwater plant: seaweed
3. an obnoxious growth, thing or person <militarism is a tough ——— to kill F.S. Oliver>
4a. Tobacco, esp. tobacco prepared for use (as a cigar or cigarette) <made the students shun both ——— and wine Time> b. slang, marijuana.
5a. something of little value; specif. an animal of poor conformation, lacking in stamina, and unfit to breed from. b. an animal that is detrimental esp. in preoccupying habitats that might otherwise harbor more desirable forms <carp forms one of the worst ——— species in some areas> or in damaging the habitat value of the land on which they live <uncontrolled deer herds may become serious ———s> (*Webster's Third New International Dictionary of the English Language*, Gove 1986).

Introduced plants often are aggressive colonizers following disturbance to native vegetation and soil. When these plants conflict, restrict, or otherwise interfere with land management objectives they are commonly referred to as weeds (MDA 1992).

Weeds are weeds only from our human egotistical point of view, because they grow where we do not want them.... A plant becomes a weed only through its relative position in relation to cultivated areas (Pfeiffer n.d.)

Weeds interfere with management goals of people (Rejmánek 1995).

In his agricultural pursuits, from the earliest periods of his existence, man has contended with certain undesirable species of plants. Such species, unwanted, nonuseful, often prolific and persistent, interfere with agricultural operations, increase labor, add to costs and reduce yields. These obnoxious plants are known as weeds (Robbins et al. 1942).

A plant growing where it is not desired (Terminology Committee of the Weed Science Society of America as cited in Radosevich and Holt 1984).

A very unsightly plant of wild growth, often found in land that has been cultivated (Thomas 1956).

b. Definition of *environmental weed*

Those species that invade native communities or ecosystems—they are undesirable from an ecological perspective but not necessarily from an economic one. Serious environmental weeds are defined as those that cause major modification to species richness, abundance or ecosystem function. Very serious environmental weeds are those that can totally and permanently destroy an ecosystem (Humphries et al. 1991).

c. Definitions of *invasive plant*

An alien plant spreading naturally (without direct assistance of people) in natural or seminatural habitats, to produce a significant change in terms of composition, structure or ecosystem processes (Cronk and Fuller 1995).

An introduced species that must be capable of establishing self-sustaining populations in areas of natural or seminatural vegetation (i.e., untransformed ecosystems) (Macdonald et al. 1989).

Invaders are spreading into areas where they are not native (Rejmánek 1995).

Invasive species are introduced species that have become pests (Usher 1988).

Invasive plants are introduced species that expand their population (and distributional range) in the new geographical location, without further human intervention (Usher 1991).

A plant that has moved into a habitat and reproduced so aggressively that it has displaced some of the original components of the vegetative community (White et al. 1993).

articles, and videos about the need to control unwanted plants invading and degrading natural areas will likely have more impact if familiar and understandable language is used consistently.

As indicated in *Webster's Third New International Dictionary* (Table 2.1a), some authors refer to unwanted animals as weeds too (Bright 1995; Harlan and deWet 1965). This usage helps emphasize that some animals are damaging and unwanted, but it may confuse many readers. Other authors label invasive nonindigenous (NI) plants, animals, and other organisms collectively as biological pollutants to make it obvious to a wide audience that these species degrade the habitats they invade (McKnight 1993). The Office of Technology Assessment (OTA 1993) used the term *Harmful non-indigenous species* in a report on introduced organisms troublesome in the United States.

Definition and Examples of Natural-Area Weeds

A natural-area weed may be defined as a plant that prevents attainment of management goals. As such, management goals for a natural area must be established before invasion occurs. Natural areas may be managed to support or increase populations of certain species, to maintain particular vegetation types or biological communities, or to restore ecosystem processes such as fire or seasonal flooding. For example, The Nature Conservancy (TNC) manages Ewauna Flat in Klamath County, Oregon, to maintain or increase the population of the rare plant Applegate's milkvetch (*Astragalus applegatei*) and associated IN grasses (Borgias 1995). On the other hand, TNC cooperates with several federal and state agencies at the Cosumnes River Preserve in central California to protect the valley oak and riparian forest community and to ensure that seasonal flooding is a regular disturbance (Reiner R, pers. com.). On other sites, the primary management goal may be to maintain green-space with a minimum of management actions or to maintain a vegetation type that was originally generated by human activities. In the United States Forest Service, land management goals and objectives, including maintenance of biological diversity and restoration of conditions that assure recovery of endangered species often fall under the label "desired future condition" (Fenwood 1992).

I use the terms *natural-area weed* and *natural-area plant pest* interchangeably to refer to IN and NI species, populations, and individual plants that interfere with management goals and objectives and are therefore unwanted. The most important consideration is the impact of the species on the site in question rather than whether it is unwanted elsewhere or indigenous to the site. A program to control or eliminate pests may properly be regarded as an integral part of an overall restoration program that has as its goal returning the site to a previously existing condition, usually that prevailing before an identifiable disturbance, stress, or invasion occurred (Lewis 1990; NRC 1992).

Most natural-area weeds are *invasive*, which is to say that they can move into an area and become dominant numerically or in terms of cover, resource use, or other ecological impacts (Rejmánek 1995b; see Table 2.1c). Impacts of these invasions can include significant alteration of ecosystem processes, competitive interactions resulting in elimination or reduced populations of IN species, promotion of invasions and population increases by NI animals, fungi, or microbes, and hybridization with IN species and consequent alteration of the gene pool. Rare species appear to be particularly vulnerable to the changes wrought by NI invaders. For example, the California Natural Heritage Database indicates that 30 of the state's 53 Federally listed endangered plant species are threatened by NI invaders (Hoshovsky M, pers. com.).

Although both IN and NI species can be invasive natural-area weeds, most species identified as such are not indigenous to the site in question—most are not indigenous to the continent (e.g., Randall 1995). A species should not be designated as a pest merely because it has been determined that it is not indigenous to the site, however, unless one of the management goals is to maintain a community entirely free of NI species (Luken 1994). As Soulé (1990) stated, control of damaging NI invaders will remain one of the most important management activities in natural areas, but blanket opposition to NI species is doomed to failure and wasteful of resources better spent on other conservation activities. Opposition to NI species is complicated in situations in which it is difficult to determine whether a species is indigenous to a site or region (see Schwartz, Chapter 1, this volume).

Populations of species known to be indigenous, but relatively rare, may invade a community, outcompeting or otherwise excluding other IN plants and animals. Invasions of this type often appear to be triggered by human-induced changes in environmental factors such as suppression of wildfires or construction of impoundments preventing seasonal flooding. Where possible, these invasions are best addressed by restoring or mimicking original conditions. For example, IN woody plants like eastern red-cedar (*Juniperus virginiana*) and a variety of oaks (*Quercus*) invade many prairie and savanna remnants in the midwest and eastern Great Plains, especially where natural fires have been suppressed (Anderson 1990; Auclair 1976; Faber-Langendoen and Davis 1995). Managers at many of these sites have implemented prescribed burn programs to suppress woody invaders and to restore pre-existing vegetation types (e.g., Faber-Langendoen and Davis 1995; Stritch 1990; White 1983, 1986).

The grass *Phragmites australis* is indigenous to the United States (Breternitz et al. 1986; Niering and Warren 1977) but is regarded as an especially troublesome weed on many preserves in the northeast and upper midwest where it invades shallow open-water and wetland habitats (Marks et al. 1994). It is not understood why it has become invasive in so many sites, but disturbances and stresses—including water pollution, alteration of natural hydrologic regimens, and increased sedimentation—apparently favor it (Roman et al. 1984). Other suggested factors include increases in soil salinity (from fresh to brackish), changes in the form of dissolved nitrogen, and introduction of a more aggressive genotype(s) from the Old World (Marks et al. 1994; McNabb and Batterson 1991; Metzler and Rozsa 1987).

There is overlap between natural-area weeds and those plants regarded as weeds in other settings, especially range lands and roadsides. For example, leafy spurge (*Euphorbia esula*), a perennial that readily establishes on disturbed sites but can invade areas subject to little or no human disturbance, is a severe pest of rangeland and natural areas in the northern Great Plains and intermountain west of North America (Watson 1985). One study indicated that invasions of leafy spurge were responsible for a $9 million reduction in net annual income for ranchers and a $75 million reduction in business activity for all sectors in North Dakota alone (Bangsund and Leistritz 1991; Leistritz et al. 1992). Leafy spurge depresses the yield of desirable forage species on range and pasturelands, and grazing animals other than sheep and goats generally avoid it (Reilly and Kaufman 1979; Watson 1985). It may cause scours and weakness in cattle that feed on it; at times, these conditions become so severe that they result in death (Kingsbury 1964). In natural areas it tends to form monospecific stands outcompeting IN plants and providing poor habitat for IN animals (Watson 1985). Belcher and Wilson (1989) found that invasion by this species was clearly related to decline in abundance of dominant IN grasses and forbs in mixed-grass prairie. Where leafy spurge was most abundant, most IN species were absent; species richness—7 to 11 outside the infestation—declined to 4 inside it.

Natural areas invaded by leafy spurge include Theodore Roosevelt National Park in North Dakota (Andrascik 1994; Everitt et al. 1995), TNC's Pine Butte Swamp Preserve along the Rocky Mountain front in Montana (Carr D, pers. com.), and other TNC preserves on the plains of western Minnesota and the eastern Dakotas (Winter B, Satrom J, and Schollett A, pers. com.). TNC's Altamont Prairie in South Dakota is so badly infested that it is no longer considered to be worth managing as native prairie and cannot be sold as cropland (Breyfogle D, pers. com.). Instead, the site is now being used to test the efficacy of sheep and goats for control of leafy spurge and restoration of prairie.

A few major cropland weeds are pests in some natural areas. Canada thistle (*Cirsium arvense*) is a widespread pest of cropland in the northern United States and southern Canada (Donald 1990). It also invades relatively pristine natural habitats, sometimes crowding out IN species, especially in riparian areas from the northern Rockies to the Cascades. The species can be found in natural areas in other regions but may have little recognizable impact on valued plants and communities. Many state, county, and local governments have declared Canada thistle a noxious weed. Laws concerning noxious weeds differ from state to state, but most require landowners, including private organizations or government agencies owning and managing preserves, to control any populations of such species on their properties. As used here, control means to prevent a population from spreading or increasing in size. Montana's law, for example, mandates that when landowners fail to control leafy spurge, a county weed agent may enter the property and treat the weeds as deemed appropriate. The landowner is charged for the treatment plus a 10% surcharge plus a $100 fine for the first such violation; additional violations bring increasingly higher fines. Thus managers of natural areas are sometimes legally obligated to control noxious weeds even if these plants are not subverting their management goals. Other cropland pests thriving on sites where soils are repeatedly disturbed, such as velvetleaf (*Abutilon theophrasti*), are rarely if ever regarded as pests in natural areas.

On the other hand, plants regarded as pests in some natural areas may be valued or viewed with indifference in other situations. Brazilian-pepper (*Schinus terebinthifolius*) is a pest in Everglades National Park and other natural areas in southern Florida and Hawaii (Doren and Whiteaker 1990a, 1990b) but is used as an ornamental landscape plant in coastal southern Califorma (Nilsen and Muller 1980a, 1980b). Similarly, Chinese tallowtree (*Sapium sebiferum*) is a serious pest in wetlands and bottomland forests in northern Florida and across the gulf coastal plain to southern Texas (Cameron and Spencer 1989; Jubinsky 1993), but it is otherwise not regarded as troublesome on cropland and rangeland in this area.

Ruderal, or r-selected, species comprise the bulk of agricultural and garden pests that most people, including many ecologists, think of as weeds (e.g., Perrins et al. 1992a). Natural-area weeds, on the other hand, have a wide variety of life histories and ecological roles. A nationwide survey of TNC stewards indicated that natural-area weeds include ferns, gymnosperms, and flowering plants; annual, biennial, and perennial herbs; floating, emersed, and submersed aquatics; and vines, shrubs, understory trees, and canopy dominants (Randall 1995). Five species were reported from 10 or more states: garlic mustard (*Alliaria petiolata*) (11 states), Japanese honeysuckle (*Lonicera japonica*) (13 states), purple loosestrife (*Lythrum salicaria*) (11 states), giant reed (*Phragmites australis*) (10 states), and black locust (*Robinia pseudoacacia*) (10 states). These widely reported species represent a range of life histories: an understory biennial herb, a vine, two wetland emergents, and a canopy dominant. Tamarisks (*Tamarix*), knapweeds (*Centaurea*), tree-of-heaven (*Ailanthus altissima*), Canada thistle (*Cirsium arvense*), Russian-olive (*Elaeagnus angustifolia*), sweet-clover (*Melilotus*), and Johnson grass (*Sorghum halepense*) were among the other widely reported species.

Having said that a species cannot be designated as a weed at a site until management goals are established, I feel compelled to add two caveats: (1) Even in natural areas where management goals do not include protection of IN plant species and communities, such as sites managed solely to furnish breeding habitat for waterfowl, invasive NI species providing food or habitat should not be planted or promoted. Using non-invasive species will help to prevent invasives from spreading to other sites. (2) Controlling or eliminating a species just beginning to colonize an area should be seriously considered if the species has been known to invade and have negative impacts on natural areas elsewhere (Hobbs and Humphries 1995). Panetta (1993) and Reichard (n.d.) found that the best predictor of whether a species will become invasive is whether it is known to be invasive elsewhere. In contrast, the taxonomic relationship of a species to a known invader has generally been a poor predictor of invasiveness (see Reichard, Chapter 15, this volume). There are some suites of invasive congeners, however. In North America, for example, several Eurasian thistles (*Cirsium*) and knapweeds and star-thistles (*Centaurea*) are now established pests. Plants in the melastome family (Melastomataceae) are so notoriously invasive on islands in the tropical Pacific that the import of any member of this family into Hawaii is now prohibited by law.

Adaptive Management of Natural-Area Weeds

Determining which plants, if any, should be controlled and setting priorities so that limited time and labor available to do so are used efficiently require a sound management strategy. My work with TNC stewards and other managers of natural areas has convinced me that an adaptive management strategy (Fig. 2.1), like that outlined by Schwartz and Randall (1995), should be used. This strategy requires managers to (1) establish management goals and objectives for the natural area (i.e., articulate what is being managed); (2) determine which plant species or populations, if any, threaten the management goals or have the potential to do so (these are natural-area weeds or pests) and assign priorities for control (see Heibert, Chapter 14, this volume); (3) determine what methods of control

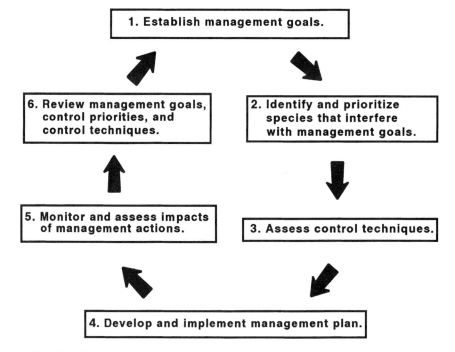

FIGURE 2.1. A flowchart for adaptive management of weeds describing management actions and decisions confronting natural-area managers. (From Schwartz and Randall [1995])

are available and, if necessary, reorder priorities based on likely impacts on target and nontarget species; (4) develop and implement a management plan based on this information and designed to move conditions toward the goals and objectives; (5) monitor and assess impacts of the management actions in terms of how well they move conditions toward the goals and objectives (see Luken, Chapter 11, this volume); and (6) reevaluate, modify, and start the cycle over again (Fig. 2.1).

The adaptive management strategy is preferred (Schwartz and Randall 1995), especially when documented in a written plan, for several reasons: (1) Establishing goals and objectives for the preserve exposes assumptions and helps ensure that means and ends do not become confused. Killing natural-area weeds is a means to the end, or ultimate goal, of promoting desirable conditions in the natural area. (2) Monitoring the results of control efforts (see Morrison, Chapter 9, this volume) helps ensure that their effects will be assessed and that modifications will be made if necessary. (3) Developing a written control plan ensures continuity of goals and actions when site managers change. (4) Revising goals, objectives, and actions ensures that actions mesh well with changing conditions or values.

Conclusions

1. A natural-area weed is a plant interfering with management goals of a site or having the potential to do so if not controlled. Management goals must be established before plants subverting them can be identified.

2. The term used to label these plants (e.g., natural-area weed, pest plant, biological pollutant, etc.) is less important than the process used to identify them as problems.

3. Indigenous and nonindigenous species can be natural-area weeds, but most species identified as weeds are not indigenous to the site in question. Plants should not be designated weeds merely because they are nonindigenous unless a management goal of maintaining the site entirely free of NI species has been established.

4. A program to control natural-area weeds may properly be regarded as an integral part of a restoration program designed to return a site to a previously existing condition.

5. Control or elimination of a species just beginning to colonize an area should be seriously considered if the species is known to be invasive in other natural areas. The best known predictor of whether a species will become invasive is whether it is invasive elsewhere; control is most effective if undertaken before a species is widespread on a site.

6. Managers of natural areas are strongly encouraged to use an adaptive management strategy. This strategy requires establishing management goals, developing and implementing control programs based on the goals, monitoring and assessing impacts of control efforts and, if necessary, modifying goals and control programs as the cycle is repeated.

Acknowledgements. I thank Mark Schwartz, Marcel Rejmánek, Larry Morse, Doria Gordon, and Ron Unger for thoughtful and productive discussions of many ideas presented here.

3
Potential Valuable Ecological Functions of Nonindigenous Plants

Charles E. Williams

Multiflora rose
(*Rosa multiflora*)

Recent popular and technical literature on nonindigenous (NI) plants is rife with examples of species run amok in natural and managed ecosystems. In North America, for example, conspicuous, well-established NI species like garlic mustard (*Alliaria petiolata*), Japanese honeysuckle (*Lonicera japonica*), kudzu (*Pueraria montana*), purple loosestrife (*Lythrum salicaria*), and multiflora rose (*Rosa multiflora*), among others, have gained considerable notoriety because of their demonstrated or presumed ability to alter the structure and function of ecosystems (Sawyers 1989; Vitousek 1990; Williams CE 1993b, 1996; Williams T 1994). Yet not all NI plants in a region are invasive pests and not all systems are equally invasible, even by the most aggressive NI species (Brothers and Spingarn 1992; Fox and Fox 1986; Hiebert 1990; Langdon and Johnson 1994; Myers 1983; Robertson et al. 1994; Vitousek 1990; Westman 1990a; Williams CE 1993a, 1993c, 1996). Moreover, in a few specific instances, NI plants may be of ecological value in some systems, playing important structural and functional roles in recovery after disturbance or as surrogates for extirpated indigenous (IN) species (Bowler 1992; De Pietri 1992; Henry 1992, 1993; Lovejoy 1985; Lugo 1988; Whelan and Dilger 1992).

Acknowledging that NI plants may be of ecological importance in some systems is generally not a popular or practical stance among many scientists and resource managers, particularly those involved in control of invasive species in natural areas where the main management goal is the maintenance or restoration of the IN biota (e.g., Coblentz 1991, 1993b; Lugo 1990, 1992; Luken 1994; Oelfke 1993; Schwartz and Randall 1995; Temple 1990). Not surprisingly, research conducted on NI plants in many regions worldwide often emphasizes development and assessment of control or removal technologies over basic studies of life history and ecological function (Lugo 1990, 1992; Luken 1994). This bias may be based in part on the belief that recognizing the ecological value of NI plants may legitimize further species introductions, weaken existing restrictions on importation and dissemination of NI species, or affect support for maintaining and restoring IN species and systems (e.g., Coblentz 1993a). Likewise, opposition to NI plants may be psychological or emotional, based in part on exposure to agricultural systems where negative weed/crop interactions are well established. Nevertheless, homogenization of regional floras by addition of NI plants continues worldwide at an increasing rate, leading some to predict that blanket opposition to NI plants will be futile in the future with management activities focused on only the most ecologically and economically damaging species (Brown 1995; Henry 1992; Soulé 1990; Westman 1990a, 1990b).

In this chapter, I provide a brief overview of the potentially valuable ecological functions provided by NI plants in three types of systems with varied management goals (Fig. 3.1):

1. *Intact natural systems*. These are natural areas and wilderness areas for which the main management goals are preserving the native biota and maintaining ecological integrity (defined as "the capability of supporting and maintaining a balanced, integrated, adaptive community of organisms having a species composition, diversity, and functional organization comparable to that of the natural habitat of the region" [Karr and Dudley 1981]). In these systems, natural disturbance regimens are relatively intact and the number of naturalized NI plants is usually low (e.g., Falk 1990). Examples include large reserves such as the Arctic National Wildlife Refuge in Alaska as well as smaller research natural areas and preserves.

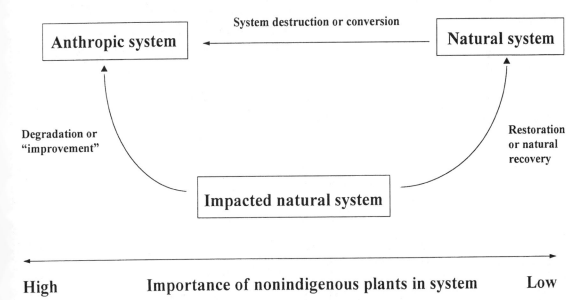

FIGURE 3.1. Relationships among anthropic systems, natural systems, and impacted natural systems relative to particular management goals or management activities.

2. *Anthropic systems*. These contain novel assemblages of IN and NI plants. Examples include utility rights-of-way, roadsides, residential areas, plantations, reclaimed strip mines, managed parks, and recreation areas, where humans strongly influence or dominate ecological processes in space and time and where management goals are varied (e.g., Luken 1990, 1994; Whitney and Adams 1980).

3. *Impacted natural systems*. These are characterized by intense or prolonged anthropogenic disturbance in the recent past and also may contain a mix of IN and NI species. Examples include severely grazed grasslands, forests subjected to intense, short-rotation harvest, and heavily used recreation areas. Impacted natural systems may be passively managed (i.e., abandoned) or they may be actively managed (i.e., "improved") through restoration and recovery leading to either an anthropic system or an intact natural system.

Assessing Potential Ecological Values of Nonindigenous Plants

The potential threat of invasive NI plants to resources of concern is often assessed or predicted by considering the life-history characteristics of the invading species, the attributes of the invaded system, and the facilitating factors that may modify system or invader attributes (Brown 1995; Ewel 1986; Fox and Fox 1986; Newsome and Noble 1986; Noble 1989; Westman 1990b; Williams CE 1996). Ecological values of NI plants could conceivably be assessed or predicted in a similar manner, considering in particular the short- and long-term management goals of the system in which the species exists.

But what is a valuable ecological function served by an NI plant species? For an NI species to be ecologically valuable, it should in some way contribute beneficially to the structure and function of a particular system (see Walker and Smith, Chapter 6, this volume). For example, an NI species could be a controller of species composition or of material and energy cycles central to system development and function (e.g., Ehrlich and Mooney 1983). The process of identifying and ultimately assessing potentially valuable ecological functions of NI plants will be guided in large part by a basic understanding of structure and function of the system of concern and by the biology of the specific NI species. Thus, ecological values will vary among NI species and systems.

The process of identifying potentially valuable ecological functions of NI plants could start with a checklist of relevant management-oriented, system-based questions such as:

1. Does the existence of the species enhance management goals of the system? Is the species noninvasive or a threat to resources of concern, particularly in adjacent systems?

2. Is the species ecologically similar to an extirpated IN species? Could it serve a similar ecological role in the system?

3. Does the species provide a keystone food resource or habitat for certain fauna without negatively affecting other system components?

4. Does the species facilitate or inhibit the regeneration of key species or the successional dynamics of the system? Is this facilitation or inhibition important in system function and development?

5. What is the role of the species in the cycling of energy and materials in the system? Would plant control destabilize the system?

6. Does the species influence disturbance regimens (e.g., fire frequency or intensity) within the system? Is this influence beneficial for system maintenance?

7. Will the ecological importance and value of the species change over time as the system ages or is affected by chronic or acute environmental stress?

These value-based questions suggest development of a research agenda for NI plants that some, particularly managers of natural areas, may deem ill-advised or misdirected. I do not argue the fact that some NI plants can cause management problems in natural systems; this point is well documented (e.g., Williams CE 1993b, 1996). What I advocate, however, is that more effort be directed at determining the ecological value of NI plants across systems, particularly in those most often ignored by ecologists: human-dominated and human-impacted systems (McDonnell and Pickett 1990; McDonnell et al. 1993).

Ecological Value: A Simple Conceptual Model

Another aid in assessing potential value of NI plants is the development of a systems-based conceptual model delineating an ecological value threshold across or within systems of concern (Fig. 3.2). I define ecological value threshold as the point at which an ecological function served by an NI plant species may be valuable or nonvaluable with respect to management goals of a particular system. As a concept, the ecological value threshold is roughly analogous to the economic threshold of insect pest management (pest density at which control measures must be undertaken to prevent economic injury to crops [Luckmann and Metcalf 1975]) and the weed-management threshold of vegetation management as applied to NI species in natural areas by Henry (1994) ("a population below a threshold that will cause native plant loss or other ecosystem degradation") except that valuable functions and not detrimental effects of NI plants are stressed.

For example, consider the conceptual model depicted by Fig. 3.2 in which the potential ecological value of an NI plant is a function of the system in which it occurs. In an intact natural system, NI plants are usually low in importance (see Fig. 3.1) either naturally or as a result of control or removal programs and are generally undesirable; thus their potential ecological value is low. At the other extreme are anthropic systems in which NI plants are desirable and dominant and so may have considerable ecological value (e.g., Westman 1990a, 1990b). In the model presented in Fig. 3.2, the ecological value threshold lies at the level of the impacted natural system primarily because of the potentially divergent management goals and development of the system as described previously (Fig. 3.1). Thus, according to the model, ecological values of NI plants generally increase as an impacted natural system is con-

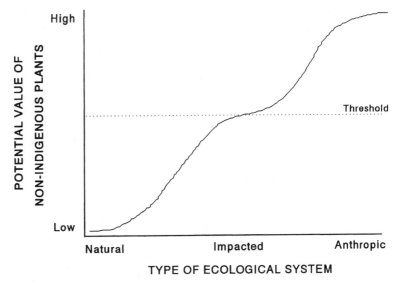

FIGURE 3.2. Conceptual model of nonindigenous plant value across natural, impacted, and anthropic systems. The ecological value threshold is indicated by the horizontal dashed line.

verted to an anthropic system, but they decrease as an impacted system is restored or recovers to a natural system. At a finer scale, a within-systems ecological value threshold, focusing on value in particular communities or habitats, could be developed in tandem with an across-systems model to better identify potentially valuable functions of NI plants at different spatial scales.

The ecological value threshold model for NI plants meshes well with the urban–rural gradient concept of McDonnell and Pickett (1990). Noindigenous plant species are usually dominant in anthropic urban/suburban systems and decrease in importance in rural (± natural) systems (McDonnell et al. 1993). This shift in importance of NI and IN species is in response to a spatially variable complex of factors that differentially influences disturbance regimens, soil characteristics, NI plant species sources, and physiological stress, selecting for one group or mix of species over another (McDonnell and Pickett 1990; McDonnell et al. 1993). Linking the urban–rural gradient responses of NI plants to a systems-based ecological value threshold model can provide a detailed picture of the shifting ecological importance and value of NI plants across large, multisystem landscapes in which system boundaries may or may not be distinct.

Developing a systems-based ecological value threshold or a similar measure or model of ecological function for NI plants may serve several short- and long-term research and management needs. An ecological value threshold model forces a broader look at the role of NI plants across systems, potentially providing insight for management of resources of concern at larger scales in a "top-down" framework (e.g., from landscapes to specific systems). This type of hierarchical perspective is central to the emerging paradigm of ecosystem-based management of natural resources (e.g., Grumbine 1994; Schroeder and Keller 1990). Likewise, development of an ecological value threshold can guide research programs by generating testable hypotheses concerning present and future ecological roles and values of NI plants across systems. Finally, an ecological value threshold can provide information useful in developing realistic, balanced, cost-benefit analyses for management of NI plants in landscapes (e.g., Brown 1995).

The Paradox of Ecological Value: Variation Within and Across Systems

Harty (1993) pointed out that most NI plants deliberately introduced into North America and other continents have been promoted for a narrow, utilitarian suite of functions including enhancing wildlife food and cover, landscaping urban and suburban areas, producing wood and fiber, controlling erosion, and improving forage production. In most anthropic systems these functions are highly valued; they fulfill specific management objectives and they are generally easy to assess. In contrast, in most natural systems, many of these functions may be irrelevant, intractable, and detrimental to system integrity, or they are fulfilled by IN species. Thus, the array of potential ecological functions and values associated with a given NI species will vary within and across anthropic, natural, and impacted systems. In particular, an ecological function deemed valuable in one system may be detrimental in another, depending on specific management goals (Table 3.1), the classic paradox of NI plants in many regions (see Whelan and Dilger 1992; Williams CE 1993b). A conflict of ecological value can also exist within a larger system, e.g., an anthropic urban nature reserve surrounded by suburban development. In the nature reserve, NI plants that escaped from cultivation may be invasive pests, but in the adjacent development they may provide valuable functions as wildlife habitat or as modifiers of microclimate.

Consider the hypothetical variation in value of ecological functions within and across systems presented in Table 3.1. Assuming "purist" management for intact natural systems in which the extant native flora provides all necessary ecological functions, NI plants are then decidedly nonvaluable and perhaps

TABLE 3.1. Some ecological functions provided by nonindigenous plants and their potential value to management of anthropic, natural, and impacted systems.

Ecological function	System type and value		
	Anthropic	Natural	Impacted
Physiognomic modification			
Habitat for fauna	+	−	+
Microclimate modification	+	−	+, −
Regeneration/successional dynamics			
Facilitate/inhibit succession	+, −	−	+, −
Material/energy cycles/soils			
Alter soil moisture/hydrology	+, −	−	+, −
Alter nutrient cycling	+, −	−	+, −
Alter soil development	+, −	−	+, −
Biotic interactions			
Food for fauna	+	−	+
Disturbance regimens			
Alter disturbance frequency or intensity	+, −	−	+, −

+ Indicates that the function is valuable to some aspect of system structure and (or) function; − indicates an undesirable, nonvaluable function; +, − indicates that the function varies in value depending on the management goals of the system.

detrimental to system integrity. In contrast, in anthropic and impacted natural systems, ecological value can vary considerably according to management goals; these systems, then, are the core of the values paradox for NI plants.

How might the ecological value of NI plants vary across anthropic and impacted systems? In general, ecological functions facilitating system change or development toward a desired management goal are those considered valuable. In contrast, functions inhibiting or interfering with attaining the desired management goal are nonvaluable. For example, in the restoration of an impacted system to a natural system, major valuable ecological functions provided by NI plants could include development of a microclimate conducive to germination, establishment and growth of key native flora typical of the region's natural habitat, and facilitation of successional change toward the desired management endpoint. These functions, valuable for restoration, may not be valuable in an impacted natural system under conversion to an anthropic system in which structural or compositional variability is undesirable.

Case Histories and Examples: Anecdotal, Hypothetical, and Otherwise

Published accounts of valuable ecological functions provided by NI plants are rare and often anecdotal or hypothetical. Detailed below are a few examples in which NI species are considered ecologically valuable beyond strict utilitarian functions. Note the emphasis on impacted natural systems, a bias generally common among ecologists who preferentially study natural and modified systems instead of anthropic systems (McDonnell et al. 1993).

Rosa rubiginosa in Argentina

Perhaps the best-documented instance in which an NI plant species is of apparent ecological value is the case of the European mosqueta rose (*Rosa rubiginosa*) in subantarctic forests of the Los Alerces National Park, Argentina (De Pietri 1992). Forests of the region have been heavily degraded owing to past overexploitation and current cattle grazing, which severely limits regeneration of woody plants. Mosqueta rose colonizes degraded areas, resists grazing, and produces an extensive shrub layer. De Pietri (1992) found that IN woody plants established almost exclusively under the rose's cover and concluded that the species functioned as a nurse plant by sheltering seedlings from cattle and by providing a favorable microenvironment for seedling establishment and growth. Importantly, seedlings of IN trees are apparently able to grow through and overtop rose canopies, eventually forming an overstory. Thus, regeneration foci for IN woody plants can develop within rose patches, eventually providing propagule sources for colonization of new favorable habitats. De Pietri (1992) concluded that mosqueta rose should "no longer be considered as a weed, but as a healing element" and suggested that the species will play a vital role in restoration of subantarctic forests in Argentina.

One should question whether the ecological value of mosqueta rose will change as grazing stress is relieved and degraded subantarctic forests are restored to a natural state. Will the species then become an invasive pest or simply become nonvaluable because it is not a natural component of the biota?

Nonindigenous Shrubs in Illinois

In many areas of the eastern and midwestern United States, remnant wood lots and forest fragments formerly or currently impacted by grazing of domestic livestock or high populations of native white-tailed deer (*Odocoileus virginianus*) are often structurally or taxonomically depauperate (Alverson et al. 1988). Commonly, a distinct shrub stratum, important as nesting habitat for native forest birds, is absent from grazed or otherwise impacted wood lots. In Illinois wood lots, Whelan and Dilger (1992) found that several species of birds, including the American robin (*Turdus migratorius*), the catbird (*Dumetella carolinensis*), the northern cardinal (*Cardinalis cardinalis*), the rose-breasted grosbeak (*Pheucticus ludovicianus*), and the wood thrush (*Hylocichla mustelina*) showed a strong preference for nesting in NI shrubs, particularly bush honeysuckles (*Lonicera*) and common buckthorn (*Rhamnus cathartica*), relative to IN shrubs. These authors speculated that this preference was due in part to the height and branch architecture of the NI shrubs, which may facilitate nest building. Importantly, IN shrub species, apparently once common in the region, were absent from the woodlots examined (Whelan and Dilger 1992). In this case, NI shrubs functioned as substitute nesting habitat for IN forest birds in the absence of a shrub stratum.

The management implications of Whelan and Dilger's (1992) study are especially pertinent for restoration of impacted systems. One of the early steps in restoration of an impacted woodlot is usually the removal of NI shrubs that may affect abundance and distribution of the IN understory flora. However, management of a system in regard to recovery of a single focal taxon or group of taxa (i.e., the IN flora) may be detrimental to other taxa (e.g., IN forest birds). Thus, Whelan and Dilger suggested that removal efforts for NI shrubs in systems similar to those they studied should be done in phases with invasion foci of NI shrubs targeted for initial treatment, coupled with active planting and restoration of native shrub populations in forest gaps and edges, thereby ensuring that adequate habitat structure for forest birds is present at all times during the restoration process.

Whelan and Dilger's (1992) suggestions have recently been dismissed as a case of "classic wildlife biology" that legitimizes use of NI plant species to promote a single focal vertebrate taxon instead of system diversity (Schwartz and Randall 1995). I strongly disagree. The take-home message of Whelan and Dilger's study is not that NI shrubs should be planted for avian food and cover (which the authors neither stated nor implied). Instead, their study clearly shows the paradox of management of naturalized NI species in human-dominated landscapes, particularly when the species may have some value in system structure or function. For example, blanket removal of NI shrubs, a potentially desirable management goal in restoring a system to a natural state, particularly the integrity of its flora, could lead to local reductions of breeding birds and thus to an overall negative impact on system integrity (see also Schiffman 1994).

Nonindigenous Plants in System Restoration

The ultimate goal in plant community restoration, particularly at heavily impacted sites, is to speed recovery of the system or to short-circuit succession. On recently disturbed sites, stabilization of soils and geomorphic surfaces is a necessary first step, usually followed by an active revegetation campaign. Establishment of a vegetation cover is particularly important in facilitating system development to the desired endpoint. In some instances, NI plants can function as facilitators of system development, particularly on heavily impacted, highly degraded sites.

Lovejoy (1985) and Lugo (1988) suggested that restoration of tropical forests on barren sites impacted by resource extraction or of former forests converted to agriculture may be initially dependent on establishment of NI plants that can tolerate adverse growing conditions. Site modification by carefully selected NI species could eventually allow IN forest plants to establish and grow, creating pockets of regeneration that could coalesce as the system develops. Lugo (1988) observed that, on mesic to wet sites in Puerto Rico, plantations of coffee (*Coffea arabica*), fruit trees, and other economically important, NI woody species provided the sheltering environment necessary for establishment and growth of IN shade-tolerant forest trees, which eventually formed a diverse understory and ultimately a secondary forest.

Similarly, Parrotta (1992) found that within 5 years of establishment, plantations of the nitrogen-fixing Asian siris tree (*Albizia lebbek*) significantly accelerated natural regeneration of IN forest species on degraded coastal pastures in Puerto Rico. An analogous example in temperate areas is the use of woody NI plants (e.g., black locust [*Robinia pseudoacacia*] and hybrid poplars [*Populus*]) in the restoration of mined lands. Establishment of a sheltering woody cover with these species can facilitate development of native-dominated, late successional stages (e.g., Aber 1987; Soni et al. 1989). Note, however, that species like the black locust have also become pests in natural systems outside of their native range (e.g., Ashby 1987).

In North America, cursory observations suggest that NI plants may play potentially valuable ecological roles in restoration of some grasslands. Nonindigenous species differ in their ability to persist in grasslands developing under management, particularly in relation to prescribed fire. For example, quack grass (*Agropyron repens*) and curly dock (*Rumex crispus*) are unable to persist more than a few years in newly restored prairie (Kline and Howell 1987). Nonpersistence in the extant vegetation is especially desirable for NI plants that could provide useful functions in grassland restoration during a discrete window of time, but would ultimately be absent from the "finished product." Certain nonpersistent, NI plants perhaps could be used to stabilize soils during the early stages of restoration. Likewise, Bowler (1992) mentioned that NI plant species such as wild artichoke (*Cynara cardunculus*), prevalent in the early stages of restoration in some California grasslands, provide critical food and cover to IN fauna and thus help to retain vertebrate species that otherwise would be lost in the absence of a food supply.

Future Ecological Values of Nonindigenous Plants in a Changing Global Environment

Focusing solely on the short-term ecological value of NI plants within and across systems of concern is nearsighted. A long-term perspective is necessary, particularly considering the specter of potential global climate change due to the greenhouse effect (Soulé 1990; Westman 1990a, 1990b). According to some climate-change scenarios, a major, rapid reorganization of ecological communities will occur with IN species shifting ranges or becoming extinct, and preadapted NI species invading the vacant niches (Westman 1990a, 1990b). In this regard, an NI plant species considered problematic today may have considerable ecological value in the future, perhaps playing key structural and functional roles in post–climate-change communities. Thus, today's NI pest could be tomorrow's naturalized keystone species. Acknowledging this possibility and incorporating the potential effects of climate change on IN and NI plants and the communities they inhabit will be crucial components in the long-term management of biological diversity and in the valuation of NI species.

Globally, human populations show little tendency to decrease in number or in rate of growth. As a result, human-impacted and anthropic systems should be expected to increase worldwide as natural systems are destroyed, converted, or degraded to meet human needs. Due to direct or indirect human actions, the ecological importance and value of NI plant species will also change across systems and scales. At a large scale, NI plants may assume valuable ecological roles in the unique cultural landscapes emerging from human endeavors across many regions (e.g., Bowler 1992; Luken 1994). Existing cultural landscapes, such as the French garrigue, are composed of a singular, dynamic assemblage of IN and NI plant species and have considerable conservation, recreation, and aesthetic value (Bowler 1992). At a more local scale, NI plants can be expected to assume important ecological functions (e.g., carbon storage, microclimate modification, resting and feeding areas for migratory birds) in urban greenways and similar anthropic systems where IN plant species grow poorly if at all. Thus, the future ecological value of NI plants will be defined in large part by the complex of spatially variable habitats arising as a result of the actions of humans, the ecological roles assumed by particular NI species or species assemblages, and the interactions between NI species and IN species.

Conclusions

1. Valuation of the ecological functions of NI plants is complex and varies within and among systems. In general, NI plants have greater ecological value in anthropic systems than in intact natural systems. Thus, as anthropic systems are restored to natural systems and natural systems are modified or converted to anthropic uses, the ecological value of NI plants will shift.

2. Ecological value of NI plants will vary most in impacted natural systems due to potentially divergent management goals (e.g., restoration to a natural system, conversion to an anthropic system).

3. Development of a conceptual model of ecological value for NI plants is vital to development of balanced cost-benefit analyses for management of NI plants within and across systems.

4. Research on NI plants has generally emphasized control over basic ecological studies. Integrated short- and long-term studies of potentially valuable ecological functions of key NI species are few and sorely needed.

5. Global climate change may cause a major shift in distribution and importance of IN and NI plants within and across systems. As a result, the ecological value of NI plants will increase in some post–climate-change communities as IN species are lost and their functions are assumed by NI species. Thus, any attempt at valuation of NI plants for the long term will require consideration of potential climate shifts, local and regional ecological dynamics, and human activities.

Acknowledgments. I thank John Knox, Dan Pavuk, and Kim Williams for helpful comments on drafts of the chapter. The interlibrary loan staff at Clarion University provided access to many key references.

Section II
Assessment of Ecological Interactions

4
Documenting Natural and Human-Caused Plant Invasions Using Paleoecological Methods

Stephen T. Jackson

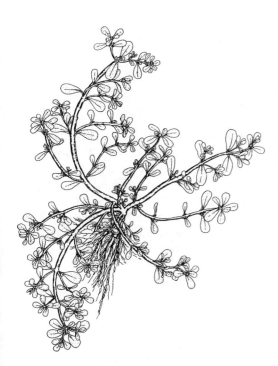

Purslane
(*Portulaca oleracea*)

Paleoecologists and ecologists are concerned with ecological dynamics, but typically at different temporal, spatial, and taxonomic scales. A key interest for both groups is documentation of species invasions and assessment of their causes and consequences. However, the terms *invasion* and *invader* have different connotations for each group owing to a differential in scale. Ecologists tend to view invasion as the introduction and expansion of a nonindigenous (NI) species within a region, typically occurring rapidly (10^1 to 10^2 years) against a backdrop of constant climate and moderate-to-intense landscape disturbance related to human activities (e.g., Elton 1958; Mooney and Drake 1986). In contrast, paleoecologists perceive invasions as spatial and population expansions of species related ultimately to secular changes in climate and other environmental factors. These "natural" invasions occur over longer time spans (10^2 to 10^6 years) and are typically detected at broad spatial scales (10^1 to 10^4 km) (e.g., Huntley and Webb 1988). Environmental changes, usually substantial at these time scales, mediate the ecological processes underlying the invasions (MacDonald 1993; Webb 1986).

In this chapter, I discuss the ways in which paleoecological methods, data, and insights can contribute to ecologists' and managers' understanding of both "natural" and "alien" invasions. The somewhat artificial boundary between ecology and paleoecology is deteriorating as ecologists, motivated in part by concerns with global climate change, become increasingly interested in long-term ecological processes and environmental change (Davis 1989a). However, the nature of paleoecological data constrains paleoecologists to spatial, temporal, and taxonomic scales only partially overlapping those conventionally studied by ecologists.

Initially, I provide a brief primer of paleoecology for ecologists and managers, emphasizing issues of scale and precision in the spatial, temporal, and taxonomic domains. These issues govern the kinds of information the fossil record can provide (Jackson 1994; MacDonald 1993; Webb 1993).

Secondly, I review selected studies illustrating the contributions that paleoecology can make in identifying local and regional invasions associated with human activities and in assessing their impacts. I focus on the New World (especially North America) and Oceania, where a variety of cultures thrived before the European colonizations of the past five centuries and where post-European plant invasions are well documented. I discuss how these disturbances and invasions are revealed in the paleoecological record and note how paleoecological studies have contributed to an understanding of which species in North America are indigenous (IN) or nonindigenous. Human land-clearance and pastoral and agricultural activities and their vegetational impacts are prominent in paleoecological records from Eurasia, especially northern Europe. I do not discuss these records, but note the existence of a rich body of paleoecological literature on human impacts in Europe (Behre 1981, 1988; Berglund 1991; Gaillard and Berglund 1988; Iversen 1941, 1973; Wiltshire and Edwards 1993).

Finally, I discuss the paleoecological record of longer-term, "natural" invasions occurring in response to secular environmental changes. In this section, I summarize four paleoecological case studies illustrating some complexities faced by resource managers attempting to restore "natural" vegetation on landscapes altered by human activity during a period of environmental change.

Nature of the Paleoecological Record

Taxonomic Scale and Precision

Paleoecologists must infer flora and vegetation from the assemblages of pollen grains and other plant remains that accumulate in sediments of lakes, wetlands, and other basins. Paleoecological inferences are necessarily incomplete and often poorly resolved compared to the floristic lists compiled by field botanists or to the plant-abundance estimates tabulated by field ecologists. Pollen assemblages tend to be heavily dominated by wind-pollinated plants, so a large component of the flora is usually unregistered in them. Furthermore, wind-pollinated plants tend to be differentially represented in sedimentary assemblages owing to differences in pollen productivity (grains produced per unit plant abundance per year), pollen dispersibility (which varies with pollen size and floral architecture), and pollen preservability (which depends on chemical composition and structure of the pollen exine and varies from nil [Juncaceae, Lauraceae] to poor [*Populus*] to excellent [Asteraceae, Poaceae, most wind-pollinated trees and shrubs]). These issues are discussed more comprehensively in recent texts (Fægri et al. 1989; Moore et al. 1991) and reviews (Jackson 1994; Prentice 1988).

The problem of differential representation of taxa in pollen assemblages has long been recognized; formal approaches to correcting for production and dispersal bias are being developed and validated (e.g., Calcote 1995; Jackson et al. 1995; Prentice 1985; Sugita 1994). These techniques can help correct the bias for many taxa but are not useful for taxa very poorly represented or unrepresented in pollen assemblages. Trace amounts of poorly represented taxa often provide positive indications of local occurrence in vegetation, and studies of assemblages of modern pollen indicate which taxa in a flora are most and least likely to be detected by pollen analysis.

A further constraint on the precision of floristic and vegetational inferences from pollen assemblages is that systematic morphological variation in pollen grains is often dampened or even absent below the level of genus or family. In fact, species differentiation for wind-pollinated taxa is the exception rather than the rule. In eastern North America, for instance, reliable species-level identifications can be made only for *Acer*, *Juglans*, *Plantago*, and *Shepherdia* among all the upland wind-pollinated taxa commonly occuring. Subgeneric splits can be made for a few genera (e.g., *Alnus*, *Cornus*, *Fraxinus*, *Pinus*, *Polygonum*, *Rumex*). In some groups, genus-level or family-level splits cannot be made routinely. For example, genus-level differentiation in Apiaceae, Asteraceae, Brassicaceae, Cyperaceae, Ericaceae, Poaceae, Ranunculaceae, and Rosaceae is rarely possible, although a few genera in each family have unique distinguishing features. Some families must even be lumped together at times (e.g., Amaranthaceae and Chenopodiaceae; Cupressaceae and Taxodiaceae).

Plant macrofossils (buds, foliage, floral organs, fruits, seeds, etc.) provide an alternative and complementary source of data on flora and vegetation. Macrofossils are preserved in a wide variety of depositional environments, including lakes, wetlands, buried soils, and packrat middens (Betancourt et al. 1990; Jackson et al. n.d.). They have a critical advantage over pollen in that they are nearly always identifiable to the generic level and frequently to the specific level.

However, macrofossils also provide a selective sample of the flora surrounding a depositional site. Macrofossil assemblages from lake sediments in temperate forested regions, for instance, consist primarily of conifer needles and seeds and *Betula* fruits, with poor representation of most deciduous trees (e.g., *Acer*, *Fagus*, *Quercus*) and understory plants (Jackson 1989). This differential representation results primarily from differential dispersibility of plant organs (e.g., wind-dispersed seeds are better represented than gravity- or animal-dispersed seeds) and from differential preservation (e.g., leaves of dicotyledons fragment and decompose more rapidly than do needles of conifers). Fluvial sediments often have much greater representation of large plant organs, especially fruits, cones, and seeds (Baker et al. 1996; Givens and Givens 1987; Jackson and Givens 1994). Macrofossil assemblages from wetland sediments provide a rich record of the local wetland flora (Jackson et al. 1988; Singer et al. n.d.; Watts and Winter 1966), but upland plants are poorly represented owing to distance from the depositional site. Packrats collect a diverse array of plant organs from within a few tens of meters from their nests (Finley 1990). However, packrat–midden floras are incomplete representations of the local flora, and different packrat species may have different plant preferences (Dial and Czaplewski 1990).

Temporal Scale and Precision

Temporal scale and precision of paleoecological records are dictated by two factors: the rate and nature of sediment accumulation and the methods used to estimate sediment ages. Sediments used in paleoecological studies vary widely in their accumulation properties. Lake sediments are preferred in most regions for studies of fossil pollen because they accumulate continuously and relatively rapidly. Physical and biological processes cause mixing of sediments at the bottoms of most lakes, imposing temporal smoothing that ranges from a few years to a few decades. For deep lakes with permanent thermal stratification, anoxic conditions at the lake bottom prevent bioturbation by invertebrates. Sediments of such lakes have annual laminations ("varves"), which allow annual and even seasonal precision in sediment records. Some temporal smoothing may still occur owing to redeposition of littoral sediments. Wetlands and small, wet hollows also provide continuous records, although sediment accumulation rates are typically slower than for lakes, and shallow basins are more liable to short- or long-term interruptions in sediment accumulation during periods of drought.

In arid and semiarid regions and in regions where natural lakes and wetlands are rare (e.g., unglaciated, nonkarstic, nonvolcanic landscapes), pollen and macrofossils are frequently recovered from other kinds of sediments, including alluvial beds and cave fill, and from packrat middens. These typically accumulate during brief episodes (10^1–10^3 years) and are usually treated as instantaneous "snapshot" samples because the interval of accumulation is within the standard error of radiocarbon dates. Landscape surfaces buried by lacustrine, aeolian, or glacial sediments also provide snapshot records. In situations where such records are abundant and span a long time period, individual snapshots can be stacked to provide a quasicontinuous record of vegetational and floristic change (e.g., Baker et al. 1996; Betancourt et al. 1990; Jackson and Givens 1994; Nowak et al. 1994). Macrofossils from fluvial sediments and pack rat middens are somewhat more subject to redeposition than those in lakes and wetlands, but careful dating can identify such problems (Givens and Givens 1987; Van Devender et al. 1985).

Age estimates of late Quaternary pollen or macrofossil assemblages are based primarily on radiocarbon (^{14}C) dates of organic-rich sediments or of plant or other organic material. Snapshot assemblages are assigned ages based upon direct ^{14}C dates of the sediments or plant materials. For lake and wetland sediments, models relating age and sediment depth are developed using linear interpolation, linear regression, or curve-fitting procedures (polynomials, splines). These age models often include the sediment surface and well-dated, regionally significant pollen events as benchmarks. Like all empirical models, age-depth models have an associated error term, which is not always specified or estimated. Webb (1982, 1993) estimated the uncertainty of Holocene age estimates from age-depth models to be 300 to 500 years.

Radiocarbon ages have traditionally been based on measurement of β-particle emission during decay of ^{14}C isotopes. Such measurement requires large amounts (usually ≥ 1 g) of carbon, which often necessitates mixing of sediment constituents or plant parts of different potential ages. Also, some aquatic vascular plants and algae use in photosynthesis HCO_3^- and $CO_3^=$, which may derive from dissolution of ancient carbonate-rich bedrock or surficial materials. Thus, conventional β-decay dates are subject to imprecision and error beyond the standard "counting" error (Olsson 1991; Pilcher 1993).

Development of accelerator mass spectrometry (AMS) as an alternative means of assessing isotopic composition of carbon samples helps circumvent some of these problems. Much smaller quantities of carbon are required, so dates can be based upon individual seeds or needles. The gain in precision as applied to age-depth models is sometimes only apparent, however, because the macrofossils used may not be exactly contemporaneous with the incorporating sediments (Bennett 1992; Webb 1993).

Ages estimated directly or indirectly from ^{14}C dates are usually presented in ^{14}C years BP (before present), rather than calendar years, owing to secular variations in ^{14}C content of the atmosphere. These variations are well understood for the past several thousand years, and models are available for calibrating ^{14}C years BP in terms of calendar years BP (Stuiver and Reimer 1986, 1993).

Age estimation of sediments spanning the period of European colonization of North America and other continents (i.e., the past two to five centuries) using ^{14}C-based models is rendered difficult by several factors. First, landscape disturbance frequently causes changes in sediment accumulation rates in lake and wetland basins, so interpolation between pre-European ^{14}C dates and the sediment surface may be problematic. Second, the precision of ^{14}C age estimates has a finite limit, typically ± 100 to 200 years except in a few unusual cases (Olsson 1991; Pilcher 1993). This is a low-end estimate of the imprecision of age estimates and does not take into account other sources of error (e.g., sediment integration, organically bound "dead" carbon, and contamination). Thus, age-depth models for the past few centuries based solely on ^{14}C dates should be used cautiously.

Other dating methods can be applied to sediments of the past few centuries. Varve counts can provide precise chronologies for lakes with annually laminated sediments (McAndrews and Boyko-Diakonow 1989). Certain short-lived isotopes that accumulate in sediments can be used for dating. Primary among these is lead-210 (^{210}Pb), which can provide reliable age models for sediments of the last 100 to 150 years (Binford 1990; Brugam 1978). Useful for very recent sediments is cesium-137 (^{137}Cs), which is produced during nuclear explosions (Pennington et al. 1973). Microscopic and macroscopic carbonaceous and metallic particles from industrial combustion can sometimes be used as stratigraphic markers (Clark and Patterson 1984; Jackson et al. 1988). In the western United States, spores of dung fungus (*Sporormiella*) can indicate the initiation of livestock grazing (Davis OK 1987). In certain depositional environments, changes in pollen density (number of grains per unit volume of sediment) can be used to refine chronologies (Brush 1989).

Finally, historically documented vegetational events can be identified in pollen profiles and used as stratigraphic markers. Some events commonly used in eastern North America include the rise in ragweed (*Ambrosia*) and other "weed" pollen following Euro-American land clearance (Brugam 1978; Russell et al. 1993), the increase in pollen of hemp/hops (*Cannabis/Humulus*) in the Midwest during early twentieth century hemp cultivation (Van Zant et al. 1979), and the decline in pollen of chestnut (*Castanea*) associated with the early twentieth century chestnut blight (Anderson 1974).

Spatial Scale and Precision

Spatial scale and precision of paleoecological inferences are determined by the dispersal properties of pollen grains and plant organs and by the nature of the depositional environment. A pollen assemblage represents a distance-weighted integration of pollen from plants growing in or adjacent to the depositional site ranging out to plants growing tens or even hundreds of kilometers away. The distance weightings vary among plant taxa according to pollen-dispersal properties and among basins according to surface area.

Quantitative models linking pollen dispersal, basin size, and pollen source area have been developed and tested (Jackson 1990, 1994; Prentice 1985, 1988; Sugita 1994; Sugita et al. n.d.). These models, along with empirical data on relationships between modern pollen assemblages and vegetation, indicate the following: (1) For a given basin size in homogeneous vegetation, pollen source areas for taxa with large, poorly dispersed pollen grains will be more locally weighted than those for taxa with smaller, better-dispersed grains. (2) For a given taxon in homogeneous vegetation, the pollen source area will become increasingly locally weighted as basin surface area decreases: more individuals of the taxon will be close to the basin center. (3) For basins of a given size, patchiness of vegetation will become increasingly less detectable (i.e., spatially smoothed) as patch size decreases. (4) For a given basin size and a given vegetation-patch size, the patch will become increasingly less detectable as its distance from the basin increases. And (5) for the smallest possible basin size (i.e., radius equal to 0 to 2 m; e.g., small, wet depressions and humus under closed canopy), more than 50% of the pollen derives from trees growing more than 50 m away.

These theoretical and empirical observations indicate that interpreting assemblages of fossil pollen is complex, but they also provide constraints that facilitate precise inferences given knowledge of basin size and pollen-dispersal properties, and given multiple sites for analysis of spatial patterns. Furthermore, the basin-size effect provides a powerful tool in design of paleoecological studies; basin size can be chosen based on the spatial scale at which the investigator seeks to identify patterns. Absolute lower limits to spatial resolution of vegetation patterns and detection of single patches or plant populations from pollen data are imposed by widespread dispersal of pollen grains in the atmosphere. An additional limit in practice is availability of sites of suitable size at sufficiently high density on the landscape to detect vegetation patterns at the desired scale.

Greater spatial precision is provided by plant macrofossils from lake and wetland sediments. These fossils are poorly dispersed in the atmosphere and tend to derive from plants growing within the basin or at or near the basin margin (Dunwiddie 1987; Jackson 1989). However, plant macrofossils are not useful for detecting plant taxa growing beyond the basin margin, especially in watersheds of gentle topography.

Paleoecological Assessments of Human-Related Plant Invasions

Pre-European Land Use and Plant Invasions in the New World and Oceania

Paleoecological detection of vegetational changes and plant invasions related to human land use before European colonizations of the fifteenth through nineteenth centuries is obscured by problems of spatial, temporal, and taxonomic scale. Small, patchy disturbances, which were probably characteristic of many native American and Polynesian cultures, are unlikely to be registered in pollen profiles unless the disturbances were immediately adjacent to the depositional site (Sugita et al. n.d.). In addition, if patches cleared for cultivation or habitation were abandoned after a few years or decades, the disturbance event might be smoothed over by sediment mixing or missed by insufficiently dense pollen sampling within a core. Finally, the taxa that might be expected to colonize the clearings (e.g., *Ambrosia*, Asteraceae, Poaceae in eastern North America) have broad ecological amplitudes, and most are not specific to human disturbances. Thus, representation of these taxa in a pollen assemblage might indicate wetland or other natural habitats, and transient pollen-maxima in them within a pollen profile might result from natural disturbances (fires, wind throw, hydrological draw down).

For these reasons it should not be surprising that unambiguous records of pre-European, human-induced vegetational changes are sparse in much of the New World (McAndrews 1988). In eastern North America, depositional basins are scarce in regions of greatest development of sedentary native populations and land clearance (e.g., the Mississippian and related cultures of the lower Mississippi Valley). The many pollen sequences from the Great Lakes region, New England, and northward show little evidence of native American land-use effects, supporting the hypothesis (Cleland 1983) that land clearance and agriculture were limited to small, shifting patches.

The clearest record of pre-European land use and local weed invasion in North America is from Crawford Lake in southern Ontario (McAndrews 1988; McAndrews and Boyko-Diakanow 1989). Crawford Lake has several attributes that make it ideally suited for registering changes in the local flora associated with land use by native Americans. First, its small surface area (1.5 ha) ensures a locally weighted pollen source area, so its pollen record should be sensitive to moderate-sized disturbances nearby. Second, the lake is permanently stratified and the sediments are annually laminated, which allows recording of short-term events and precise chronology. Third, Iroquoian villages were established within a few hundred meters of the lake between 1360 and 1660 A.D. Archeological evidence indicates that the Iroquoian people at the site cultivated beans (*Phaseolus*) and maize (*Zea mays*).

A pollen diagram from Crawford Lake (Fig. 4.1) clearly indicates local land-use effects (McAndrews and Boyko-Diakonow 1989). Pollen of Poaceae is common throughout the period of Iroquoian occupation and rare before and after. This probably represents annual and other grasses that colonized fields and other disturbed areas. Maize pollen also occurs.

4. Documenting Natural and Human-Caused Plant Invasions Using Paleoecological Methods

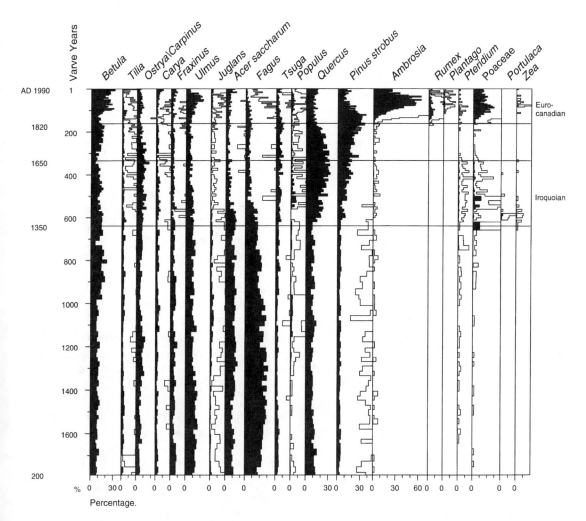

FIGURE 4.1. Pollen percentage diagram from Crawford Lake, southern Ontario, showing selected pollen types. Age estimates are from varve counts and range from 212 to 1990 A.D. *Open silhouettes* represent 10× exaggeration of pollen percentages. Note occurrences of Poaceae (= Gramineae), *Portulaca*, and *Zea* pollen during the Iroquoian occupation period (ca.1350–1650 A.D.), and increases in *Ambrosia*, Poaceae, *Plantago*, and *Rumex* following European land clearance in the mid-nineteenth century. Note also the mid-twentieth century decline of elm (*Ulmus*) pollen, recording widespread killing of elms by Dutch elm disease. (From McAndrews [1988] and McAndrews JH, pers. com. [1995]. Reprinted from McAndrews [1988] by permission of Kluwer Academic Publishers)

Particularly noteworthy at Crawford Lake is the occurrence of pollen and seeds of purslane (*Portulaca oleracea*) in the Iroquoian period (Byrne and McAndrews 1975). This species has been regarded by North American botanists as a European invader, although it was reported in Hispaniola in the early sixteenth century and in eastern Canada and New England in the early seventeenth century. The Crawford Lake record clearly indicates that purslane is an IN North American species.

Paleoecological studies have led to identification of other putative North American "aliens" as IN species. Seeds of carpetweed (*Mollugo verticillata*) and purslane have been found in Holocene archeological sites in Tennessee (Chapman et al. 1974) and in Pleistocene wetland sediments in Alabama (Delcourt 1980). Seeds of bugseed (*Corispermum*), often regarded as an NI invader from Eurasia, have been dated using AMS to the late Pleistocene and Holocene in Alaska, western Canada, and the southwestern United States (Betancourt et al. 1984).

These occurrences suggest caution in identifying plants as nonindigenous in the absence of clear historical or botanical documentation of their entry into a continent or island. Corispermum spp., *Mollugo verticillata*, and *Portulaca oleracea* in North America are all IN taxa that have opportunistically expanded their populations and geographic ranges following the intense landscape disturbances of the past four centuries. Until the eighteenth and nineteenth centuries, botanists and plant collectors were usually not among the first Europeans to enter newly "discovered" regions, and hence most botanical records postdate extensive land-clearance activities.

A few other North American sites besides Crawford Lake show paleoecological evidence of native American land clearance and agriculture (e.g., Betancourt and Davis 1984; Betancourt and Van Devender 1981; Burden et al. 1986; McAndrews 1988; Whitehead and Sheehan 1985). Pollen of plants that colonize disturbed sites today (*Boerhavia* type, *Cleome*, *Euphorbia* type, *Kallstroemia*, *Oenothera* type, *Sphaeralcea*, *Tidestromia* type) occurs at many archeological sites in the southwestern United States together with evidence of plant cultivation (Gish 1991; Martin and Byers 1965).

An oft-cited example of native American land clearance is the pollen and macrofossil record from Tuskegee Pond in eastern Tennessee (Delcourt et al. 1986). Archeological and paleoethnobotanical studies near the site provide clear evidence of at least local land clearance and agriculture in the past 2000 to 3000 years. Based on the age model provided by Delcourt et al. (1986), the Tuskegee Pond pollen record is consistent with this evidence, with high representation of herb pollen during the past 1500 years. In particular, percentages of ragweed pollen range between 20% and 50% in the pre-European period, indicating land clearance on a broad regional scale. Comparable percentages of ragweed pollen occur today only in portions of the midwestern and southeastern United States where forests are sparse and agricultural lands are widespread (Peterson 1978; Webb 1973; Webb and McAndrews 1976). However, the age model at Tuskegee Pond is based entirely on ^{14}C dates. Bedrock in the region is carbonate-rich (Delcourt et al. 1986), and hence the ^{14}C dates may be erroneously old. The entire Tuskegee Pond pollen profile may be post-European.

Pollen records from Central America and South America are few, and little direct evidence exists of pre-European land use and impacts. A pollen sequence from a lake in the Petén lowlands of Guatemala provides good evidence for extensive Mayan deforestation starting 3000 years BP (Leyden 1987). Pollen assemblages in the Mayan period are characterized by early-successional trees and by ragweed and other Asteraceae. Maize pollen also occurs consistently. Extensive soil erosion is indicated by the clayey sediments deposited in the lake. The Spanish invasion 400 years ago led to abandonment and reforestation of the region, which is still not complete (Leyden 1987). Other studies in Central America show similar impacts of Mayan, Aztec, and other cultures (Bush et al. 1992; McAndrews 1988; Watts and Bradbury 1982).

Evidence for pre-European land clearance and plant introduction in Oceania is sparse for the same reasons as for the American continents. Pollen records indicate extensive deforestation in New Zealand after Polynesian colonization (McGlone 1983, 1989), and paleontological and archeological studies indicate introductions of animals (lizards, mice, pigs) and cultivated plants and extinction of indigenous fauna throughout Oceania following Polynesian arrival (Dye and Steadman 1990; McGlone 1989; Steadman 1995). Pollen evidence for land clearance and species introduction is particularly clear at Easter Island because of the island's small size, high insularity, and low species diversity. Polynesian colonization 1200 years BP was followed by destruction of the native palm forests and the introduction of NI Asteraceae (Liguliflorae), Caryophyllaceae, *Plantago*, and *Rumex* (Flenley et al. 1991). Increases in pollen percentages of Poaceae probably resulted from expansion of IN grasses and colonization by NI species. Flenley et al. (1991) also demonstrated that totora (*Scirpus californicus*) has occurred on the island for at least the past 36,000 years and therefore was not introduced by human colonizers from South America as asserted by Heyerdahl (1971).

Post-European Land Use and Plant Invasions in the New World and Oceania

The earliest documented colonization of the New World by Europeans was the tenth-century Norse settlement of southern Greenland. The settlements lasted three to five centuries and were supported by grazing and cultivation. Pollen and plant macrofossil records from the vicinity of the settlements show introduction of NI annual plants from Iceland and Scandinavia, including annual blue grass (*Poa annua*), common chickweed (*Stellaria media*), knotweed (*Polygonum aviculare*), sheep-sorrel (*Rumex acetosella*), shepherd's purse (*Capsella bursa-pastoris*), and yarrow (*Achillea millefolium*) (Fredskild 1978, 1988). Populations of several IN species also expanded in cultivated and grazed areas. Most of the IN and NI species are important weeds in pastures and settlements of Greenland today (Fredskild 1988). Norse colonization of North America was limited to small, ephemeral settlements. The best documented of these, L'Anse aux Meadows (Newfoundland), lacks pollen evidence of land clearance or plant introduction (Davis et al. 1988). Any such activities or effects must have been spatially restricted.

Spanish colonization of the New World began in the late fifteenth century and was rapid and widespread in the Caribbean region, Central America, and South America during the next three centuries. This region was not intensively studied paleoecologically until very recently; hence, few pollen records of post-European land clearance and vegetation changes exist. Paleolimnological studies in Haiti indicate extensive European land clearance and soil erosion (Binford et al. 1987).

English, French, and Dutch settlement of North America was concentrated along the Atlantic Coast in the seventeenth and eighteenth centuries. Extensive settlement and land clearance in the interior of the continent did not begin until the late eighteenth and early nineteenth centuries. Replacement of forest by croplands and pastures over broad areas is well documented by decreases in tree pollen and increases in pollen of annual herbs (ragweed, other Asteraceae, plantains [*Plantago lanceolata*, *P. major*], Poaceae, *Rumex acetosella*) from the Atlantic Coast to the western Great Lakes region (Figs. 4.1 and 4.2) (McAndrews 1988). Comparison of pollen chronologies based on varve counts or ^{210}Pb-dating with historical records shows clearly that increases in these taxa accompanied regional land clearance and that the increases were time-transgressive westward (Brugam 1978; Swain 1973). Pollen of *Plantago lanceolata*, *P. major*, and *Rumex acetosella* is absent from pre-European sediments in North America, confirming that these species were not indigenous to the continent.

The Pacific and Rocky Mountain regions of North America were not extensively cleared or settled until the early to mid-nineteenth century. Many of the NI plants that accompanied Europeans (e.g., *Acacia*, *Erodium cicutarium*, *Eucalyptus*, *Plantago lanceolata*, *Podocarpus*, *Rumex acetosella*, *Salsola*) are well recorded in pollen assemblages (Davis 1992; Davis and Turner 1986; McAndrews 1988; Mudie and Byrne 1980).

Oceania provides several good records of plant invasions associated with European colonization during the eighteenth and nineteenth centuries. A lake-sediment core from Tahiti shows extensive land clearance by Polynesians, followed by forest recovery in

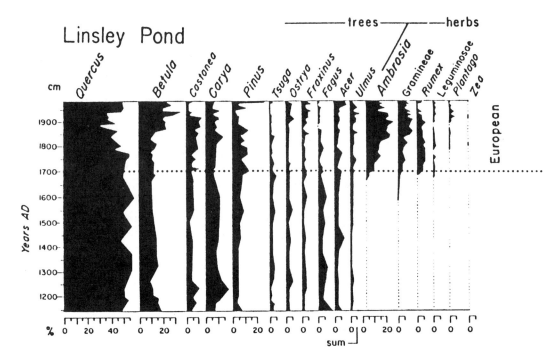

FIGURE 4.2. Pollen percentage diagram from Linsley Pond, southern Connecticut, showing selected pollen types. Age estimates are from ^{210}Pb and ^{14}C dating. Note increases in *Ambrosia*, Gramineae (= Poaceae), Leguminosae (= Fabaceae), *Plantago*, and *Rumex* following European occupation and land clearance in the late seventeenth and early eighteenth centuries. (From McAndrews [1988]; modified from Brugam [1978]. Reprinted from McAndrews [1988] by permission of Kluwer Academic Publishers)

the early nineteenth century when the local Polynesian society collapsed owing to introduced diseases and social disruption (Parkes et al. 1992). The secondary forests included a substantial component of invasive species (*Ageratum*, *Miconia*, *Psidium*) not indigenous to the island (Parkes et al. 1992). Pollen records from Easter Island document nineteenth century invasion by NI Asteraceae (Liguliflorae), *Casuarina*, *Psidium*, and Urticaceae and Moraceae (including *Broussonetia*, *Ficus*, *Morus*) (Flenley et al. 1991).

Much potential exists for using paleoecological techniques to further document invasions of NI plants in the New World and Oceania. Spatial networks of pollen and macrofossil sites with varve count or ^{210}Pb dating control would permit mapping of invasion patterns and determination of rates of spread for taxa reliably registered in paleoecological records.

Long-Term Records of Plant Invasions and Environmental Change

Vegetational dynamics occur at time scales ranging from decades to tens of thousands of years. Terrestrial environments of the earth have undergone fluctuations at these time scales during the past million years owing to variations in earth's orbital geometry, atmospheric chemistry, ice-sheet volume, sea-surface temperature, and ocean circulation (COHMAP Members 1988; Kutzbach and Webb 1991; Webb and Bartlein 1992). Plant communities and ecosystems have changed in response to these variations. Depending on rates and magnitude of the environmental changes (which partly depend on the time scale considered), these responses have ranged from local population changes to shifts along regional environmental gradients (elevation, soil) to continental-scale shifts in range boundaries (Prentice 1992).

At sufficiently broad time scales (typically ≥ 10^3 to 10^4 years), most plant taxa in any given region or habitat can be viewed as invaders in that they have not been continuous constituents of the vegetation. Examples of such "natural" invasions include (1) appearance of North American taxa (*Alnus*, *Quercus*) in the Andes after Pliocene joining of North America and South America and subsequent Pleistocene cooling (Hooghiemstra 1984; Hooghiemstra and Sarmiento 1991); (2) colonization of glaciated regions of North America following late Wisconsinan ice retreat (Davis 1976; Jackson et al. n.d.; Webb 1988); (3) Holocene expansion of species' ranges along edaphic, elevational, and climatic gradients spanning 10^1 to 10^2 km (Davis MB 1987; Gaudreau et al. 1989; Jacobson 1979); and (4) appearance of new species in local forest and wetland communities during the late Holocene as climate changed toward modern regimens (Bradshaw and Hannon 1992; Davis et al. 1994; Singer et al. n.d.). A key concern in Quaternary paleoecology is understanding the timing, mechanisms, and consequences of these long-term invasion events, all of which are related in some way to long-term environmental changes.

Apparent persistence of plant populations in a region for more than 10,000 years has been documented in a few cases, including beech (*Fagus grandifolia*), black walnut (*Juglans nigra*), and red-cedar (*Juniperus virginiana*) in the Tunica Hills of Louisiana and Mississippi (Delcourt and Delcourt 1977; Givens and Givens 1987; Jackson and Givens 1994) and Utah juniper (*Juniperus osteosperma*) in the Painted Hills of western Nevada (Nowak et al. 1994). However, in both these regions other species have come and gone in the past 20,000 years, and the broader-scale geographic ranges of these locally persistent species have undergone substantial shifts (Delcourt and Delcourt 1987; Thompson 1988; Webb 1988). Migrations of tens to thousands of kilometers appear to be the norm for time-scales of more than 5000 years in most continental regions (Huntley and Webb 1988). At broad time scales and narrow spatial scales, invasion is a routine ecological phenomenon, representing responses of plant species to environmental changes of high magnitudes and low frequencies.

Paleoecologists have long been interested in the ecological consequences of plant invasions recorded in pollen sequences (Watts 1973). Various attempts have been made to fit spatial diffusion models, single-species population-growth models (exponential, logistic), and the Lotka–Volterra two-species interspecific competition model to pollen data. These efforts, reviewed recently by MacDonald (1993), require spatial, temporal, and taxonomic precision not always available in the fossil record (see discussion above and in MacDonald [1993]). Application of these kinds of models also assumes that the environment has remained constant during the period of population growth or interaction. This assumption was long considered reasonable by many paleoecologists for the Holocene. Rapid climate change as the late Pleistocene ice sheets melted, followed by a 10,000-year period of relative constancy, would provide an ideal "natural laboratory" for studying a variety of phenomena associated with ecological invasion, including intrinsic dispersal rates, disturbance, establishment, population growth rates, density compensation, habitat displacement, and community "resistance" to invasion.

The "constant-climate" view of the Holocene has been largely superseded in the past 15 years by a perspective assuming that ecologically significant climatic changes have occurred at frequencies of less than 2000 years throughout the Holocene. Three key lines of evidence and argument have led to this transition (Webb 1986). First, advances in theoretical climatology and paleoclimate modeling indicate that such climate changes have occurred throughout the Holocene (Kutzbach and Guetter 1986; Kutzbach and Webb 1991). Furthermore, these climate changes have been complex, consisting of changes in seasonal temperatures and precipitation as well as in annual means. Second, paleoclimates simulated from general circulation models (GCMs) are sufficient to explain most of the spatial and temporal variation in Holocene pollen assemblages at frequencies of 2000 years and greater (COHMAP Members 1988; Huntley and Prentice 1993; Webb et al. 1993). Also, paleoclimates inferred from pollen data for selected taxa successfully predict pollen abundances of other, independent taxa (Prentice et al. 1991). Third, the paleoecological observations used to justify the constant-climate viewpoint (different rates and directions of migration, simultaneous southward and northward expansions of different taxa, vegetation with no modern analogues) can all be explained equally well or better by climate change (Davis 1989c; Graumlich and Davis 1993; Jackson et al. n.d.; MacDonald 1993; Overpeck et al. 1992; Webb 1986; Webb et al. 1983).

Thus the paleoecological record is poorly suited for studying "pure" effects of species invasion in the absence of environmental change, at least at time scales greater than 2000 years. Even at finer time scales it is difficult to differentiate between ecological consequences of species invasion and environmental change (MacDonald 1993). Most North American pre-Columbian ecological invasions not caused by human activity were linked ultimately to environmental changes and were probably directly constrained or mediated by environmental changes. The most likely exception is the local response to the rapid mid-Holocene decline in hemlock (*Tsuga*) populations in eastern North America, which evidently resulted from a widespread pathogen outbreak (Allison et al. 1986; Bhiry and Filion 1996; Davis 1981a). The high and rapid mortality of hemlocks, which were widespread and abundant in the region at the time, led to density compensation and habitat expansion among other tree species. Forest stands were undoubtedly invaded by species from other communities. Dynamics of these local invasions could be studied by using networks of small-hollow sites for pollen and macrofossil analysis, although this has not been attempted to date (but see Foster and Zebryk 1993).

A question of topical interest in paleoecology is the role of natural disturbances (fire, blowdown, canopy-gap formation) in mediating environmentally induced plant invasions at fine spatial (patch, stand, landscape) and temporal (less than 10^3 years) scales. Such invasions can be accelerated or retarded by disturbances, and the establishment of the invaders can in turn create new disturbance regimens (Björkman and Bradshaw n.d.; Bradshaw and Hannon 1992; Bradshaw and Zackrisson 1990; Grimm 1983). Charcoal particles preserved in sediments are valuable in assessing changing fire regimes (Clark 1988; Patterson et al. 1987). However, differentiating cause and effect among disturbance patterns and vegetation composition in the fossil record is often difficult.

Paleoecological records from eastern North America and other regions show nearly continual flux in taxon ranges and abundances owing to climatic variation and disturbances (Huntley and Webb 1988). A key element of this flux, especially at time scales of greater than 2000 years, is aggregation and disaggregation of plant communities. As the environment changes, species recombine to form new local communities (Bradshaw and Hannon 1992; Davis et al. 1994), new landscape vegetation patterns (Jackson and Whitehead 1991; Spear et al. 1994), and new regional or subcontinental "formations" (Huntley 1990; Overpeck et al. 1992; Webb 1988). These community recombinations are a consequence of recombinations of environmental factors and gradients. Complex changes in environment induce complex responses in vegetation.

Interactions Between Environmental Change and Human Disturbance

Resource managers are often charged with maintaining vegetation in a "natural" state, which in North America is frequently defined as the condition existing immediately before intensive Euro-American disturbance and land use. The desired natural state may be a static community or a dynamic but stationary mosaic of communities of different developmental stages. Invasive species, whether indigenous or nonindigenous, pose obvious potential impediments to this goal. Many management programs focus on "surgical removal" of the invading species, which is presumed to be followed by restoration of vegetation to its predisturbance state by natural processes.

Restoration of predisturbance vegetation may be an elusive goal unless intensive and continuing intervention by managers is practiced. First, time has not stood still since inception of land use by European cultures. Paleoecological and paleoclimatic studies indicate climate change of sufficient magnitude during the past few centuries to cause changes in vegetation composition in the absence of human disturbance. Thus, the natural state of vegetation pursued by managers may be a moving target. Second, land-use practices and disturbances can have long-term or even permanent effects on landscapes and ecosystems, which may be rendered unsuitable for the communities that existed before human disturbance. A final problem is accurate determination of predisturbance vegetation composition and even physiognomy. These are frequently inferred from presumed relict patches or advanced secondary vegetation, which may not be accurate representations of original vegetation owing to undocumented disturbance effects, climate changes, or both.

These problems are becoming increasingly recognized and discussed by ecologists and managers. Paleoecology can contribute to this discussion by providing direct records of predisturbance vegetation, rates and patterns of postdisturbance change, and long-term environmental trends. I present selected examples of these contributions.

Widespread invasion of red-cedars into semidesert grasslands in Arizona during the past 150 years has been variously attributed to climate change, overgrazing, and fire suppression (Hastings and Turner 1965). A pollen sequence from Pecks Lake in central Arizona (Fig. 4.3) reveals that this historic expansion is superimposed on a trend of increasing red-cedar populations during the past 2000 years (Davis and Turner 1986). The long-term trend is related to gradual regional cooling; that trend was amplified since European occupation by intensive cattle grazing (Davis and Turner 1986). A key remaining question is to what extent grazing and fire suppression altered the natural climate-forced red-cedar trajectory. Would red-cedars be as abundant today in the absence of past or present grazing?

Abandonment of pastures and cultivated fields in New England during the past century has led to extensive reforestation (Russell et al. 1993). Paleoecological studies indicate that even the oldest and least-disturbed secondary forest stands differ from presettlement (European) forests (Foster et al. 1992) and that some putatively natural plant communities (e.g., coastal heathlands) are artifacts of human land-use practices (Dunwiddie 1989). Pollen profiles from closed-canopy settings in Harvard Forest, where land-use history is particularly well documented, show continuing impacts of human activity on forest composition (Fig. 4.4). Presettlement mixed hardwood-conifer forest consisting mainly of sugar maple (*Acer saccharum*), birches (*Betula*), beech, red oak (*Quercus rubra*), and eastern hemlock (*Tsuga canadensis*) was cut over, leading to expansion of chestnut (*Castanea dentata*) from sprouting (Fig. 4.4) (Foster et al. 1992). Continued use of the stand as a woodlot for poles and firewood maintained the high chestnut populations. Late nineteenth century cessation of most cutting activities and early twentieth century extirpation of chestnut by the chestnut blight have, however, not led to restoration of forest resembling the presettlement condition (Foster et al. 1992). It is unclear whether this is a result of climate change, soil alteration, inadequate seed sources (e.g., for sugar maple and beech), insufficient time since last disturbance, or a combination of these factors.

A broad-scale synthesis by Russell et al. (1993) showed that forests throughout northern New England were undergoing composi-

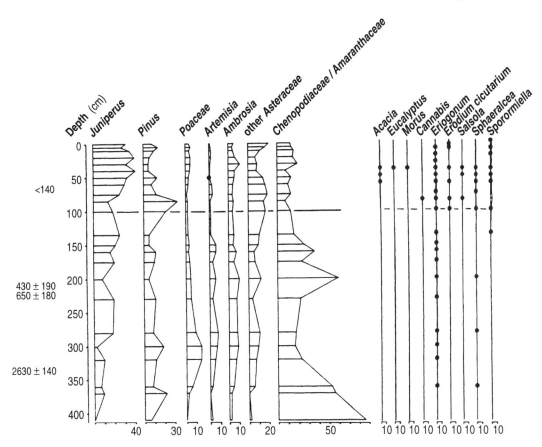

FIGURE 4.3. Pollen percentage diagram showing selected pollen types from Pecks Lake, central Arizona. Dates on left side of diagram are uncalibrated ^{14}C dates (year BP) with standard errors. *Solid circles* represent percentages ≤ 2%. Note steadily increasing trend of *Juniperus* pollen, both before and after the horizon of European land-use inception (horizontal line at 100-cm depth). Note pollen occurrences of indigenous weeds (*Eriogonum*, *Sphaeralcea*), nonindigenous annuals (*Cannabis*, *Erodium cicutarium*, *Salsola*), and cultivated trees (*Acacia*, *Eucalyptus*, *Morus*) above 100-cm depth. *Sporormiella* spores derive from a fungus that grows on dung of large herbivores. (Modified from Davis and Turner [1986])

tional change in the centuries before European colonization. These changes were probably related to cooling associated with the Little Ice Age (Bernabo 1981; Gajewski 1988; Russell et al. 1993). Russell et al. (1993) also showed that secondary forests throughout the northeastern United States have not returned to their pre-European composition owing to post–Little Ice Age warming, altered fire regimens, site alteration, introduced plant pathogens, and direct disturbance effects (e.g., selective removal of species).

Wetlands are particularly sensitive to hydrological changes induced by human activities (Wilcox 1995). Pollen and macrofossil studies by Jackson et al. (1988) demonstrated that landscape modification in the Lake Michigan dunes region led to rapid and dramatic changes in composition of vegetation in a shallow wetland (Fig. 4.5). Several important species (*Bidens cernua*, *Brasenia schreberi*, *Cyperus* spp., *Eleocharis* spp.) were extirpated from the marsh, and other species (all indigenous) invaded the site (*Cephalanthus occidentalis*, *Equisetum* cf. *fluviatile*, *Sparganium*, *Proserpinaca palustris*, *Ranunculus flabellaris*, *Typha*). The hydrology of the marsh has changed not only because of direct disruption

4. Documenting Natural and Human-Caused Plant Invasions Using Paleoecological Methods

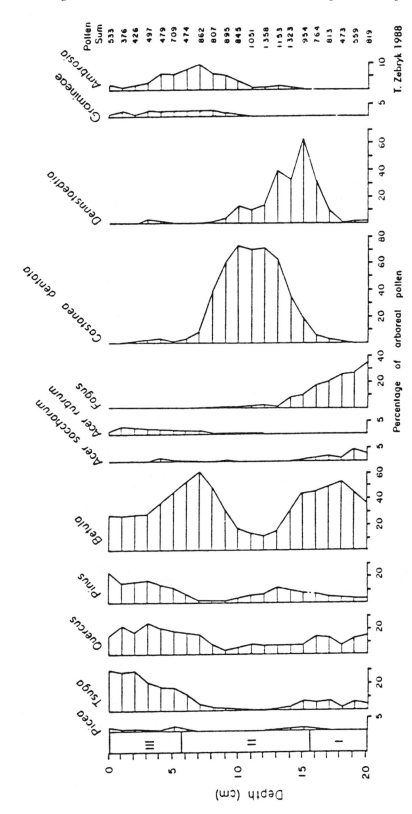

FIGURE 4.4. Pollen percentage diagram from forest-floor humus at Harvard Forest in central Massachusetts. Note (1) abundance of *Acer saccharum* and *Fagus* at bottom of profile (presettlement), (2) declines in percentages of all tree taxa except *Castanea* in Zone II (local and regional harvesting of trees by Euro-American settlers), (3) transient maximum of *Castanea* during woodlot period, (4) maxima of *Ambrosia* and Gramineae (= Poaceae) owing to regional land clearance, and (5) development of local forest dominated by *Pinus* and *Tsuga* and lacking *Acer saccharum* and *Fagus* in Zone III (reforestation during the past century). (From Foster et al. [1992]. Used by permission of Blackwell Scientific Publishing)

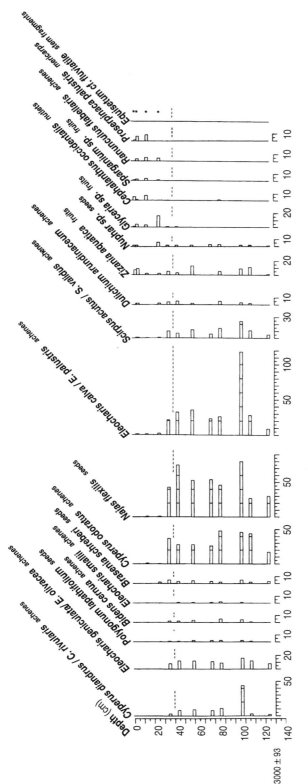

FIGURE 4.5. Plant-macrofossil concentration (number of specimens/115 cm³ sediment) diagram from Miller Woods Pond 51 in Indiana Dunes National Lakeshore, northwestern Indiana. *Dashed horizontal line* represents regional Euro-American land clearance (mid-nineteenth century) as indicated by increase in *Ambrosia* pollen. Note declines in abundance of most wetland plant species following human disturbance, and invasions of the site by other indigenous wetland species. Date at left side of diagram is an uncalibrated ^{14}C date (year BP) with standard error. (From Jackson et al. [1988]. Used by permission of The Ecological Society of America)

of drainage by cut-and-fill for railroad grades but also because rapid accumulation of organic sediment following the disturbance has substantially reduced basin depth. Restoration of the original vegetation would require intensive and continuous intervention. Because the species dominating the marsh vegetation today are all elements of the indigenous regional flora, it is not obvious that vegetation of the marsh is not in its original, presettlement condition. Paleoecological documentation of the actual presettlement flora of the marsh was required to recognize the extent of floristic and vegetational change (Jackson et al. 1988).

Apparent relict or remnant stands, recognized by high species diversity and occurrence of indicator species, are often used as models for restoration of natural vegetation in nature preserves in Europe and North America. Paleoecological studies of small hollows or humus profiles can test the assumption that such stands have not been heavily disturbed by human activities. Segerström et al. (1994) showed that an apparently undisturbed swamp forest in northern Sweden was in fact heavily disturbed before the eighteenth century by slash-and-burn agriculture (Fig. 4.6). The "old-growth indicator" species must have colonized the site following reforestation or, alternatively, must have survived locally in spatially shifting undisturbed patches. In a similar study, Edwards (1986) demonstrated past human disturbance in four "relict" woodlands in northern Wales.

A study of pollen in a small hollow by Björkman and Bradshaw (n.d.) confirmed the old-growth nature of an apparent relict beech/

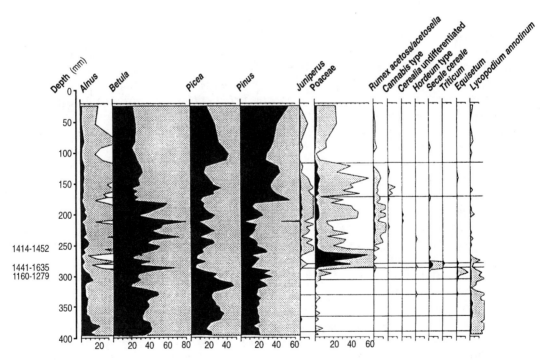

FIGURE 4.6. Pollen percentage diagram showing selected pollen types from forest peat within the Långrumpskogen nature reserve, northeastern Sweden. *Gray shading* represents 10× exaggeration of pollen percentages. Note high percentages or occurrences of *Cannabis*, Poaceae (= Gramineae), *Rumex*, and cereals (*Hordeum* [barley], *Secale cereale* [rye], *Triticum* [wheat]) between 125- and 275-cm depth, indicating local land clearance and cultivation between 1500 and 1700 A.D. Dates at left side of diagram are ^{14}C dates calibrated in calendar years A.D.). (Modified from Segerström et al. [1994] with kind permission from Elsevier Science Ltd, The Boulevard, Langford Lane, Kidlington 0X5 1GB, U.K.)

spruce stand in southern Sweden. However, the study also showed that stand composition has changed substantially within the past few centuries as a result of local fires and regional climate change. Beech and spruce invaded the stand 950 and 175 years ago, respectively, and attained dominance within the past two centuries at the expense of several hardwood species (Björkman and Bradshaw n.d.).

Cowles (1901, p. 81) observed that vegetational change could be described as "a variable approaching a variable rather than a constant." He was speaking of climate and physiography as the variable(s) being pursued by the vegetation. His perspective is confirmed by paleoecological studies indicating continual climate change and vegetational response on landscapes, regardless of the occurrence or nature of human land use. To climate change we must add human land use because it may also permanently alter the trajectory of the natural vegetation. Paleoecology can indicate the nature and trends of vegetation before human land use and the rates and magnitudes of land-use effects. Such information will aid vegetation managers in determining what is possible and desirable in natural and disturbed landscapes.

Conclusions

1. Paleoecology can serve as a powerful tool in assessing natural and human-caused plant invasions. Application of paleoecological data is constrained by limits to taxonomic, temporal, and spatial resolution of the records. Despite these limitations, paleoecological studies provide unique information and perspectives not available by other means.

2. Paleoecological studies in North America provide scanty evidence for extensive land clearance and plant invasion associated with presettlement (i.e., native American) cultures. In some cases, this is a function of a scale differential between human impact and the fossil record (i.e., application of a coarse-grained sensor to fine-grained phenomena).

Also, paleoecological site coverage is sparse in the regions of greatest development of sedentary, cultivation-based populations. The few available sites in Oceania indicate extensive land clearance (New Zealand, Easter Island) and plant invasions (Easter Island) related to Polynesian occupation and land use.

3. Intensive and extensive land clearance by European colonizers of the New World and Oceania are well recorded by paleoecological data. Population expansion of IN annuals (Amaranthaceae/Chenopodiaceae, Asteraceae, Poaceae, ragweed) and invasion of Eurasian annuals (*Erodium cicutarium*, *Plantago lanceolata*, *P. major*, *Rumex acetosella*, *Salsola*) during the past three centuries are clearly evident in pollen sequences from North America. Paleoecological records from Oceania also record postsettlement plant invasions. Much unexploited potential remains for documenting patterns and rates of spread of selected post-settlement invaders.

4. Pollen and macrofossil data indicate that some presumed nonindigenous species (*Corispermum*, *Mollugo verticillata*, *Portulaca oleracea*) occurred in North America thousands of years before European colonization.

5. Paleoecological studies spanning the 20,000 years that have elapsed since the last glacial maximum indicate continual invasion and extirpation of species owing to secular changes in climate. At sufficiently broad temporal scales, most constituents of a local flora can be viewed as invaders. Composition of vegetation from the arctic to the tropics undergoes continual flux at time scales greater than about 2000 years owing to environmental change.

6. Long-term environmental change presents a moving target for vegetation and for managers seeking to restore or maintain "natural" vegetation. Since European colonization of the New World, the environment has changed in many regions because of natural trends and variations in climate and because of long-term effects of human land-use practices on landscapes and ecosystems. Restoration of vegetation to a presettlement

ideal will frequently require intensive and continuing intervention.

7. Paleoecological studies can aid vegetation managers by providing direct records of vegetation and land-use history within a site or region, by documenting invasions of IN and NI species, and by indicating rates and patterns of long-term changes in vegetation and environment.

Acknowledgments. Preparation of this paper was facilitated by grants from the National Science Foundation and the National Biological Service. I thank Leif Björkman, Jim Luken, Jock McAndrews, Susie Smith, and Tom Webb for comments and discussion. Jock McAndrews provided an updated pollen diagram for Crawford Lake.

5
Community Response to Plant Invasion

Kerry D. Woods

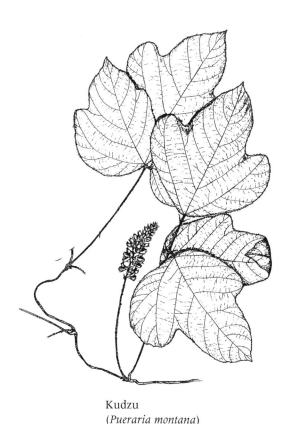

Kudzu
(*Pueraria montana*)

The potential effects of plant invasion on community structure are easily listed. Altered competitive interactions may produce changes in species composition, either by loss or reduction of indigenous (IN) species or, conceivably, by allowing entry of other species previously absent. Diversity—either species richness or dominance patterns—may be modified. Physical structure of the community may be changed. Phenology may be shifted. These changes may, in turn, lead to novel disturbance regimens and new successional paths.

However, it is much more difficult to link unambiguously such changes to invasion. Invasion appears in many instances to be associated with novel, usually anthropogenic disturbance regimens (Huenneke et al. 1990). Thus, it is often difficult to assign causes to subsequent changes in community properties. Are they ecological consequences of invasion, or do they simply follow introduction of the disturbance? Distinguishing between these possibilities ultimately requires careful analysis of experimental or natural controls. Even when no novel disturbance is evident in association with an invasion, it is difficult to be certain of the absence of such. Unfortunately, there are relatively few studies addressing community-level consequences of invasion. Most studies of invasion deal primarily with demographic and biogeographical properties of the invading species itself, and many of these do not resolve the potential confounding of effects of invasion with effects of disturbance.

Resolving these issues is of obvious importance in dealing with nonindigenous (NI) species in community-level management or conservation. Efforts at direct management of invading species by herbicides or mechanical removal may be of little avail in maintaining native communities if invasion and community change are parallel results of disturbance or subtle anthropic influences (see Luken, Chapter 11, this volume). In such instances, addressing management efforts directly to the disturbance may be the most practical means of reaching management goals.

The possibility that community properties can be nonequilibrial or historically contingent (i.e., that there are alternative stable states for communities, or that successional pathways depend on chance or historical circumstance) (Niering and Egler 1955; Niering et al. 1986) adds another dimension to the problem of understanding and managing the effects of invasion on plant communities. For example, classical community theory would suggest that, if a novel disturbance has facilitated invasion and subsequent community alterations, removal of the disturbance should initiate a successional return to the "natural" equilibrium state. However, newer models of community dynamics suggest that, even if an invasion is permitted by anthropogenic disturbance, removal of that disturbance may not lead to a return of the community to original conditions. Even if late successional communities converge toward preinvasion conditions, invading species may continue to influence early successional response and so be of concern to managers and conservationists. For example, although New England old fields eventually give way to indigenous forest, some invasive species, indigenous or not, may continue to play a role in response to natural disturbances in these forests, altering community properties and adversely affecting IN species (Niering and Egler 1955; Niering et al. 1986). Very little is known of the successional impact of invading species or of the potential for community recovery following release from the factors initially permitting invasion, but management concern should not be limited to effects on late-successional communities.

Well-targeted community management, then, requires not only determining the causal role of invading species in cases of community change, including effects on patterns of disturbance and succession, but also understanding of likely successional response of the community to management efforts. In this chapter I review documented instances of community effects due to invasion in light of these issues. I will not dwell on nonindigenous communities that obviously result from intense anthropogenic disturbance (e.g., old-field communities), but will consider largely "native" communities altered as a result of invasion but without wholesale removal of IN species.

Management needs, as well as ecological understanding, call for appropriate predictive generalizations. Toward this end I will attempt to address four general questions:

1. Are there relationships between the characteristics of invasions or invasive species and the likelihood of community effects?
2. Is the nature of community alteration predictable from characteristics of invasive species?
3. Are communities likely to be severely affected by invaders recognizable by some community attribute?
4. Do invasions facilitated by anthropogenic disturbance have more or less likelihood of producing community change than those occurring in the absence of such disturbance?

Potential Mechanisms of Community Effects

Introduction of new competitive interactions and the alteration of niche structures constitute the most direct avenue of influence when invasion occurs, but these may be particularly difficult to verify. Addition of a species to a community may not lead to any significant change in community structure or composition (aside from the addition of the species to the flora). The fossil and pollen records appear to document cases of species entrance into a community without obvious change in the abundances of other species (e.g., Vermeij 1989 [regarding animals]; Webb 1988; Woods and Davis 1989). Modern instances of plants entering communities without apparent consequence may also be found where, for example, NI species remain sparsely distributed (e.g., the helleborine orchid [*Epipactis helleborine*] in forests of eastern North America). Noninindigenous species may even have positive effects on some IN species, for example, where NI species of distinct phenology help maintain higher levels of pollinator activity (Parrish and Bazzaz 1978). However, invasive species established in dense populations must generally have negative effects on some IN species and thus have consequences for community composition through competitive interactions. This will occur except in the unlikely event that such a population could be limited by particular resources not otherwise fully exploited (i.e., that there are "unfilled niches").

Competitive effects might involve interactions between species of similar life history, in which case the consequences of establishment of an NI species may be restricted to reduction or displacement of one or a few directly competing species, that is, the invading species plays a community role similar to that of the displaced IN species. In other cases invading species may have broad competitive effects on species of different guilds; for example, an invasive tree or shrub may cause population changes in most or all species of lower stature or in seedlings of other trees and shrubs. In this latter case, competitive effects might lead, beyond simple floristic replacement, to alteration of community structure and decreased overall diversity. This might be especially likely where new competitive interactions alter successional dynamics. In either case, definitely establishing that observed changes have been induced by competition is difficult, requiring either appropriate experimental work or extremely careful interpretation of field data.

Intergrading with, but sometimes distinct from direct competitive effects, invading species may alter natural disturbance regimens. The most dramatic of such changes might be if different species characteristics increase or decrease the vulnerability of a community to fire, but it is also plausible that frequency and intensity of wind throw, erosion, or herbivory impacts could be influenced by invading species. In any of these cases, community changes could be dramatic and of any of the types previously defined.

A final class of community effects might follow from alterations in mineral or hydrological cycles due to particular properties of invading species. Again, there is intergradation with competitive effects. However, effects on soil chemistry or water budgets may mediate community change through induced habitat change without actual resource competition between species if, for example, establishment of an invader resulted in acidification of soils, elevation of water tables, or deposition of toxic materials.

Possible Instances of Effects on Community Composition and Structure

This review of community changes that may be induced by invasion is organized around the mechanisms postulated in the previous section. Arguments for effects of invading plants on community properties grow from several kinds of data: field observation of existing patterns without manipulation, extended observation of changes occurring concurrently with the process of invasion (including paleoecological "observation"), and experimental manipulations specifically testing hypotheses about effects of invaders. These approaches are, of course, progressively more powerful in unambiguously establishing the relationships between community change and invasion; however, they are also progressively more difficult to execute and, not surprisingly, progessively more rare in published literature.

Competitive Effects

Competition for light is most frequently suggested as a predominant mechanism for community effects of invaders, perhaps because the precondition for such competition—shading by invasive plants—is particularly easy to observe. However, few studies provide well-controlled experimental support for light competition as the causal factor for changes in indigenous vegetation. Although competition for other limiting resources should have the same potential for community alteration, there are only a few instances where such competition is suggested or supported.

Replacement of individual IN species by ecologically analagous invaders may not be seen as a "community-level" effect of invasion. Cases of such restricted impact appear to be few, but some possible examples are documented, including replacement of golden wattle (*Acacia longifolia*) by *Chrysanthemoides monilifera* in Australia (Weiss and Noble 1984). More significant, from a community perspective, is the replacement of several members of an ecological guild by a single invader of similar life form, and reduction of community diversity, as in the spread of Monterey pine (*Pinus radiata*) into mixed eucalypt (*Eucalyptus*) forests in Australia (Burdon and Chilvers 1977; Chilvers and Burdon 1983). Invasive tree species may alter late-successional forest composition by displacing IN species from the pool of suppressed saplings or potential canopy successors, in effect competing for the occupation of canopy gaps. Norway maple (*Acer platanoides*) dominates the understory of hardwood forests in northern New Jersey at the expense of seedling abundance of beech (*Fagus grandifolia*) and sugar maple (*Acer saccharum*) (Webb and Kaunzinger 1993).

Of even more concern, in terms of community properties, is the possibility that invading species can produce dramatic and potentially permanent alterations in community structure, function, and composition. In many apparent cases of such changes, the invader is of a life form, habit, or phenology not previously abundant in the indigenous community. Maritime pine (*Pinus pinaster*) invasion can convert diverse fynbos shrublands of South Africa to essentially monotypic forests (Kruger 1977). (The structurally similar chaparral of southern California, on the other hand, may be relatively resistant to such invasion and conversion [Burns and Sauer 1992].) *Rhododendron ponticum*, in establishing an "almost impenetrable" shrub layer in oak-holly forests of southern Ireland, appears to produce dramatic decreases in diversity and cover in the ground layer with all vascular species decreasing (Kelly 1981). Through suppression of seedlings of durmast oak (*Quercus petraea*) and English holly (*Ilex aquifolium*), *Rhododendron* may be, in effect, replacing *Ilex* and inhibiting canopy regeneration (Cross 1982; Usher 1986). A parallel situation has been suggested for shrubby honeysuckles (*Lonicera tatarica* and related taxa), which establish an unusually dense shrub stratum in some forests of northeastern United States and appear to reduce diversity of herbs and woody seedlings, potentially leading to partial failure of canopy replacement (Woods 1993). These invading shrubs establish a nearly closed stratum (previously lacking or sparse in these forests) that apparently produces decreased diversity and cover of the ground layer and may lead to opening of the canopy. In both of these cases, claims of invasion impact were made on the basis of "chronosequence" studies—comparisons of areas of high invader cover with areas lacking the invader—with the assumption that such areas were similar prior to establishment of the invader.

Mimosa pigra may have a similar effect in *Melaleuca* forests in tropical Australia, converting them to shrubland as the canopy trees die without replacement (Braithwaite et al. 1989). *Mimosa pigra* also invades seasonally flooded herbaceous vegetation, converting it to shrubland with reduced diversity. This invasion has probably been facilitated by disturbance caused by NI water buffalo (*Bubalus bubalis*), but the change in community structure is likely to be permanent even with reduction or removal of the animal.

Invasive lianas have also been implicated in competitive effects on indigenous flora with resulting changes in community structure. English ivy (*Hedera helix*), Japanese honeysuckle (*Lonicera japonica*), and kudzu (*Pueraria montana*) are all perceived as aggressive and noxious invaders in the southeastern United States. Thomas (1980) described invasion of honeysuckle and ivy in forests of Theodore Roosevelt Island in the Potomac River, District of Columbia, and argued that both species kill host trees by shading, potentially producing a more shrubby and less arborescent vegetation. The ivy, an evergreen, may also inhibit recruitment of herbaceous species. These hypotheses and predictions, however, remain untested by experiment or follow-up study (as is the case in most studies cited). *Chromolaena odorata* apparently behaves similarly in South Africa (MacDonald and Frame 1988).

Even invading herbs may alter community structure dramatically. A tradescantia (*Tradescantia fluminensis*) inhibits tree regeneration in New Zealand forests and may lead to an opening of forest canopies (Atkinson and Cameron 1993).

Invasion of nonarborescent vegetation by trees can also have far-reaching community effects. A variety of species is known to invade shrubby vegetation in New Zealand (Atkinson and Cameron 1993) and South Africa (Kruger 1977; MacDonald and Richardson 1986), with far-reaching effects on community structure.

Aquatic communities appear to be subject to similar effects. Invasion by Eurasian water milfoil (*Myriophyllum spicatum*) has been associated with reduction of cover by indigenous plants and decline in the number of IN species (including other species of *Myriophyllum* with similar habit) (Coffey and McNabb 1974; Madsen et al. 1991b; Smith and Barko 1990). Communities dominated by *M. spicatum* may be of different physiognomy as compared to communities of IN macrophytes. Madsen et al. (1991b) did not actually observe the community prior to invasion. Their study, however, documented both the invasion of an apparently healthy indigenous community and the changes associated with that invasion. They suggested that community conversion was a consequence of shading. Various species of invasive floating plants (e.g., salvinia [*Salvinia molesta*], water-hyacinth [*Eichhornia crassipes*], and water-lettuce [*Pistia stratiotes*] in South Africa and India) can form "canopies" on previously open bodies of water, leading to large changes in subjacent indigenous macrophyte communities (MacDonald and Frame 1988; Thomas 1981).

The peculiar vulnerability of islands to alteration of communities by invasion has been much discussed for animals but less documented for plants. Competitive effects on islands, as elsewhere, may be most significant in preventing regeneration of IN plants, with consequent changes in diversity and structure. Sanders et al. (1982) suggested that *Aristotelia chilensis* has replaced several species indigenous to the Juan Fernandez Islands by usurpation of regeneration niche. Several species in the Hawaiian Islands appear to spread spontaneously through competitive displacement of IN species, generally reducing diversity and sometimes changing community structure (Cuddihy and Stone 1990; Wester and Wood 1977). Imported invaders include lianas (e.g., banana poka [*Passiflora mollissima*]) and trees (e.g., strawberry guava [*Psidium cattleianum*]), whose effects probably involve light competition, and a variety of herbs and grasses, whose establishment appears to reduce diversity of IN herbs and to prevent establishment of tree seedlings (in some cases converting forests to herbaceous vegetation). The Galápagos Islands have experienced similar effects (Schofield 1989).

New Zealand's flora contains a very large proportion of NI species, some of which have actively invaded and altered natural vegetation (Atkinson and Cameron 1993). Indigenous herbs and grasslands have apparently been displaced by two species of hawkweed (*Hieracium*) and by a bent grass (*Agrostis capillaris*). In the case of *Agrostis*, a 27-year study monitoring invasion has provided an unusually firm cause-and-effect relationship. Other herbs appear to inhibit tree regeneration, altering stand structure in indigenous forests. Some northern hemisphere tree species invading nonarborescent communities simply tolerate more severe cold or drought than the IN species.

Finally, it is important to recognize that there have been few cases where competition from invaders has been shown unambiguously to be responsible for significant alteration of communities. Most of the extensive literature suggesting such effects is based on correlative studies, historical records, or anecdotes. Impressions thus gained can be misleading. Anderson (1995), for example, cast doubt on the frequent assertion that purple loosestrife (*Lythrum salicaria*) displaces IN plants in North American wetlands; his literature review shows no clear evidence of reduction of diversity or total biomass of IN plants.

Altered Disturbance Regimen

Disturbance regimens are a function of interaction between flora and physical environment. Changes in biota can have dramatic feedback on community structure and composition through alteration of disturbance dynamics. Any change in nature or frequency of disturbance may be expected to induce changes in dominance, successional status, landscape pattern, and other community properties.

Invaders have been reported to alter native communities by altering stream flow and flooding regimen as in the case of tamarisk (*Tamarix*) invasion of riverine woodlands in Australia (Griffin et al. 1989) and western North America (Graf 1978; Turner 1974). Effects on geomorphological dynamics of beaches and dunes have also been noted in the invasion of European beach grass (*Ammophila arenaria*) on the west coast of North America, where IN species were replaced and diversity was reduced (Mooney et al. 1986), and of Australian-pine (*Casuarina equisetifolia*) in subtropical coastal regions of North America (Barbour and Johnson 1977). Lianas can change the likelihood of windfall by binding trees together and weighting their canopies (Thomas 1980).

Most dramatic and perhaps most significant, however, are the community effects of fire-encouraging invaders in communities not normally subject to frequent fire. Often this effect is a consequence of phenological differences, such that fuel load is higher during dry seasons, as in the takeover of diverse indigenous dry grasslands by invading grass species in Australia (MacDonald and Frame 1988), North America (Anable et al. 1992; Bock et al. 1986; Kincaid et al. 1959), and tropical America (Parson 1972). However, fire encouraged by invading plants can also play a role in the conversion of forests to savanna or grassland (MacDonald and Frame 1988), or of Mediterranean-climate shrubland to grassland. Rye grass (*Lolium*), used to stabilize soils in southern California chaparral following fire, provides fuel load favoring rapid recurrence of fire and ultimately leading to grassland conversion (Zedler et al. 1983).

Again, this mode of community alteration seems to be particularly significant on islands. Loope et al. (1988) suggested that island floras have generally not been subjected, evolutionarily, to the intensity of disturbance by fire and herbivory experienced by continental floras. If so, introduction of species tolerating and encouraging fire could have particularly dramatic effects.

Community Effects Due to Changes in Nutrient Cycling and Chemical Effects

Community alteration has been ascribed to effects of NI plants on various chemical cycles and hydrology. These effects may involve either allelopathy or changed availability of resources.

A number of invasive species appear to have altered hydrology or soil water conditions to the extent that IN species are displaced and the nature of the community is changed. Yellow iris (*Iris pseudacorus*), in a Potomac River marsh, facilitated conversion of marshland to mesic forest when its rhizome mat provided a raised seed bed favoring ashes (*Fraxinus*) over willows (*Salix*) (Thomas 1980). Conversely, broom-sedge (*Andropogon virginicus*) establishment in Hawaii led to development of boggy areas, potentially causing loss of forest cover and also erosion in rain forests (Mueller-Dombois 1973). Dramatic changes in hydrology and in native communities are attributed to invasion of riverine forests by tamarisk (*Tamarix*) in Australia and North America (Blackburn et al. 1982; Graf 1978; Griffin et al. 1989; Turner 1974; see Walker and Smith, Chapter 6, this volume).

Alteration of other mineral cycles is likely to affect communities but is difficult to document. It is possible, for example, that the loss of cryptogam cover documented with invasion of northwestern North American grasslands by spotted knapweed (*Centaurea maculosa*) (Tyser 1992) mediated the observed reduction in diversity and cover of IN species through changed nutrient cycling. More apparent cases involve establishment of nitrogen-fixing species in nitrogen-limited systems. In Australian grasslands, intentionally introduced nitrogen-fixing clovers (*Trifolium*) may foster establishment of other invaders with consequent decline of IN bunchgrasses through alteration of the nitrogen cycle. Fayatree (*Myrica faya*) in Hawaii, through enhancement of nitrogen availability, appears to alter successional development with potentially long-term effects (Vitousek 1990; Vitousek et al. 1987).

Some invaders appear to have toxic effects. Iceplant (*Mesembryanthemum crystallinum*) in California appears to dominate coastal grasslands to the exclusion of IN annuals by actively taking up salt from soils and depositing it on the surface in plant residues, producing a near-monotypic vegetation (Vivrette and Muller 1977). A similar replacement of annual pasture communities has been documented in Australia (Kloot 1983), although the community being replaced there is itself nonindigenous.

The Confounding Effects of Anthropogenic Disturbance in Invasion and Community Change

There are many instances where the association between dramatic change in indigenous plant communities and extensive invasion by NI species is difficult to interpret due to simultaneous imposition of novel anthropogenic disturbance. In such cases, the cause of community change is difficult to assign without careful monitoring of the processes of change or experimental manipulation. The problem can be aggravated when the changes occur over large geographical scales and are of long historical standing (note Darwin's [1909] observations of replacement of Argentinian grasslands by cardoon [*Cynara cardunculus*]). For example, the effect of invasion of savanna woodland in Australia by lantana (*Lantana camara*) is associated with community changes, but its invasion is facilitated by pig rooting, and it is unclear whether the indigenous community would recover in the absence of pigs (Fensham et al. 1994).

Perhaps the most frequently cited cases where invasion is linked to anthropogenic disturbance involve conversion of diverse grassland communities, in the presence of heavy grazing by domestic livestock, to radically different taxonomic composition and phenology or even to wholly different physiognomy. The grasslands of semiarid North America, for example, have been severely altered by a number of NI grasses, particularly cheat grass (*Bromus tectorum*) (see, for example, Mack 1981, 1986, 1989), but in many instances these changes appear to have occurred in association with heavy grazing by cattle or sheep (the situation is complicated by the fact that these conversions often occurred very quickly and without much careful documentation). In such cases it may be impossible to establish the importance of grazing in mediating invasion and community change. Grazing exclusion experiments can explore whether natural communities can be reestablished without imposed disturbance, but failure of reestablishment in such cases does not preclude a critical role for grazing in initiating and maintaining the change; indigenous communities may require a "native landscape" and associated fire regimen, herbivory patterns, and seed rain; restoration of these properties is not generally feasible.

Invasion and alteration of native grasslands without the influence of grazers has been documented in the case of love grass (*Eragrostis lehmanniana*) in Arizona (Anable et al. 1992; Bock et al. 1986; Kincaid et al. 1959; McClaran and Anable 1992) and of timothy (*Phleum pratense*) and spotted knapweed (*Centaurea maculosa*) in Montana (Tyser 1992; Tyser and Key 1988; Tyser and Worley 1992). In both cases the changes initiated appear to be self-maintaining. *Centaurea* and *Phleum* reduce both vascular plant diversity and cover of the cryptogam crust typical of indigenous bunch grass prairies of the western United States. *Eragrostis* invasion reduces diversity by displacing IN species and leads to as much as a quadrupling of aboveground biomass (and a potential positive feedback with increased fire frequency). However, it remains unclear whether these local invasions were enabled by the massive regional presence of these species that might, in turn, have been a consequence of grazing.

Grazing has been associated in a number of other cases of aggressive invasion with substantial community change. Again, islands without large IN grazers may be particularly vulnerable (Atkinson and Cameron 1993; Cuddihy and Stone 1990; Schofield 1989). Wagner (1989) suggested a similar phenomenon in the replacement of indigenous grassland in California's Central Valley by a low-diversity nonindigenous community, arguing that California lacked large grazers during most of the Holocene. *Eupatorium adenophorum* is a widespread invader throughout the tropics, but Mahat et al. (1987) noted that in Nepal the main infestations are associated with heavy grazing, logging, or fire. Despite its local name of "forest killer" and the suggestion by Mishra and Ramakrishnan (1983) that it alters the course of succession following agriculture, it is unclear whether *Eupatorium* invades successfully without such disturbance or whether it will be replaced if disturbance ceases. Invasion of western Australian woodland and tall shrub communities is associated with lowered diversity in herbaceous strata and conversion to an annual herbaceous flora (Bridgewater and Backshall 1981), but it is not known whether reduced diversity is a consequence of invasion or if both are a result of altered fire regimen.

The apparent negative impact of plant invaders in human-disturbed communities can, in fact, be deceptive. Scotch broom (*Cytisus scoparius*) and gorse (*Ulex europaeus*) both establish dense stands on postagricultural land in New Zealand, but both can act as "nurse plants" for indigenous trees that replace them; neither species invades or persists in closed forests (Atkinson and Cameron 1993). The "altered communities," in this case, appear to be ephemeral with removal of the initiating disturbance (although it is unclear whether the NI shrubs will persist, through seed banks, in novel successional communities). However, managers should not dismiss the potential for long-lasting consequences of the presence of NI species in early succession and should be particularly cautious about intentional use of such species in revegetation. Wilson (1989) noted the apparent suppression of IN prairie species, over at least 8 years, in sites where mixes of NI species were seeded to stabilize soil. Although Egler (1942) suggested that NI species colonizing disturbed areas in Hawaii will eventually be replaced by IN species, subsequent studies are not so sanguine (Cuddihy and Stone 1990).

The preceding studies indicate a need for clear distinction between correlation and causation. Any consideration of community impacts of invasion should address the role of disturbance in initiating and maintaining invaders or in causing the putative "invasion effects." It is also important to understand whether removal of anthropogenic disturbances will allow reversion of communities to prior states; changed species pools may permanently alter successional pathways (Hobbs and Huenneke 1992). It is probably impossible, however, to isolate completely the effects of disturbance and invasion because they are too intricately intermingled and because the anthropogenic "disturbances" may be practically irreversible.

Invasion by Indigenous Species and the Paleoecological Record

Natural instances of invasion occur continuously with changes in species ranges. Although range changes have been extensively documented by modern observers and especially through paleoecological methods, there are few careful considerations of community effects. The mechanisms by which such invaders might effect community changes are the same as those discussed above, but it is not clear whether the effects are comparable in frequency or intensity. Most such "natural" invasions bring together species that have had previous contact. Perhaps of greater importance is that plants undergoing natural range expansion are less likely to experience the loss of predators or pathogens that may permit NI invaders to expand explosively and dominate communities. If this is the case, fewer community changes might be anticipated.

Hobbs and Mooney (1986) offered one of the few direct observational records of such an invasion. Coyotebrush (*Baccharis pilularis*) periodically invades grasslands in northern California. In a chronosequence of plots of different invasion history, Hobbs and Mooney noted a dramatic change in many community parameters with canopy closure by the invading species. Repeated observations support the causal connection. The community effects appear to be due both to direct competition, probably for light, and fostering of herbivore populations by coyotebrush.

Human disturbance may facilitate similar phenomena. Relatively stable shrubby or herbaceous communities can apparently become established in normally forested habitats following agricultural abandonment (Niering and Egler 1955; Niering et al. 1986) or logging (Horsley 1977a, 1977b). This phenomenon, which may be a result of either chemical or physical interference with tree regeneration, apparently occurs where a novel form of disturbance permits unusual dominance by species or life forms normally of limited occurrence during succession. Potential for exploitation of such phenomena in vegetation management has already been recognized (e.g., Bramble and Byrnes 1976; Hill et al. 1995; Luken 1990; Niering and Goodwin 1974; Putz and Canham 1992).

Sedimentary pollen provides a record of invasion by a variety of species, but paleoecologists have seldom used the pollen record to evaluate community response to these invasions. Such interpretation is difficult because of the limits on precision with which community composition can be reconstructed, the taxonomic incompleteness of the pollen record, and the low temporal resolution of most sedimentary records (see Jackson, Chapter 4, this volume). Furthermore, many paleoecologists regard community composition as determined solely by climatic response of individual species, disregarding the possible roles of species interactions.

Various conclusions have been reached by paleoecologists attempting to find a community response to invasion. Woods and Davis (1989) noted that invasion by beech in the Upper Great Lakes region of North America was associated with no apparent further change in regional canopy composition, even though this species is generally understood to be a late-successional dominant. However, in a fine-scale study of eastern hemlock (*Tsuga canadensis*) in the same region, Davis et al. (1991b) concluded that invasion was followed by significant change in ecosystem and community properties. Invasion by hemlock was coincident with increased frequencies of pollen of northern hardwood species (e.g., sugar maple [*Acer saccharum*], yellow birch [*Betula alleghaniensis*], and basswood [*Tilia americana*]) at the expense of pines (*Pinus*) and oaks (*Quercus*). Their interpretation is that hemlock changed the flammability of the forest, altering fire frequencies, and may also have changed nutrient cycling rates, microclimate for seedling establishment, deer browse patterns, and landscape heterogeneity. Although simultaneous climate change likely favored these changes as well, the invasion probably accelerated them.

Generalizations

No strong or detailed generalizations concerning the questions posed at the outset emerge from an examination of available studies of effects of invasion on community properties. This may be simply a consequence of the scarcity of studies focused on these issues and of the near total absence of carefully controlled or experimental analysis. I suspect, however, that even if such studies were available, a great deal of variation in effects would remain; the individuality and complexity of communities and species interactions may mitigate against precise prediction. However, some patterns may be suggested by available information; potentially valuable avenues for further research do exist.

Which Invaders Will Have the Greatest Consequences?

Vitousek's (1990) generalization—that invaders are likely to alter ecosystem properties when they differ substantially from IN species in resource acquisition or utilization, alter trophic structure, or alter disturbance frequency or intensity—can be seen as covering most of the instances discussed here. Unfortunately, it is probably impossible to say in most cases, before the fact, whether a particular species, as an invader, will have major effects on a particular community (this is quite aside from the difficulty of predicting whether any given species will, in fact, successfully invade a particular community). Consistent with Vitousek, available studies suggest that, in a wide array of situations, community alterations are particularly marked when an invading species is of a previously absent or scarce growth form. In most cases, community changes appear to be the consequence of intensified light competition leading to diminution and reduction of diversity in community components shaded by the new species. Thus, invading shrubs in forests where shrub cover is normally modest leads to diminished cover and diversity of herbs and tree seedlings. Shrubs or trees invading nonwoody vegetation have similar consequence, as do floating aquatic plants or lianas.

In a number of cases, invaders are of life forms already present in the community, but they achieve cover or abundance never reached by IN species. There are IN species of honeysuckle (*Lonicera*) in the forests invaded by *L. japonica* and *L. tatarica*. Eurasian water milfoil invades lakes that already harbor several congeners. The aggressive growth of the invaders, presumably due to release from population controls present in their native habitat, may have detrimental effects on their ecological analogues in the new habitat. Community effects are more far-reaching as a consequence of increased abundance of a particular growth form or habit, which may cause changes in overall community structure and have long-term consequences for community development.

Similarly, abundant invaders with novel phenological patterns may be particularly likely to change community patterns by altering both community structure and dynamic community properties (Huenneke and Mooney 1989). Phenological differences in productivity can, for example, influence fire patterns (frequently seen in changes in grassland communities) and other disturbance phenomena. Unique phenological characteristics in opportunistic species might lead to altered successional response; Young and Evans (1989) suggested, for example, that phenological properties of invasive grasses in California grasslands lead to a series of changes in successional processes and to altered end results.

Assessment of the hypothesis that alteration of community physiognomy or phenology is an important determinant of whether an invader will lead to significant community alteration will be difficult. It is possible that instances of significant establishment of invasives without such changes could be found and compared. The more critical management issue, however, concerns prediction of which potential invaders are likely to cause such alteration; this remains problematic.

It is probably even more difficult to predict which invaders are likely to change functional attributes—disturbance regimens, successional dynamics, or mineral cycling—of communities. Changes in fire frequency and successional patterns might, in some cases, be foreseen when invaders are fire adapted and particularly flammable. Invasive grasses or herbaceous plants without herbivores in the invaded community might be expected to produce high fuel loads. Invaders whose patterns of ground-water utilization differ from those of IN species might be anticipated to change local water budgets with ramifications at the community level.

All of these generalizations are consistent with the notion that community properties are individualistic manifestations of unique combinations of species' life histories (Moore and Noble 1990). The introductions most likely to change community properties are those introducing the greatest change to the assemblage of life histories.

What Community Effects Can Be Anticipated?

The same reservations apply here as in the previous section. Two rather broad generalizations emerge from the collection of studies and observations discussed. Cover and diversity of particular community components or strata are often reduced by competition for light when an invader alters community structure. In extreme situations, community structure may be altered, as in cases where dense cover by NI shrubs or herbs impedes tree regeneration in forests.

Changes in disturbance regimen induced by invasive species typically have far-reaching effects, but their nature is less predictable. Diversity of IN species tends to decrease with altered disturbance regimen, but this may be partly compensated by increased diversity of weedy species (Hobbs and Huenneke 1992; Myers and Henry 1979). More obviously, in terms of community effects, such invasions may alter community structure in dramatic ways, with far-reaching consequences on subsequent community dynamics.

In fact, the complex nature of communities may be such as to make strong generalizations impossible. As traditional ideas about community dynamics, typically based on notions of equilibrium and stability, are challenged by "nonequilibrium" models, the apparent potential for such generalization may become more remote. The complexity of interactions within communities leads to many interlocking feedback cycles, and the indirect effects of perturbations can therefore be many and extremely difficult to predict.

What Communities Are Most Likely to Be Affected?

Several suggestions have been made in the past concerning vulnerability of communities to invasion, and it is probably reasonable to expect that communities more vulnerable to invasion are also more likely to be significantly altered by invasion. It has been stated (Huenneke et al. 1990; Ramakrishnan and Vitousek 1989) that communities subject to anthropogenic disturbance are more vulnerable to invasion, and there is little doubt that this is true (although the extreme claim that pristine communities are never invaded is falsified by a number of the studies considered here). As discussed earlier, it is difficult to assign causality to changes observed in strongly human-influenced communities; invasion may either cause or be permitted by some of the community changes observed. Thus it is unclear whether it is more appropriate to say that disturbed communities are more likely to be changed by invasion or that community-level changes induced by human disturbance increase likelihood of invasion. Lodge (1993a) suggested that more diverse communities suffer fewer successful invasions, but it is unclear what this might mean with respect to community-level effects.

It is possible that environmental characteristics are correlated with frequency of invasion. MacDonald and Frame (1988), for example, concluded that drier parts of the tropics have experienced fewer plant invasions. However, such generalizations are currently not well supported and, even if they are strong, it is unclear whether frequency of invasion is a strong predictor of degree of alteration of community properties. Humans may act as dispersal agents, so that the nature of human movement through indigenous communities may influence the frequency of invasion (MacDonald et al. 1989; McDonnell and Pickett 1990), but, again, it is unclear how such patterns are related to changes in community characteristics.

Similarly, it is a strong generalization that island communities, or communities otherwise long isolated from potential invaders, are particularly prone to invasion. It is not so well documented that these communities are particularly likely to experience changes in community properties upon invasion, but available data tend to suggest this. This possibility might be explored by comparison of the magnitude of community alteration in island and mainland communities with similar proportions of NI species (if appropriate examples could be found).

The character of early-successional communities may also be more frequently changed by invaders, as suggested by the generally higher frequency of NI plants among early-successional species. However, this is difficult to assess since many successional communities are anthropogenic in some degree and, again, it is difficult to isolate the effects of human activity from those of invasive species. It is not clear whether any generalization can be made about the long-term effects of changes induced during succession; in some cases, NI species may be replaced during the course of succession.

Does Anthropogenic Disturbance Facilitate Invasion-Induced Community Change?

It is possible that invasive species whose establishment has little or moderate impact on community properties in relatively natural vegetation will have substantially larger impact in the presence of increased rates or novel types of disturbance. These impacts should be seen as a consequence of invasion only if community properties are altered more in the presence of both invader and anthropogenic disturbance than by the disturbance alone. This scenario is highly plausible, especially in the potential for altered successional sequence. However, it remains very difficult to isolate the direct and indirect effects of disturbance in terms of community-level change; few studies have done so.

Conclusions

1. It is clear that both direct and indirect effects of invasive species can be powerful factors in determining floristic, structural, and dynamic community properties.

2. Community effects of invasive species range from simple competitive replacement of one or more species, to loss or reduction of whole guilds or strata within the community with consequent changes in community diversity, to wholesale conversion of community structure and organization.

3. In addition to direct and indirect competitive effects, alteration of disturbance patterns appears a particularly powerful factor in such conversions; changes in mineral cycling or water regimen can also play a significant role.

4. Unfortunately, very few predictive generalizations about these effects can be elaborated or strongly supported given available studies and data. It remains unclear whether there is any effective way of predicting which invaders are likely to have significant impacts on the host community or what the nature of such impacts would be in a particular community.

5. The paucity of generalizations regarding community effects of invaders presents a substantial dilemma for community managers. Nonindigenous species are most easily controlled or eliminated by concentration of efforts in the initial phases of establishment, but it may be impossible to tell a priori which invaders pose serious threats to native communities.

6. Prior experience with a given species is most informative of its potential effects in

new areas, so it is important that community effects of invasion be better documented and that managers be well acquainted with that documentation. Without prior experience, management efforts can be guided only by ecological judgments that must remain tentative.

7. Evaluation of potential effects of an unfamiliar invader should take into account as many aspects of its life history as possible. Introduction of novel structural or life-history traits to the community may be suggestive of particular threat.

8. It is also important to distinguish between community-level effects of invasion and interacting effects of anthropogenic disturbance. Community change and invasion by NI species may both be direct results of disturbance but not themselves causally related. Even when alteration of communities is a consequence of establishment of NI species, the invasion may be a consequence of anthropogenic disturbance. In such instances, it is important to know whether modification of the disturbance regimen can lead to restoration of the indigenous community. Management of invasions may eventually focus on management of anthropogenic disturbance.

9. Range expansions of regional IN species constitute invasions as well, but these events may have less dramatic consequences for invaded communities if invaders continue to be limited by natural enemies and so fail to assert dominance. Nonetheless, some such invasions are likely to induce significant community change.

10. Unless and until better understanding is gained, the foregoing guidelines and suggestions should be used only in the setting of priorities for control, not for designation of some invaders as of no concern. Establishment of any NI species should be seen as carrying the threat of extensive community disruption.

6
Impacts of Invasive Plants on Community and Ecosystem Properties

Lawrence R. Walker and Stanley D. Smith

Salt-cedar
(*Tamarix* sp.)

Invasive indigenous (IN) or nonindigenous (NI) plant species can alter various properties of plant communities, including species diversity, primary productivity, interactions between species, stability, and rates or pathways of successional recovery of a community following disturbance (Ramakrishnan and Vitousek 1989; Versfeld and van Wilgen 1986; Vitousek 1990; Vitousek and Walker 1989; Walker and Vitousek 1991). They can also alter the actual disturbance regime (Breytenbach 1986; Hughes et al. 1991; Vitousek 1990). Studies of ecosystem processes are usually not focused on a particular organism but on the linkages between organisms and their environment. Examination of the flow of energy, water, or nutrients through an ecosystem can provide a sensitive measure of the degree to which invasive plants alter indigenous ecosystems. Successful management of communities and ecosystems containing invasive species involves (1) assessing whether the invaders have significantly altered the ecosystem from its preinvasion condition, (2) recognizing and measuring specific community and ecosystem properties potentially being altered by the invader, and (3) developing strategies that return communities and the associated ecosystem processes to the preinvasion state (if such is deemed desirable by management goals).

The susceptibility of a community to invasion is thought to be related to successional status, species density or diversity, and the extent of disturbed ground within a community (Crawley 1987; Rejmánek 1989). Plants can invade ecosystems that are only occasionally disrupted by natural disturbance, but more commonly they invade ecosystems subjected to various degrees of natural or anthropogenic disturbance. A typical early-successional invader is able to monopolize recently disturbed ground due to its rapid dispersal to the site, followed by rapid growth and reproduction in the absence of significant competition for space and nutrients. However, separating the impact of the invader from the impact of the disturbance can be difficult (Vitousek 1986), ideally requiring long-term removals of the invasive species.

For a successful invader to alter community or ecosystem properties it must have an impact on energy, nutrient, or water flow, on the disturbance regime, or on the community response to the disturbance regime (Fig. 6.1). This "impact" often occurs out of proportion to the invader's relative dominance in the community (Vitousek 1986). Primary productivity (and associated aspects of flow of energy through vegetation, including vegetation structure, composition, growth, and species diversity) can be altered directly by an invader or indirectly through effects of an invader on disturbance frequency, nutrient retention and turnover, or site water balance. Soil or plant nutrient levels and cycling of nutrients and water can be modified by invasive species. The disturbance regime (frequency, magnitude, severity, type) can be altered by invasive species that increase fire frequency or decrease erosion, for example. Invasive plants can also alter species interactions—for example, competitive dominance, facilitation, mutualism, and herbivore resistance—and community responses to disturbance—for example, stability or successional pathways. The likelihood and nature of an invasion are in turn affected by these impacts.

In this chapter we examine the impacts of terrestrial invasive plants on many of the community and ecosystem properties listed above, illustrating the impacts with examples from several geographical areas and biomes. For each property we discuss how the impacts can be measured and how to evaluate various management options. Although some properties of communities are easily measured (e.g., species diversity), others are measured with some difficulty (e.g., species interactions, suc-

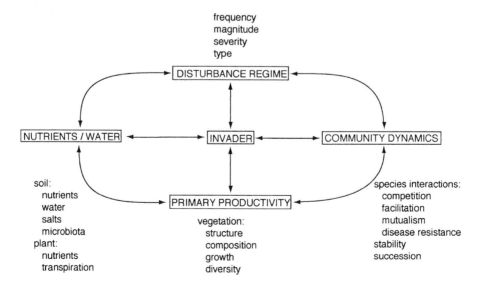

FIGURE 6.1. Effects of an invasive plant species on community and ecosystem processes. Arrows indicate direction of influence. Note that the invader is affected by the processes that it modifies.

6. Impacts of Invasive Plants on Community and Ecosystem Properties

cessional change) or with great difficulty (e.g., stability). Some of the difficulty is in defining each property; indeed the very existence of communities with emergent properties is still being debated (e.g., Wilson 1994). Similarly, ecosystem properties differ in ease of measurement (e.g., in order of increasing difficulty: plant and soil water and nutrient status, plant gas exchange, nutrient cycling, ecosystem gas and energy fluxes) and in the confidence with which one extrapolates from detailed measurements to generalizations about the ecosystem. Managers will want to rely on properties that are easy to define, relatively straightforward to measure, and relevant with regard to issues of scale (O'Neill et al. 1986).

Primary Productivity

Concepts

Primary productivity is a useful measure of the impact of invasive plants on communities and ecosystems because it can be directly measured with relatively objective criteria and because it integrates several community and ecosystem properties. Primary productivity (we refer to net primary productivity defined as gross primary productivity minus respiration) can also be compared across communities and ecosystems, allowing one to measure the relative importance of an invader on more than one site.

Effects of plant invasions on primary productivity can be positive, negative, or neutral (Table 6.1). Positive effects (i.e., increases in overall community productivity) can occur if the successfully established invader fills a new niche (ecological role) in the community, adding its productivity to the preinvasion total. To have such an increase, any detrimental effects of the invader on IN species would have to be offset by the greater production of the invader. Examples of previously unoccupied roles could include (1) new life forms, for example, trees invading a grassland, as mesquite (*Prosopis*) in Texas (Archer et al. 1988); (2) new phenological patterns, for example,

TABLE 6.1. "Positive" (+) and "negative" (−) effects or feedbacks of an invader on community and ecosystem attributes.

Primary productivity
 + Exploits vacant structural, functional, or successional niche.
 + Exploits underutilized resource in a new way.
 + Outcompetes indigenous species.
 − Sequesters nutrients.
 − Grows more slowly than displaced indigenous species.
 − Promotes disturbance.
Nutrient dynamics
 + Fixes nitrogen.
 + Adds nutrient-rich litter.
 + Increases weathering rates.
 − Increases nutrient uptake.
 − Sequesters nutrients.
 − Adds nutrient-poor litter.
 − Adds recalcitrant litter.
 − Accumulates salt.
Soil moisture content and site water balance
 + Lowers leaf transpiration rate.
 + Lowers leaf area.
 + Increases infiltration and stemflow.
 − Increases leaf transpiration rate.
 − Increases leaf area.
 − Increases run-off (litter repellancy or surface disturbances).
 − Accesses unused moisture sources.
Disturbance regimen
 + Increases fire frequency.
 + Increases erosion.
 + Decreases herbivore resistance.
 + Increases susceptibility to wind damage.
 − Decreases fire frequency.
 − Controls erosion.
 − Increases herbivore resistance.
Community dynamics
 + Increases species diversity.
 + Increases facilitation.
 + Increases rate of succession.
 + Increases primary productivity.
 − Decreases species diversity.
 − Increases competition.
 − Decreases rate of succession.

annuals invading a perennial-dominated community, as cheat grass (*Bromus tectorum*) invading shrub steppe in the western United States (Mack 1981); (3) new modes of resource acquisition, for example, a nitrogen fixer, as acacias (*Acacia*) in South Africa (Versfeld and van Wilgen 1986) and the faya tree (*Myrica faya*) in Hawaii (Vitousek and

Walker 1989); or (4) a new successional niche, for example, late successional, shade-tolerant tree species, as beech (*Fagus grandifolia*) and hemlock (*Tsuga canadensis*) in the Great Lakes region (Davis MB 1987). This increase in overall productivity occurs in early stages of plant succession, particularly during primary succession where primary productivity is initially absent or low. Alternatively, the community may have a naturally low productivity due to harsh climatic conditions (e.g., desert, alpine). Then the invader may add to productivity by utilizing a resource unavailable to the IN plants, for example, salt-cedar (*Tamarix*) tapping soil water not previously utilized by IN species along desert riparian corridors (Busch et al. 1992). Finally, increased productivity can result from the invader competitively displacing slower-growing or smaller IN species, for example, pines (*Pinus*) invading protea (*Protea*) shrublands in South Africa (Rutherford et al. 1986). This scenario could occur at any stage of succession. However, greater size or faster rate of growth are not, by themselves, enough to predict the probability of a successful invasion.

Negative or neutral impacts of an invading plant species on overall productivity (Table 6.1) can occur if the invader grows more slowly than, or at a rate similar to, the species it replaced (e.g., similar growth among IN thicket-forming species and invasive trees in South Africa (Rutherford et al. 1986). The successional analogue would be a decrease in net productivity of late successional stages as more of the nutrients become tied up in biomass or undecomposed organic matter (Chapin et al. 1994). Under such nutrient-poor conditions, only certain stress-tolerant species (sensu Grime 1977) can establish. Although stress-tolerant species are often part of the local flora and are adapted to the particular constraints of the habitat, some invaders can be successful here as well. For example, nutrient-poor bogs and mature forests in England have fewer invasive species than disturbed riverbanks or abandoned pastures with higher levels of available nutrients, but each does have invaders (Crawley 1987). An indirect method of reducing primary productivity could occur if the invader promoted a more frequent or severe disturbance regime (see below).

Measurements

Plant productivity, measured by increments of plant growth over a chosen time interval, is typically expressed in terms of $g\,m^{-2}\,yr^{-1}$ or $kg\,ha^{-1}\,yr^{-1}$. For small plants (e.g., some annuals), measures of total, aboveground, or below-ground biomass can be made by harvesting whole individuals and obtaining fresh and dry mass. For small perennials, shoot extension or leaf number or biomass can be used. For larger perennials (e.g., trees), diameter or height increments can be used to estimate productivity. For many trees and some shrubs, regressions exist to calculate biomass or biomass increment from simple measures of height or trunk diameter. Additionally, flower, fruit, or seed production and species density or cover can be used to estimate productivity.

The best way to measure the impacts of an invader on productivity is to have measurements at the same site before, during, and after an invasion (see Morrison, Chapter 9, this volume). Such data are rare. Second best, and more realistic, is to compare nearby control sites (similar vegetation, soils, geology, climate, and land-use history) with invaded sites. The best control site is one that is identical to the invaded site but into which the invader has not yet dispersed. A third approach, to complement the above two approaches, or if no uninvaded sites are available, is to conduct removal experiments (Aarssen and Epp 1990). The invader may be totally removed or partially removed (i.e., selective thinning or cutting back of individuals) once or continuously. Removal experiments are imperfect because one usually removes only the aboveground portions of the invader, leaving the roots intact. However, the damage to mycorrhizal populations and soil structure that would occur if root removal were attempted may be too large to be worthwhile (McClellan et al. 1995). Also, the biotic community that remains (or the one that is colonized following removal of the invader) may bear little resemblance to the preinvasion community. Removal experiments actually measure residual effects of the invader, rather than an unaltered community, unless the removals are begun prior to widespread establishment of the invader.

Addition experiments (Aarssen and Epp 1990) are another way to evaluate impacts of an invasive plant on primary productivity. For the evaluation of impacts they may be preferable, because they best mimic the potential invasion, particularly if plants are added at the appropriate life stage (generally as seeds or seedlings), densities, and time of year. The drawback of this approach is that it may take several years for effects to be manifested. With addition experiments, one is also able to vary densities or other factors thought to be important. Further experiments, including field or greenhouse studies that control all but a few environmental or biotic variables and that manipulate resources, are needed if one wants to identify particular mechanisms by which an invader alters primary productivity (Aarssen and Epp 1990; DiTommaso and Aarssen 1989).

Management

Management goals will direct any experimental investigations. Some questions to consider: Is an invader acceptable if it does not alter primary productivity of indigenous vegetation? Are IN species currently being used for commodity production? If so, then an invader will generally be unacceptable. If commodity production is not important, of what value is the primary productivity to the indigenous system (e.g., aesthetic, erosion control, energy base for higher trophic levels)? Another potential value of primary productivity is as an indicator of species diversity, but this cannot be assumed for all ecosystems. Can the invader serve as an acceptable replacement for IN species?

A key problem in using primary productivity as a parameter to assess the impacts of an invader is the difficulty in obtaining estimates of belowground production. This is particularly important because some invaders may invest more biomass below ground, particularly if nitrogen or water is more limiting than light. Research programs attempting to measure effects of invasive species but failing to tackle the more difficult measurement of belowground productivity may not be addressing the full range of impacts the invader is having on the ecosystem.

Soil Nutrients

Concepts

Invasive plants may have no effect on levels of a particular nutrient if that nutrient is not limiting or if it is utilized at the same level by IN and invasive species. For example, total soil nitrogen levels did not change when grassland converted to pine forest in Swaziland (Morris 1984, cited in: Versfeld and van Wilgen 1986) or when indigenous evergreen forest converted to nonindigenous grassland in Hawaii (D'Antonio CM and Vitousek PM, pers. com.). Invasive plants may also increase levels of nutrients through nitrogen fixation or increased weathering rates, leading to addition of nutrient-rich leaf litter or to retention of nutrients in early successional soils (Ramakrishnan and Vitousek 1989; Versfeld and van Wilgen 1986; Vitousek and Walker 1989) (Table 6.1). Increased levels of available nutrients can increase growth, the rate of leaf production, leaf turnover, and litter decomposition of IN and invasive species alike, perhaps favoring those that have rapid growth but are not efficient in converting nutrients to biomass (Ramakrishnan and Vitousek 1989). Changes in rooting patterns by invasive plants can increase exploitation of soil nutrients (Caldwell et al. 1987; Elliott and White 1989); changes in species of mycorrhizae can also alter patterns of nutrient use (Lodge et al. 1994).

Invasive plants can lower soil nutrient levels directly through competition or indirectly through additions of nutrient-poor or recalcitrant litter from, for example, pines (Versfeld and van Wilgen 1986) and eucalypts (*Eucalyptus*) (Robles and Chapin 1995) or via accumulations of salts that decrease nutrient availability or dramatically alter soil pH, such as the ice plant *Carpobrotus edulis* (D'Antonio 1990a). Invaders that increase fire frequency or severity can indirectly cause increased volatilization of nitrogen (D'Antonio CM and Vitousek PM, pers. com.) and perhaps a decline in soil nitrogen (Ramakrishnan and Vitousek 1989). Changes in nutrient levels in plant tissues due to invasive plants are therefore mediated through changes in soil nutrient levels.

Measurements

Soil nitrogen and phosphorus are the nutrients that most often limit plant growth. Each is found in soils in a range of forms, from fractions readily available for plant uptake (but susceptible to erosion) to fractions bound within organic compounds and thus completely unavailable for plant growth (Binkley and Vitousek 1989; Chapin et al. 1986). Available nitrogen and phosphorus can be extracted with salt solutions or weak acids; total nutrient pools are determined by digestion in strong acids; field or lab incubations are used to measure potential mineralization; and lysimeters measure leaching below the root zone. Samples are collected typically with small corers, but the use of quantitative soil pits allows for calculation of bulk density and extrapolation to soil volume. Seasonal changes in availability can be great because the microbial populations regulating nutrient availability fluctuate depending on the amount of water, carbon, nitrogen, and other nutrients available in the soil (Lodge et al. 1994). Some invaders may lower levels of other nutrients, for example, potassium, calcium, and magnesium (D'Antonio 1990a; Versfeld and van Wilgen 1986), to degrees that hinder growth of IN species.

Pre- and postinvasion measurements on the same soils are, as noted above, the most direct evidence of the influence of invasive plants. However, effects on soil nutrients can be detected with addition or removal studies. Appropriate spatial scales and temporal sampling regimes must be determined to address management concerns, and the high variability typical of soil nutrient content must be accounted for in the sampling program. Sampling of soil nutrients can give managers of plant invasions a sensitive tool to predict eventual species change, thus lengthening the time to formulate adequate responses to the invasion.

Management

Levels of soil nutrients can have a strong impact on species composition, species diversity, primary productivity, and other community and ecosystem properties; therefore they can affect management decisions. First, one must establish that the invasive plant is altering availability of one or more soil nutrients. Correlative evidence may substitute for experimental evidence if effects of a particular invader have been documented elsewhere. Second, does the change in soil nutrients promote or hinder management goals? For example, if one is managing for maximum richness of IN species, an invader that reduces levels of an essential nutrient below levels at which these species can survive will be undesirable. However, if one is managing for increased plant cover and soil enrichment, an invader providing dense ground cover and increasing soil nutrient levels may be desirable. Finally, how does one encourage or discourage the invader? If the source of nutrient addition or depletion is through rapid growth and decomposition, the invader's growth must be curtailed (or enhanced). If the source is through nitrogen fixation, then soil aeration or soil phosphorus levels can be manipulated to improve nitrogen fixation, if that is the desired goal. Alternatively, flooding or acidification of soil to decrease phosphorus availability can reduce nitrogen inputs. In some instances, purposeful reduction of nutrients from cropping or burning is used to halt successional change and to promote recovery of indigenous communities (Luken 1990).

Soil Water and Salinity

Concepts

Invasive plants can strongly impact soil water content and landscape water balance of indigenous communities, with both positive and negative effects (Table 6.1). The most direct negative impact is to use more water than the IN species they are replacing, or to increase overall community water use in a water-limited habitat. This can happen as a result of the invasive species having intrinsically higher transpiration rates per unit leaf area or as a result of maintenance of higher total leaf area (in the event that intrinsic transpiration rates are similar). For example, salt-cedar is known to be capable of desiccating wetlands in desert areas, apparently by maintenance of high leaf area, not by higher intrinsic water-loss rates per unit leaf area (Sala et al. 1996). In contrast, broom-sedge (*Andropogon virginicus*), through its reduced canopy-level transpiration rates, promotes the formation of swampy areas in Hawaiian forests that it invades (Mueller-Dombois 1973).

A second way that invasive plants can influence landscape water balance is if they effectively change the surface characteristics of the habitat. If invasive species alter infiltration processes due to different canopy architecture or to the production of more hydrophobic litter layers, they may significantly influence soil water balance as they increase in dominance in the community. There are apparently no data sets indicating that invasive plants increase or decrease water repellency beneath their crowns, although invasive plants may contribute to increased litter layers. There is better documentation of the influence of architectural differences between IN and invasive species on landscape water balance. For example, juniper (*Juniperus*) trees are IN invaders of grasslands and sagebrush steppe in the Great Basin (Blackburn and Tueller 1970). Juniper woodlands intercept large amounts of rainfall and snowfall before these reach the soil surface, resulting in high evaporation and sublimation losses directly back to the atmosphere from wetted canopies; this phenomenon has been estimated to reduce winter soil moisture recharge by more than 50% in dense juniper stands (Eddleman and Miller 1992). Additionally, if invasive species result in greater surface disturbance, this may have a negative impact on surface water balance by increasing run-off after large rainfalls.

A third important way that invasive species may alter soil water balance is through shifts in phenological schedules. If a plant invades a seasonal climate to which it is not pre-adapted, it may have a seasonal pattern of canopy formation and physiological activity differing from the IN species in the community. Although inherent phenotypic plasticity may contribute to acclimation in the new climate, there could still be significant differences between the invader and the IN species in timing of leaf production, litter fall, or patterns of maximum growth rate. This could result in periods of growth for the invader when IN species are dormant, thus releasing the invader from competition for certain periods of its life cycle. For example, in tropical dry forests the first heavy rains after the dry season result in synchronous leafing out of the forest (Borchert 1994a), but trees with access to deep soil-moisture sources or stem water storage may be able to leaf out earlier (Borchert 1994b) and thus have access to high light levels prior to full canopy development.

One example of the importance of phenological differences is illustrated by the invasion of Great Basin shrublands by an NI annual, cheat grass (*Bromus tectorum*). The Great Basin is characterized by cold, wet winters in which snow melt recharges the soil profile, followed by a predictably dry spring and summer season in which the moisture stored during winter recharge is depleted after the vegetation breaks dormancy. Cheat grass can germinate and exhibit root elongation at soil temperatures below which sympatric shrubs or perennial grasses can grow roots (Harris and Wilson 1970). As a result, the grass is able to commence growth and to deplete soil moisture prior to breaking of dormancy of the IN perennials, which gives it a significant competitive advantage in this cold, semiarid environment (Harris 1967). The observation that perennial shrubs and grasses growing next to cheat grass exhibited greater physiological stress (Melgoza et al. 1990) and reduced total root length (Melgoza and Nowak 1991) than plants without it as a neighbor strongly suggests that the grass has a significant competitive effect on IN species and that its enhanced competitive ability may be traced to its accelerated developmental schedule at the onset of the growing season.

Finally, an invasive species may significantly alter the water balance of an ecosystem if the species gains access to water sources previously unavailable to, or lightly utilized by, the indigenous vegetation. Primary examples of this are deep-rooted phreatophytes that invade herbaceous wetlands or shallow-rooted annuals that invade shrublands. If the invasive species then utilizes a previously unused moisture source to invade the habitat successfully, its presence may effectively increase overall biomass and leaf area of the ecosystem, thus increasing landscape-level evapotranspiration and therefore potentially resulting in a long-term decline in site water balance.

The effects of invasive plants on soil salinity have not been so well studied as the effects on soil water. Clearly, one mechanism by which plant invasion can increase site salinity is through invasion of halophytes (salt-tolerant species) into communities dominated by glycophytes (salt-sensitive species). One example is the invasion of flood plain environments in the western United States by halophytes after river systems have been regulated, thus eliminating annual floods that tend to leach salts out of the soil profile. In addition to salt-cedar (see below), a number of IN shrubs or trees such as mesquite (*Prosopis*), pickleweed (*Allenrolfea*), and saltbushes (*Atriplex*) can become dominant in many regulated riparian environments and internal drainage basins in the West. Most halophytes are able to take up salts and to compartmentalize them where they are not physiologically damaging, but this process concentrates salts in the shoot, which results in deposition of high-salt-content litter to surface soils over time. A shift in species composition on salinized sites toward dominance by halophytes is particularly prevalent in arid climates, where relatively little water moves vertically through the soil, thus allowing salts to accumulate within the soil profile (due to water extraction by roots) or at the soil surface (due to litter decomposition) (Virginia and Jarrell 1983). Once halophytes salinize the ecosystem, it may rapidly become uninhabitable for seedlings or adults of the formerly dominant glycophytes such as cottonwood (*Populus*) and willow (*Salix*) (Busch and Smith 1995).

Measurements

Soil water can be measured at various levels of sophistication, providing researchers and managers with various levels of accuracy and precision. Because of the complex nature of soils and the nonuniform distribution of plant roots, it is not a trivial task to decide what is an ecologically relevant measure of soil moisture status and to decide how to measure that parameter appropriately. For example, agronomists and ecologists routinely take soil moisture readings at one or more standard depths (e.g., 10 and 30 cm) without quantifying the vertical distribution of the root systems of the plants of experimental interest. As a result, those 10- and 30-cm soil moisture readings may not be representative of the regions of the soil where plants are actually taking up water, nor would they be relevant over time if plants seasonally shift their zones of maximum moisture absorption, which can vary up to a vertical meter or more in some perennials (Caldwell and Richards 1986).

Prior to determining soil water content, the researcher needs to determine if the relevant parameter is percent soil moisture (mass of water per unit mass of soil) or soil water potential (a measure of the energy status, and therefore "extractability" of water in the soil). If plant water potentials are being collected, then soil water potentials are often desirable for correlative purposes, although percent soil moisture data are also useful.

There is a variety of methods for assessing soil moisture in ecosystems; these were outlined by Rundel and Jarrell (1989). Techniques used to measure percent soil moisture are (1) gravimetric analysis, (2) neutron probes, and (3) time domain reflectometry (TDR) (Topp and Davis 1985). The last two are quite accurate but require expensive equipment, whereas gravimetric sampling can be performed by most field biologists. Unfortunately, it is very difficult to get representative and accurate samples of soil moisture with gravimetric sampling due to sampling and coring difficulties, errors in handling, and general destruction of physical structure of the soil. These problems can be minimized by always taking intact soil cores with a standard auger and then immediately placing them in sealed soil tins in a cool, dry container to eliminate evaporation errors during sampling and handling. Another drawback of gravimetric sampling is that a single location cannot be measured repeatedly over time. The methods used to obtain soil water potential (or matric potential, a measure of the energy required to extract water adsorbed to soil particles) are (1) tensiometers, (2) resistance blocks, and (3) soil psychrometers (Rundel and Jarrell 1989). Tensiometers are useful only in very wet soils and so are not applicable to dry (or seasonally dry) ecosystems. Resistance blocks and soil psychrometers, both of which require expensive equipment, give only a localized measurement with an individual probe. Therefore, in intact ecosystems with heterogeneous substrates, a large number of probes (and associated data-logging equipment) must be used to adequately assess horizontal and vertical variability in moisture content. Because of that, many ecologists today utilize techniques such as neutron probes and TDR, which can integrate soil moisture across vertical depth gradients. Additionally, neutron probes and TDR are very amenable to making repeated measurements at depth over time. They are easily integrated with microcomputers (e.g., Hulsman 1985) to obtain relatively continuous data sets; in the case of TDR, they can accurately measure variation in soil water content over small spatial scales (Amato and Ritchie 1995). However, neutron probes and TDR are both expensive; the researcher requires training in use of these techniques before they can be successfully implemented in the field.

Because soils, and thus soil moisture, are horizontally and vertically heterogeneous, it is often far easier to measure plant water status directly than it is to obtain accurate estimates of soil moisture status in the active rooting zone. Although relative water content of plant tissues can be measured, the preferred method of measuring this parameter is through the use of a pressure chamber to measure plant water potential (Turner 1981; Waring and Cleary 1967). Because most plants close their stomata at night and thus reequilibrate with soil moisture, most field ecologists record a plant water potential before dawn and then also record water potential at midday to determine the effects of transpiration on plant water status. Pre-dawn water potentials of perennial plants correlate well with soil water potentials in coniferous forests (Fahey and Young 1984) and in desert scrub ecosystems (Fonteyn et al. 1987).

One new technique allowing investigators to determine moisture sources in vegetation is the use of stable isotopes of water extracted from xylem sap to track moisture sources of the vegetation (Ehleringer and Dawson 1992). This technique has been used to determine the relative contribution of ground water versus soil water in plants (Busch et al. 1992; Dawson and Ehleringer 1991), because ground water is isotopically lighter than is soil water. It has also been successfully used to determine seasonal utilization of soil moisture by vegetation based on the fact that winter and summer rains have different isotopic signatures (Ehleringer et al. 1991; Flanagan et al. 1992). The technique could be used to determine if invasive species are capable of differentially utilizing seasonal moisture sources that IN plants may not exploit.

The methodologies available to measure landscape water balance accurately are extremely difficult due to the complexities involved and to the sophisticated instrumentation required. For accurate calculation of landscape water balance, it is necessary to measure (1) precipitation, (2) infiltration of water into the soil, (3) run-off, (4) evaporation of water from the soil and other wet surfaces, (5) plant transpiration, and (6) internal transport of water in the soil, including potential deep losses to the ground water (Rundel and Jarrell 1989). Only large, well-financed research groups are able to conduct this array of measurements for intact ecosystems. As a result, there are few comparative studies in the literature that have addressed the question of whether invasive species alter landscape water balance.

A variety of parameters can be measured that are related to ecosystem salinity, depending on the investigator's interest. Standard parameters measured by soil scientists include (1) electrical conductivity (EC), which is determined from soil extracts by use of an electrical conductivity probe; (2) total dissolved salts, which is determined by measuring individual cations by atomic absorption on aqueous soil extracts; and (3) direct measurement of salinity with an appropriate sensor such as TDR (Dalton et al. 1984). It is particularly important to measure salinity along vertical depth profiles in the soil because salts accumulate on the soil surface at many sites. Thus the concentration of salts in the surface crust, and the depth of salt deposits, may be extremely important when considering processes such as plant recruitment and nutrient cycling. Other salinity-related parameters that investigators should consider include salinity of groundwater sources in the event that ground water rises during certain times of the year or that the vegetation includes phreatophytes; and boron content of the soil, particularly surface layers, since many saline soils are also sodic, and it may be the sodicity of the soil that is the most ecologically important parameter when considering plant performance and survival.

Management

When a researcher considers the potential impacts of invasive species on water balance in ecosystems, several questions may help determine subsequent management strategy. Does an invasive species significantly alter mean soil moisture content and landscape water balance of the ecosystem? If so, how does the species accomplish this? Is the shift in water balance deleterious to long-term survival of IN species that dominate (or previously dominated) the ecosystem? Simple visual analysis of an impacted habitat may provide useful information, for example, in situations when an invasive species dries up a spring or wetland. However, such observations are not quantitative, so most management programs need quantitative data of soil moisture content in the rhizosphere (discussed previously) or perhaps discharge rates from a wetland.

The question of how an invasive species may potentially alter landscape water balance is related to intrinsic water loss rates and production of transpirational surface area of the invasive species, as discussed above. Invasive species tend to be fast growing, and so they would be expected to have high water loss rates on a surface area basis and to produce a high leaf area when they become established in the ecosystem. If they accomplish high water use primarily because of high leaf area, then management programs (e.g., periodic burning or clearing) that keep the invasive species in an early successional stage may be successful. Of course, such programs must be compatible with the indigenous vegetation being managed. Attempts have been made in the past to spray high-water-use vegetation with antitranspirants, but these efforts have been costly and largely unsuccessful.

It was suggested above that an invasive species, by potentially utilizing resources not previously exploited, could conceivably increase the overall carrying capacity of the ecosystem without having deleterious effects on IN species. However, this is usually not the case. In the event that belowground competition remains minimal due to the invasive species utilizing soil moisture and nutrients at a depth (or time of year) not being exploited by other species, the invasive species may nevertheless crowd out or overtop the previous dominants in the community. However, over time there would almost certainly be overlap in root systems of the invasive and IN species, with belowground competition increasing through time.

With regard to soil moisture, this competitive effect can be determined in a number of ways. In concert with quantitative estimates of soil moisture content by depth interval, researchers could also directly determine the effects of an invasive species on indigenous vegetation by determining the level of water stress among IN species as soil moisture seasonally declines. This can either be accomplished with paired sites in which the invasive species is common or absent, or it can be experimentally tested by selectively removing the invasive species from around individuals of the IN dominants (e.g., Busch and Smith 1995; D'Antonio and Mahall 1991; Fonteyn and Mahall 1978). If the IN species exhibit higher plant water potentials and leaf area production in the absence of the neighboring invasive species, it can be concluded that belowground moisture supplies are probably more plentiful in the rooting zone of the IN species than they were when the invasive species was present. More mechanistic experiments (Mahall and Callaway 1992) are necessary if the exact nature of competition between the IN and the invasive species is to be determined.

Disturbance Regimes

Concepts

The impact of invasive plant species on disturbance regimes can be the most pervasive impact of all, with secondary impacts on primary productivity, nutrient cycling, and community dynamics. An invasive species may increase or decrease the susceptibility of an ecosystem to disturbance (Table 6.1).

Invasive species can affect fire frequency and severity by altering the structure and composition of fuel loads. Invasion of arid shrubland by grasses generally increases fire frequency and may increase fire severity as well due to increases in flammable litter (D'Antonio and Vitousek 1992; Hughes et al. 1991; Mack 1986; Smith 1985; Young and Evans 1978). The introduction of fire to a system with little or no fire history can result in high mortality of the IN shrubs (Brown and Minnich 1986; Callison et al. 1985; Young and Evans 1978) and a physiognomic shift from dominance by shrubs to dominance by fire-adapted grasses. Similarly, invasion of desert riparian areas by salt-cedar (discussed previously) has precipitated an increase in wildfires in an ecosystem that had limited historical exposure to fire (Busch n.d.). Invasion of grasslands or shrublands by trees may decrease fire frequency but, depending on water content of the accumulating litter, may promote hotter, more severe fires that occur only occasionally (Versfeld and van Wilgen 1986).

Erosion is another disturbance that can be affected by invasive species (Versfeld and van Wilgen 1986). Many invaders, through rapid growth, canopy interception of moisture, and extensive root or rhizome development, reduce erosion in recently disturbed habitats. Seasonal erosion might increase in the case of annuals replacing perennials, especially if the annuals die back in a rainy season or if scattered trees replace dense understory species, increasing the amount of exposed ground. If the trees are shallow rooted or more susceptible to burning than the species they replace, erosion may also be promoted (Versfeld and van Wilgen 1986).

The susceptibility of forest trees to wind damage is often a function of species characteristics such as height, diameter, wood density, and root morphology or of foliar characteristics and stand characteristics, including density, age distribution, and topographical location (Brokaw and Walker 1991). Invasive (or planted) tree species such as pines are generally more susceptible than IN hardwoods (Tanner et al. 1991).

Invasive plant species are also likely to alter plant/herbivore interactions through changes in plant density, foliar chemistry, and seed production. However, wide fluctuations in normal herbivore populations, particularly following a disturbance, make it difficult to determine the effect of the invasive plant (Breytenbach 1986).

Measurements

Disturbances occur at many spatial and temporal scales, so it is critical to match the scales with the expected impacts of the invasive species. Examining the respective impacts of algal blooms, annual herbs, perennial shrubs, or long-lived clonal shrubs and trees on disturbance regimes requires different measurements. For long-lived species invading as seeds or seedlings, their impact on the local disturbance regime will change as they grow. Long-term observations can place individual measurements in a more appropriate context, avoiding generalizations from a limited data set (Magnuson 1990). Choice of measurements will reflect what one expects to be altered by the invader (e.g., erosion rates). Research on invaded and control sites will accumulate correlative evidence, but better evidence of a causal relationship between invader and disturbance would involve experiments where the disturbance regime and such aspects of the invasion as invader density are manipulated.

Management

Again, a series of questions must be formulated to help focus management policy. What benefits does the altered disturbance regime provide? Perhaps increased fire frequency would be useful to suppress undesirable species or to deter insect pests; increased erosion might allow more run-off to accumulate in flood-plain croplands. To what degree is disturbance important for maintaining the desired ecosystem? If one wants to return the altered system to its original state, is disturbance really the factor that one must alter? Or should management concerns be focused on primary productivity and altered species composition?

How much effort will it take to return the disturbance regime to pre-invasion levels or frequencies? Will removal of the invader return the system to its pre-invasion disturbance regime, or will additional manipulations be needed? Invasive plant species often alter many properties of the invaded ecosystem, so simple removal may not be adequate to restore pre-invasion ecosystem structure and function. Damage to indigenous vegetation (e.g., by increased fire frequency, erosion, herbivory), introduction of additional invasive species, or changes in soil water or nutrient levels could require extensive manipulations to restore the initial disturbance regime.

Community Dynamics

Concepts

Community dynamics are inevitably altered by species that impact primary productivity, nutrient cycling, or the disturbance regime. An aggressive, productive invader is probably a good competitor for space, nutrients, and light. Although it may have a neutral or positive effect on species diversity, an invader is more likely to reduce diversity (Table 6.1) by its monopoly of resources, while simultaneously increasing primary productivity or biomass (Breytenbach 1986). Reduction in diversity can come from competitive exclusion via shading or from nutrient or water depletion, and may affect any life stage of the IN species, from seedling to adult to propagule (Busch and Smith 1995; Walker and Vitousek 1991).

Successional trajectories can also be altered by a species that changes the abundance of critical resources. Invaders that form dense thickets and reduce ambient light levels to the point where few species can survive or regenerate in the deep shade might arrest succession at least until the thicket-forming species dies (Luken 1990; Niering et al. 1986; Vitousek et al. 1993; Walker 1993, 1994). Nitrogen-fixing species may locally increase available nitrogen (Vitousek and Walker 1989), but the availability of the nitrogen to other species depends on density and life form of the thicket-forming species. Small, herbaceous N-fixers may increase the rate of succession by contributing nitrogen while not forming inhibitory thickets. Large, thicket-forming N-fixers may not only arrest succession by their capture of light but also recapture whatever nitrogen they fix (Walker 1993). Nitrogen-fixers may also alter succession by favoring invasion of other, more nitrogen-demanding species. For example, the N-fixing fayatree may promote establishment of other NI species in Hawaii (Aplet et al. n.d.; Mueller-Dombois and Whiteaker 1990; Vitousek and Walker 1989). Facilitation of later successional (indigenous) species by early successional N-fixers is presumed to be particularly important in primary succession where nitrogen levels are initially low, but experimental evidence is rare (Walker 1993).

Another way that succession can be altered is by an invader that increases the fire frequency and that is fire tolerant or able to colonize quickly after fire (Vitousek 1990). Clements (1928) termed this kind of arrested succession a fire disclimax. Important fire disclimaxes include the annual grasslands of California's Central Valley, parts of the Great Basin, and flood plains dominated by salt-cedar in the arid southwestern United States.

Measurements

As noted earlier, some community properties are relatively easy to measure (e.g., species diversity, primary productivity, plant cover or growth) but others are more difficult to measure (e.g., species interactions, successional change, stability). Most of these properties are dynamic and best measured over a period of at least several years so that influences of the invader can be monitored. Causal mechanisms for changes in community properties may best be found by measuring environmental factors (e.g., light and nutrient levels) in conjunction with community properties and experimental studies. The properties to measure will again be determined by effects hypothetically associated with the invader. Actual measurements vary and standard methods can be obtained from several sources (e.g., Hairston 1989; Magurran 1988; Mueller-Dombois and Ellenberg 1974).

Management

Do invasive species reduce species diversity? Maintaining biodiversity is often one of the primary goals of ecosystem management. Reductions in diversity may destabilize trophic dynamics, alter wildlife populations (by affecting food supplies or shelter), and change nutrient cycles or decomposition rates. If the goal of management is to restore pre-invasion diversity levels, is that a feasible goal and how is it best accomplished? In some cases, "restoration" of the original system is not realistic; "reclamation" of some functions of the original ecosystem, perhaps with an alternate species composition, is more practical.

If pre-invasion successional trajectories have been changed, what are the consequences for the organisms of interest? In Hawaii, invasion of the fayatree reduced the importance of the IN tree species during succession. Whether invasive species alter rates of successional change, either arresting succession or speeding it up, or alter successional trajectories, management scenarios must address the temporal dynamics of the altered system (see Luken, Chapter 11, this volume). Is an undesirable shrub, currently outcompeting IN species, a long-lived plant that will arrest succession, or is it short lived? In the latter case, what will characterize the next successional stage—surviving IN species or a new set of species adapted to the altered environment? In the case where the invader increases the disturbance frequency (e.g., a fire disclimax), can one rearrange the succession to favor IN species? Perhaps one must settle for an alternate suite of IN species that are also tolerant of frequent fires.

Finally, the temporal dynamics of plant communities require study to determine how long one should monitor the system to be sure the desired effect has occurred. Variation in species life cycles, in rapidity of nutrient cycles, and in return time of disturbance types suggests that management goals must specifically address these temporal issues.

Case Studies

In this section we present two case studies to illustrate the range of possible impacts of invasive species on natural ecosystems. These examples will also illustrate how impacts on vegetation, soil and water cycling, disturbance regime, and community dynamics are related. The first example addresses the impacts of the invasive faya tree on soil nutrients and succession. The second example addresses the impact of salt-cedar (*Tamarix*) on soil water and salinity. We compare the two case studies and discuss how one would investigate what is not known about each species (e.g., effects of faya tree on soil water or of salt-cedar on soil nutrients). In addition, we address both the management history and the future management strategies for these important invaders.

Fayatree

Fayatree, medium-sized (to 20m), has nitrogen-fixing symbionts in root nodules. Indigenous to Macaronesia (Madeira and the Canary Islands), it was introduced to the Hawaiian Islands in the late 1800s by Portuguese settlers, then planted widely in Hawaii in the 1920s and 1930s to control erosion in pastures (Lutzow-Felling et al. 1995; Vitousek and Walker 1989). At low elevations on the island of Hawaii it now occurs in several dense, monospecific stands as a mature forest with no regeneration of IN or NI species in the understory (Whiteaker and Gardner 1985). In the uplands, fayatree is actively invading low-nutrient volcanic soils where indigenous forests are either sparse or damaged. Again, dense monospecific stands develop with no regeneration of any species in the understory (Walker and Vitousek 1991).

Invasion of the uplands around Hawaii Volcanoes National Park (HAVO) was very rapid. Only isolated trees were reported in 1961; by 1977, 609 ha were covered with fayatrees, despite the removal of 92,000 individuals between 1973 and 1977. By 1985, fayatrees covered 12,200 continuous ha (Whiteaker and Gardner 1985); by 1992, 15,100 ha were covered (Loh et al. n.d.). In 1985, removal efforts were restricted to 3000 ha of special ecological study areas comprising less than 25% of the total range of the species in HAVO (Tunison n.d. (a)).

Primary productivity, soil nitrogen, and earthworm populations increased dramatically in the upland volcanic soils when fayatree invaded, while growth and regeneration of IN trees were inhibited (Aplet 1990; Vitousek and Walker 1989; Walker and Vitousek 1991). The primary disturbance regime—episodic volcanic eruptions—was not altered by invasion of the fayatree, but it is possible that the denser forests created by the species have altered wind dispersal of propagules. Primary succession was definitely altered. Prior to the tree's arrival, forest development took several centuries (Aplet and Vitousek 1994). In contrast, faya forests developed within 30 years (Vitousek and Walker 1989) and no IN species survived in the understory or overstory. The life span of the fayatree is not known, but stands more than 35 years old in the lowlands of Hawaii, still appearing vigorous (Whiteaker and Gardner 1985), and mature faya forests are abundant in Macaronesia (Lutzow-Felling et al. 1995).

The failure of removal efforts, by physically uprooting young fayatrees or cutting them at ground level and applying herbicides to prevent stump sprouting (Tunison n.d.(b)), was not due to resprouting but to extensive and rapid reinvasion. Vitousek and Walker (1989) calculated that one stand of 21 fruiting trees would produce 152 new fruiting trees within 6 years. There were simply not enough resources to control the invasion. HAVO budgets about $75,000 annually on manual and herbicidal removal of fayatrees just from the 3000 ha of ecological study areas (Tunison n.d.(a)). Currently, biocontrol efforts are underway, but several releases of leaf-eating insects imported from Macaronesia have had limited success (Lutzow-Felling et al. 1995). Experimental removals of fayatrees suggest that IN tree ferns may benefit from partial thinning of faya stands by girdling (fayatrees die slowly). When faya stands were clearcut, the principal invaders were NI species such as anemone (*Anemone*) and blackberry (*Rubus*) with some seedling regeneration of fayatrees (Aplet et al. n.d.).

Fayatrees have completely altered the soil and aboveground environments where they have invaded. Responses to the improved nutrient status are positive for some IN species only if the fayatree is removed. As long as it remains, it outcompetes IN species for light and recycles the nutrients it has added to the system. Its impact on long-term forest dynamics is still being evaluated, but it has essentially replaced the IN forest trees throughout most of its invaded range (Walker and Vitousek 1991; Whiteaker and Gardner 1987).

Future management of fayatree in Hawaii depends on success of the biological control effort, on continued removals in select study areas, and on unplanned events such as a recent dieback of fayatrees apparently caused by a leafhopper (Yang et al. n.d.). Dieback may also be aggravated by winter drought (Tunison et al. n.d.[a,b]), suggesting the importance of knowing more about the water relations of the tree. The tree appears to establish best in dry, open grasslands and open stands of the indigenous forest. Its invasion is slowest in the dense, closed forests where the soils are very wet (Lutzow-Felling et al. 1995; Vitousek and Walker 1989; Walker and Vitousek 1991). Lipp (1994), noting that fayatree was most likely to occur in forest gaps, demonstrated its tolerance to high light conditions. Vitousek and Walker (1989) found that fayatree seedlings grew best under high light, but this study did not measure tree growth under different light conditions. Better understanding of physiological response to light and water might clarify population dynamics and recent diebacks (Lutzow-Felling et al. n.d.).

Salt-cedar

Salt-cedar is an NI thicket-forming species that occurs in dense, often monospecific stands along many perennial watercourses in arid and semiarid regions of western North America (Brock 1994). It occurs in almost all western states of the United States, occupying well over 500,000 ha of primarily riparian flood plain habitat (Robinson 1965). The history and spread of salt-cedar into the southwestern United States has been well documented (Horton 1964; Robinson 1965). Briefly, salt-cedar was introduced into the United States as an ornamental and windbreak species more than 100 years ago. Its most rapid increase has occurred in the past 50 years, primarily in response to anthropogenic disturbances associated with regulation of river flows that accompanied construction of dams, reservoirs, and large diversion projects (Everitt 1980). Once established along major drainages in the southwestern United States, salt-cedar successfully invaded less-extensive wetland habitats such as isolated marshes, springs, and ephemeral watercourses via its windblown seeds and occasionally through deliberate plantings. The extent to which salt-cedar assumed dominance in each habitat to which it was introduced was a function of initial plant community structure, seasonal water availability, and water table depth (Everitt 1980; Graf 1982).

The taxonomy of salt-cedar is complex, but several species with previously disjunct distributions across the Eurasian continent now freely hybridize into a single naturalized species complex, which may help explain its apparent extreme adaptability and tolerance of a wide range of environmental conditions (Brotherson and Field 1987). Everitt (1980), in a review of the literature, concluded that little is known about stand demographics or habitat requirements of salt-cedar. As a result, the natural successional processes of southwestern riparian communities experiencing potential invasion by salt-cedar have until recently been only speculated upon, making it difficult to implement management programs for habitats impacted by salt-cedar.

Recent research has more clearly documented the role of salt-cedar in desert riparian ecosystems and has found that salt-cedar invasion can profoundly alter species composition of riparian communities (Crins 1989) and can alter fundamental processes in riparian ecosystems (Busch and Smith 1995; Vitousek 1990). Effects of salt-cedar on primary ecosystem attributes include increased evapotranspiration and potential desiccation of flood plains (Davenport et al. 1982), increased salinity of soil and water, increased frequency of fire (Busch n.d.), declines in numbers and diversity of faunal communities (Hunter et al. 1988), and probably changes in litter decomposition and nutrient cycling processes.

One of the primary biological attributes of salt-cedar is its alleged prodigious rate of water use, which has been the focus of much research and management effort. Early reports of extremely high rates of water loss in salt-cedar (Gatewood et al. 1950; Robinson 1965) led to management efforts that focused on clearing salt-cedar in favor of replacement phreatophytic vegetation (Culler 1970) or spraying salt-cedar stands with antitranspirants (Davenport et al. 1978). However, recent research efforts have found that salt-cedar exhibits effective stomatal control of water loss when exposed to high temperatures or low humidities (Anderson 1982) and that transpiration rates of salt-cedar are similar to sympatric phreatophytes (Busch and Smith 1995). Thus the extremely high water losses previously attributed to salt-cedar may not occur under many conditions (Graf 1982). In addition to climatic factors, total canopy leaf area or volume (Gay and Fritschen 1979; Hughes 1972) and subsurface conditions such as salinity and depth to ground water (van Hylckama 1974) are of importance in determining transpiration in salt-cedar. Leaf area may be of greatest importance; Sala et al. (1996) found that dense stands of salt-cedar may reach a Leaf Area Index (LAI; leaf area per unit ground area) of 4.0, which is probably higher than the LAI attained by stands of IN phreatophytes. Under such conditions salt-cedar stands can lose 50% more water than predicted potential evapotranspiration (i.e., theoretical maximum water loss from a uniform, wet surface). However, ground water depth and salinity may also be important, as a decline in water table or an increase in subsurface salinity will tend to increase hydraulic resistance to water flow in the soil-plant-atmosphere continuum. As a result, sites with shallow water tables and low salinities should exhibit the highest community transpiration rates. These sites are also the best candidates to reclaim from the early stages of salt-cedar invasion, as salt-cedar stands tend to depress water tables and increase soil salinities as they increase in dominance and in leaf area.

Salinization of flood-plain habitats may be the most important single way that the invasion of salt-cedar fundamentally alters ecosystems. Sites dominated by salt-cedar are much more saline (both soils and ground water) than are adjacent sites still supporting indigenous cottonwood and willow forest (Busch and Smith 1995). This is because salt-cedar takes up salt from the subsurface and stores it in salt glands in the leaves prior to secreting it onto the exterior of the leaves (Decker 1961).

Being deciduous, salt-cedar adds a new layer of salt-rich litter to the soil surface each year. As a result, even after salt-cedar has been mechanically cleared, burned, or sprayed with herbicide, it still leaves behind a saline ecosystem that may not be conducive to revegetation efforts, particularly if those efforts include IN glycophytic species. Burning of salt-cedar stands can also result in a dramatic increase in surface salts immediately after the fire (Busch and Smith 1993). Therefore, in many ecosystems being reclaimed from salt-cedar invasion, only a return of annual floods, which leach the soil of salts, will allow the ecosystem to be revegetated with former dominants such as cottonwood and willow. Often the more practical approach may be to attempt to revegetate with IN salt-tolerant grasses such as alkali sacaton (*Sporobolus airoides*) or saltgrass (*Distichlis spicata*) or both and salt-tolerant trees such as mesquite, particularly if the site in question is adjacent to a regulated river that precludes the return of historical flooding events.

A final, but potentially critical, ecosystem attribute that salt-cedar invasion may alter is nutrient cycling. However, there is essentially no information on nutrient cycling processes in stands dominated by salt-cedar. In one comparative study, Busch and Smith (1995) found that sites dominated by indigenous cottonwood and willow forest did not have higher soil concentrations of organic matter, nitrogen, or phosphorous than did nearby sites dominated by salt-cedar. However, such comparisons do not necessarily mean that the two vegetation types exhibit similar nutrient cycling processes. Important attributes such as high soil salinity and the potential for "hydraulic lift" (cf. Caldwell and Richards 1989; the tendency for deep-rooted plants to "lift" water from deep sources and deposit it in shallow soils, which may stimulate nutrient cycling when surface soils are otherwise dry) could be important in distinguishing nutrient cycling processes in indigenous ecosystems versus those dominated by salt-cedar.

In conclusion, deciduous salt-cedar tends to significantly alter ecosystems it invades, primarily by salinizing surface soils and depressing water tables due to its high transpiration rate. Once salt-cedar attains dominance, it controls most ecosystem processes and creates an environment in which the formerly dominant IN species have little probability of reestablishing without human intervention. As a result, efforts to control salt-cedar in the southwestern United States have tended to concentrate on localized infestations of the plant that can be economically cleared, leaving vast tracts of desert flood plains with essentially 100% salt-cedar cover. As water supplies become more scarce in the desert Southwest, the "salt-cedar problem" will probably become one of the most important management issues in the region. Managers taking an ecosystem approach to the problem, rather than single-factor solutions, should achieve the greatest success in converting salt-cedar stands to communities of more desirable IN species.

Conclusions

1. Invasive plant species are most likely to alter community and ecosystem properties when these species occupy a new niche or alter the disturbance regime.

2. Alterations in primary productivity, nutrient cycling, landscape water balance, and species interactions are inevitably linked. Alter one parameter and most other parameters are likely to change, including the likelihood of future invasions. As a result, measurement of impact of an invasive plant is in one sense arbitrary, primarily dependent on what parameter is of interest to or most readily measured by managers.

3. In some instances, impacts may be desirable or at least tolerable. For example, invaders may increase primary productivity and thereby reduce erosion or increase the rate of successional return to IN species.

4. Because management of undesirable impacts may be limited by practical or financial constraints, prioritization should be given either to management strategies having the broadest possible effect or to those focusing on enhancement of a particular community or ecosystem attribute.

5. To alleviate undesirable impacts, some degree of removal of an invader is often the best strategy because this reduces competition with IN species for resources and space.

6. Altering the disturbance regime can also be an effective management tool. For example, renewal of annual floods through controlled releases from dams on rivers in the southwestern United States may be the most practical way to permit reestablishment of IN riparian trees in areas subject to salt-cedar invasion.

7. An understanding of the intricate interaction between invasive plant and invaded ecosystem is critical to successful management. Managers can further their cause by engaging ecologists in studies of such interactions. The study of how species alter ecosystem properties (Vitousek 1996), how biodiversity and ecosystem processes interface (Silver et al. 1996), and how biodiversity and ecosystem processes interact with landscape dynamics (Turner et al. 1995) is a current, exciting topic in ecology with relevance for both management of invasive species and reconstruction of damaged ecosystems (Jordan et al. 1987; Luken 1990).

Acknowledgments. We thank C. D'Antonio, W. Silver, K. Woods, and the editors for comments. Ideas presented here were developed while the first author was supported by NSF grants BSR 84-15821 and BSR 87-18003 to Stanford University and the University of Hawaii, and NSF grants BSR 88-11902 and DEB

94-11973 to the Terrestrial Ecology Division, University of Puerto Rico, and to the International Institute of Tropical Forestry (USDA Forest Service) as part of the Long-Term Ecological Research Program in the Luquillo Experimental Forest. The second author was supported by NSF and DOE EPSCoR grants and research funding from the Las Vegas Water District.

7
Animal-Mediated Dispersal and Disturbance: Driving Forces Behind Alien Plant Naturalization

Paula M. Schiffman

Wild oat
(*Avena fatua*)

Nonindigenous (NI) plants did not evolve in the ecosystems they colonize. Therefore, relationships that develop between NI plants and indigenous (IN) animals are evolutionarily and ecologically novel and may have unexpected community- and ecosystem-level consequences (Brown 1989). This problem of uniqueness suggests that development of generalizations about animal/NI plant interactions, potentially useful for habitat management and conservation purposes, may be difficult and impractical. It also points to the importance of studying the ecology of NI plants on a species-specific basis (Wagner 1993). Yet, NI species pose serious ongoing problems that must be dealt with in a timely manner (Lubchenco et al. 1991); and the species-specific approach can be slow and costly.

Perhaps if NI plants are viewed specifically within the ecological framework that enables them to integrate into new environments, useful generalizations regarding animal/NI plant interactions will become apparent. Though a complex process, this integration can be divided into two distinct phases, *introduction* and *naturalization*, that may in turn be dissected and evaluated in detail. Interactions with animals are often central to both phases (Fig. 7.1). Moreover, evaluation of animal/NI plant interactions in the context of this conceptual framework should be helpful in point-

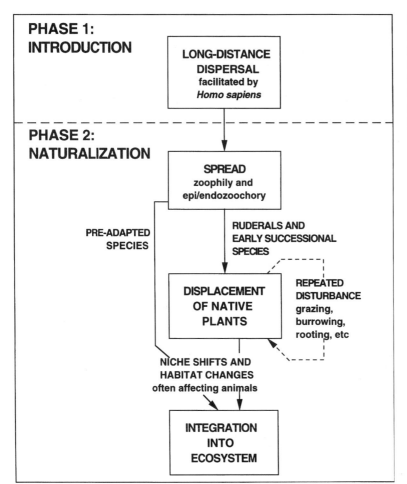

FIGURE 7.1. Diagrammatic representation of the processes that integrate nonindigenous plants into ecosystems. The role of animals is emphasized. (Modified from Bazzaz [1986])

ing research efforts in productive directions, leading to improved management of ecosystems affected by NI plants.

The first phase of the integration process, introduction (Fig. 7.1), includes the means by which plants are transported to and initially colonize new regions. Since most terrestrial angiosperms are mobile only as seeds, introductions of NI plants nearly always occur via long-distance seed dispersal. A variety of vectors can disperse seeds (Sauer 1988; Willson et al. 1990), but most modern introductions of ecologically problematic NI plants have been due to the movements and activities of a single common animal species, *Homo sapiens* (Baker 1986; Chaloupka and Domm 1986; Cronk and Fuller 1995; Heywood 1989; Rejmánek et al. 1991; Wester 1992). These introductions can be either accidental (as seed contaminants in ship ballast, crop seed, and nursery stock or through adherence to livestock fur or automobiles) or purposeful (for forage, fiber, ornamental, medicinal uses, erosion control, and timber production) (Baker 1986; Lonsdale and Lane 1994). While only a fraction of all such long-distance dispersal events results in successful colonizations by NI plants (Groves 1986; Lodge 1993b), this can drastically modify regional floras. For example, since the arrival of Europeans in Hawaii and California about 200 years ago, 813 and 1025 vascular plant species, respectively, have been successfully introduced (Rejmánek and Randall 1994; Wester 1992).

Naturalization is the second phase of the integration process; it is more ecologically complex and immediately problematic than the introduction phase (Fig. 7.1). Therefore, control and eradication of NI plants are dependent upon understanding the processes that have facilitated naturalization. The naturalization phase begins with the reproduction and subsequent spread of plants to parts of the landscape beyond the locations of initial colonization (Fig. 7.1). While some NI species reproduce primarily by asexual means (Barrett 1989; Lodge 1993a, 1993b), population growth of many other species is dependent on some form of pollination, including zoophily (pollination by animals, usually insects, birds, or mammals). Because zoophilous NI plants are often pollination generalists (pollinated by a variety of IN and NI animals), they readily reproduce and spread in the habitats they colonize. In California, for instance, wild radish (*Raphanus sativus*), a widespread NI plant native to Mediterranean Europe, is visited (and presumably pollinated) by a variety of Diptera, Hymenoptera, and Lepidoptera (Ellstrand and Marshall 1985). Interestingly, the honeybee (*Apis mellifera*) is among these. This common pollinator of NI and IN plants is native to southern Asia (Hogue 1993) and is also the primary pollinator of many problematic NI plants, including eucalyptus (*Eucalyptus* spp.), fennel (*Foeniculum vulgare*), horehound (*Marrubium vulgare*), and field mustard (*Brassica rapa*) on Santa Cruz Island in Channel Islands National Park, CA (Thorp et al. 1994). Eradication of feral honeybees from the island appears to have reduced reproduction of NI plants and to have had no perceptible negative impact on IN plant reproduction (Thorp et al. 1994; Wenner and Thorp 1994).

Following successful reproduction, animal-mediated seed dispersal (Herrera 1995; Howe 1986; Howe and Smallwood 1982; Willson et al. 1990) is often the means by which NI plants spread throughout a new habitat. Nonindigenous plants originally from habitats ecologically similar to those they colonize may be sufficiently pre-adapted to become rapidly integrated into the ecosystem (Bazzaz 1986). More often, however, the integration of NI plants into new ecosystems is facilitated by disturbance (Hobbs and Huenneke 1992; Ramakrishnan and Vitousek 1989) (Fig. 7.1). Although human-caused disturbances can be important (Heywood 1989; Lodge 1993b), other animal species are often more directly responsible for producing the disturbances that enable NI plants to become integrated into new ecosystems (Hobbs and Huenneke 1992). The role of animals in the naturalization of NI plants is discussed in more detail below.

Naturalization

Animal-Mediated Spread

Lodge (1993a) noted that invasive organisms typically exhibit high rates of dispersal, enabling them to spread readily throughout newly colonized regions. Although wind and water sometimes disperse seeds of NI plants into the surrounding landscape (Sauer 1988), many of the most important and widespread NI plants produce fruits and seeds with adaptations for dispersal by animals (nearly always birds or mammals). White and Stiles (1991) found that NI plants constituted 33% of the bird-dispersed species in central New Jersey forests. Interestingly, the proportion of invasive species dispersed by vertebrates in South Africa was nearly identical, 32% (Knight 1986). Clearly, if managers are to limit or prevent the spread of NI plants, relationships with potential dispersal agents must be considered and manipulated.

By moving seeds away from parent plants, which are competitively superior, animal dispersers are likely to increase the probability of seedling survivorship (Herrera 1995; Howe and Smallwood 1982). In addition, many animals that disperse seeds around the landscape maintain a certain amount of habitat fidelity. Therefore, animal-dispersed seeds often spread among environmentally similar sites (Janzen 1983; Sorensen 1986). This can increase the likelihood that NI plants will reach conditions necessary for successful establishment (Howe and Smallwood 1982) and can be particularly important for naturalization in fragmented landscapes.

One common mode of animal-mediated dispersal, endozoochory, occurs when animals consume seeds and later regurgitate or defecate them in other locations (Hoffmann and Armesto 1995; Howe and Smallwood 1982; Janzen 1983). Birds usually consume and disperse brightly colored arillate seeds, drupes, or berries that have no scent and are relatively high in nutritional value. Mammal-dispersed seeds tend to occur within large, nutritious arils or fruits that are more aromatic and duller in color than the fruits preferred by birds (Howe 1986). It should be noted, however, that many animals are opportunistic and may disperse seeds from fruits that do not match the above descriptions. For example, the NI deciduous shrub Amur honeysuckle (*Lonicera maackii*), an invader of concern in the eastern United States, produces fleshy fruits with low protein and lipid contents. Despite this relatively low nutritive value, birds and mammals consume these fruits (Ingold and Craycraft 1983; Williams et al. 1992) and disperse the seeds.

In addition to benefiting from animal-mediated transport, germination of some invasive species with thick seed coats (e.g., acacia [*Acacia cyclops*] in South Africa) may be enhanced by gut scarification (Glyphis et al. 1981). Establishment of defecated seeds can also be improved because they are deposited along with a dose of organic fertilizer (Ramakrishnan and Vitousek 1989).

Epizoochory is another important means by which animals facilitate the dispersal of NI seeds into newly colonized landscapes (Milton et al. 1990). It occurs when hooks or sticky substances on the surfaces of seeds, fruits, or associated structures inadvertently attach to animal skin, fur, or feathers. Dispersal occurs when the animal moves to another location and the seeds detach (Howe and Smallwood 1982; Leishman et al. 1995; Sorensen 1986). Because seeds often attach to pelage of grazing animals, grassland and rangeland communities frequently include many epizoochorous invaders, for example, burclover (*Medicago polymorpha*), horehound (*Marrubium vulgare*), and wild oat (*Avena fatua*) in California and spiny cocklebur (*Xanthium spinosum*) in Australia (Baker 1986; Groves 1986; Milton et al. 1990; Sauer 1988). Epizoochorous invaders can be as problematic as endozoochorous ones despite the fact that epizoochory is a relatively rare mode of dispersal in angiosperms (Sorensen 1986). In one California grassland, 21 naturalized NI plant species produced seeds or fruits with morphologies consistent with epizoochory. These species included the most successful plants in the community: NI annual grasses accounting for nearly 100% of the total vegetation cover through competitive displacement of IN grasses and forbs (Schiffman, unpublished data).

There are hundreds of examples, of NI plants that produce seeds spread by animals (Cronk and Fuller 1995). Sometimes these NI plants become tightly allied with their animal dispersers. Such close relationships are particularly problematic because they can lead to the efficient spread and establishment of NI plants. In Natal, South Africa, for example, the Rameron pigeon (*Columba arquatrix*) feeds on bugweed (*Solanum mauritianum*), an NI pest plant, because timber harvesting has eliminated the pigeon's food sources. It has come to depend on bugweed and is now largely responsible for dispersal of this plant (Oatley 1984). A somewhat similar relationship has developed in California grasslands where giant kangaroo rats (*Dipodomys ingens*) occur. The diet of this small granivorous mammal has shifted to accommodate NI plants. The seeds of NI species, for example, filaree (*Erodium cicutarium*), red brome (*Bromus madritensis*), and others, are significantly larger in size than the seeds of IN species and are now preferentially consumed and dispersed by the kangaroo rats (Schiffman 1994).

Invasive organisms usually have generalist life histories (Lodge 1993b), so it is not surprising that naturalization of a single NI plant can be facilitated by several different animal dispersers. Unfortunately, multiple animal dispersers are likely to magnify the difficulties inherent in managing NI plants. The situation can become even more complex if the suite of animals that disperses a particular plant includes NI as well as IN species. Many well-documented cases of this sort occur in Hawaii. The fayatree (*Myrica faya*), for example, is a widespread and aggressive endozoochorous alien. Its seeds are used by several IN birds including 'amakihi (*Hemignathus virens*), 'apapane (*Himatione sanguinea*), and 'oma'o (*Phaeornis obscurus*). Yet, two fruit-eating NI birds, the house finch (*Carpodacus mexicanus*) and the Japanese white-eye (*Zosterops japonica*) are primarily responsible for its dispersal (Cronk and Fuller 1995; LaRosa et al. 1985; Vitousek 1992; Vitousek and Walker 1989). Pigs (*Sus scrofa*) are also very important NI dispersers of invasive plants in Hawaii. They have been implicated in the dispersal of Koster's curse (*Clidemia hirta*) and numerous other NI plant species (Vitousek 1992; Wester and Wood 1977). Additional NI dispersers of Koster's curse include the Japanese white-eye and the mongoose (*Herpestes auropunctatus*) (Wester and Wood 1977).

Controlling the naturalization of NI plants can also be complicated when seeds are dispersed by counterintuitive means. For example, Malo and Suárez (1995a, 1995b) found that grazing animals (cattle [*Bos taurus*], fallow deer [*Dama dama*], red deer [*Cervus elaphus*], and rabbits [*Oryctolagus cuniculus*]) inadvertently consumed and dispersed seeds of 107 Mediterranean species. Moreover, they estimated that a single cow was capable of dispersing 300,000 viable seeds per day. From these impressive numbers and the fact that many of the dispersed species are invasive in California and other parts of the world (Malo and Suárez 1995b), it is clear that endozoochory by animals generally perceived to be foliage-eaters (Janzen 1983, 1984) can rapidly and dramatically alter community composition.

Given the unique and often unpredictable nature of ecological interactions involving NI plants (Brown 1989; Wagner 1993), one might initially expect the management of an NI plant species to be particularly complicated when its animal disperser is also introduced. However, just the opposite may actually be true. Control or eradication of an NI animal that disperses seeds of an NI plant should also negatively affect the plants derived from those seeds. For example, if cattle are found to disperse seeds of NI plants in a California grassland, cessation of grazing would likely lessen the problem. Of course, if an NI plant has multiple dispersers, the ones that remain extant may compensate for the loss of an NI disperser. In such a case, there might be little net effect on the NI plant's ability to spread.

Animal-Mediated Disturbance

The relationship between disturbance and the naturalization of NI plants is widely discussed in the literature (Hobbs 1989; Hobbs and Huenneke 1992; Lodge 1993b; Ramakrishnan and Vitousek 1989). In fact, most NI species are ruderals or they colonize early-successional communities (Heywood 1989; Hobbs and Huenneke 1992; Lodge 1993b) and exhibit a variety of traits, for example, short generation times, high fecundities, and rapid growth rates, enabling them to exploit disturbed habitats (Lodge 1993b). Likewise, the open space and altered resource availabilities characteristic of disturbed habitats promote NI plant establishment (Heywood 1989; Hobbs 1989; Orians 1986). Invasions by NI plants most likely occur at sites that experience disturbances differing from the natural or historical disturbance regimens. In particular, changes in disturbance type and increases in disturbance intensity and/or frequency tend to enhance NI plant invasions by creating new niches that IN species may be less capable of occupying (Hobbs and Huenneke 1992).

Although changes to natural disturbance regimens are virtually always anthropogenic at the core (Heywood 1989), other animal species are often more directly linked to disturbance and subsequent NI plant invasion. Two general disturbance types directly involve animals and typically lead to NI plant invasions: (1) soil disturbances caused by burrowing or rooting and (2) foliar disturbances caused by grazing or other forms of herbivory. Of course, both of these would have occurred in the absence of humans, but they would not have led to NI plant establishment. It is the interplay between human impacts and animal behavior that frequently results in the naturalization of NI plants in disturbed areas.

For example, rooting by feral pigs destroys vegetation and leaf-litter cover and significantly alters soil characteristics (Singer et al. 1984). In Hawaii, pigs were among the earliest human-caused introductions; many aggressive NI plants, including Koster's curse, strawberry guava (*Psidium cattleianum*), passionflower (*Passiflora*), and others, are now naturalized at sites disturbed by pigs (Vitousek 1992; Wester and Wood 1977). Soil disturbances caused by pigs and other feral animals have facilitated NI plant naturalization in many other regions as well (Brockie et al. 1988; Coblentz 1980). Another example can be found in California where grasslands include large numbers of NI plants introduced from the Mediterranean region by humans. The grassland vegetation that becomes established on the disturbed soil of giant kangaroo rat burrows has significantly greater diversity and cover of NI plants than does vegetation growing in relatively undisturbed areas located between burrows (Schiffman 1994). It is both interesting and troubling to note that not only do feral pigs and IN giant kangaroo rats create disturbed sites where NI plants can establish, but they both further enhance the naturalization process by dispersing seeds to the sites they disturb (Schiffman 1994; Vitousek 1992).

Herbivory can also promote integration of NI plants into ecosystems. Schierenbeck et al. (1994) found that NI Japanese honeysuckle (*Lonicera japonica*), a common resident of roadsides, fencerows, and pastures in the southeastern United States (Radford et al. 1968), was better able to withstand herbivory by IN insects and mammals than coral honeysuckle (*L. sempervirens*), an early successional species and congener indigenous to the southeastern United States. Because of its compensatory response to herbivory, the NI honeysuckle would be expected to survive, spread, and become further integrated into the ecosystem, possibly at the expense of IN species.

Sometimes the effects of herbivory are massive, causing large-scale shifts in community composition, such as when changes in disturbance type and intensity associated with introduction of large herds of domesticated livestock result in naturalization of large numbers of NI plants (Aschmann 1991; Groves 1986; Lidicker 1989; Wagner 1989). Prior to the onset of livestock grazing in California, for example, grassland vegetation was dominated by IN perennial grasses and annual forbs, such as California poppy (*Eschscholzia californica*), goldfields (*Lasthenia californica*), needlegrasses (*Nassella*), and pine bluegrass (*Poa secunda*). This vegetation was altered by NI species, for example, brome grasses (*Bromus*), filaree (*Erodium cicutarium*), and oats (*Avena*) when widespread and intensive grazing by cattle and sheep (*Ovis aries*) became a more important form of disturbance than the smaller scale and less intense disturbances (patchy, intermittent herbivory and burrowing) caused by the IN fauna (Keeley 1990).

Niche Shifts and Habitat Change

The successful integration of NI plants into natural ecosystems ultimately requires that the invaded ecosystem somehow change to accommodate the presence of the new species. As noted above, animal-mediated seed dispersal and site disturbance are important driving forces behind this. Animals cause niche shifts, thereby enabling NI plants to fit into the system. The changes to existing habitats resulting from this naturalization process are sometimes far-reaching and problematic from a management perspective.

For example, the branch structure of several NI shrubs, such as Amur honeysuckle, buckthorn (*Rhamnus cathartica*), and *Euonymus*, in the eastern deciduous forest serves as nesting habitat for a large number of IN bird species (Whelan and Dilger 1992; discussed by Williams, Chapter 3, this volume). Similarly, the horizontal cover and increased forage associated with NI submerged hydrilla (*Hydrilla verticillata*) in a Texas lake has led to increased populations of several waterfowl species, including American coot (*Fulica americana*) and pied-billed grebe (*Polilymbus podiceps*) (Esler 1990). Likewise, the availability of food (fruits) associated with spread of the NI multiflora rose (*Rosa multiflora*) has enabled the mockingbird (*Mimus polyglottos*) to expand its range in the northeastern United States (Stiles 1982). In all three of these cases, the naturalization of NI plants has caused avian niche shifts. Eradication of the NI plants would probably negatively affect these bird populations (Whelan and Dilger 1992).

Still, it is important to note that many other invasive NI plants do not enhance bird habitat. For example, in the northern United States and Canada, purple loosestrife (*Lythrum salicaria*), a European native, has aggressively invaded wetlands and, in so doing, has destroyed waterfowl habitat (Hight 1993; Thompson et al. 1987). Moreover, the bird species benefiting from invasions by Amur honeysuckle, buckthorn, *Euonymus*, hydrilla, and multiflora rose are common and widespread and probably would not be driven to extinction by plant eradication efforts. In fact, eradication of these NI plants might actually enable other, less conspicuous IN species, such as invertebrates, plants, and fungi, to reestablish. It is possible that declines of many such less conspicuous species were directly or indirectly caused by NI plant naturalization (Ramakrishnan and Vitousek 1989) but went unnoticed by researchers focused on the increasing bird populations.

Sometimes, however, niche shifts can lead to IN animal/NI plant relationships with very high levels of interdependency. In such situations, the impacts of NI plant eradication can be very uncertain. Interdependencies can be particularly problematic when an IN animal species is already threatened with extinction. South Africa's once-threatened Rameron pigeon and California's endangered giant kangaroo rat are two species that currently depend greatly upon NI plants for food. Research suggests that strong facultative mutualisms have developed between Rameron pigeons and bugweed (Oatley 1984) and between giant kangaroo rats and Mediterranean grasses (Schiffman 1994). If populations of these NI plants are reduced or eradicated, their dependent animals might either go extinct or, alternatively, could shift again to other food sources (possibly produced by recovering IN plants that had been suppressed or displaced by NI competitors). The vast difference between these two management outcomes indicates that the potential for extreme consequences must be carefully considered when strong animal/NI plant interdependencies are apparent.

Conclusions

Because ecologial processes involving invasive NI plants are complex, far-reaching, and often devastating, these species lead to changes now known as "biological pollution" (McKnight 1993). Effective strategies for controlling or eradicating NI plants, and hence stemming the tide of biological pollution, depend on first identifying and understanding the factors enabling these plants to become integrated into ecosystems. To this end, NI species should be viewed expressly within the context of the biotic processes that facilitate naturalization. For many NI plants, naturalization involves interactions with animals. Seed dispersal and habitat disturbance, two types of processes frequently mediated by animals, are key to ensuring the spread and maintenance of NI plants in natural areas.

Interspecific interactions involving NI plants are evolutionarily and ecologically novel; the ecological roles of species in their IN habitats may not be accurate predictors of their roles when introduced into similar but foreign environments, for example, most Mediterranean grasses in California (Jackson 1985). Despite the uncertainties that NI plants pose, a few potentially important and useful generalizations can be drawn regarding animal-mediated seed dispersal, habitat disturbance, and the processes by which NI plants become naturalized.

1. Many NI plants produce seeds or fruits (or related structures) with morpholo-

gies consistent with endozoochory or epizoochory. Initially, identities of animal dispersers may not be known. However, a knowledge of extant indigenous and naturalized fauna should make it possible to identify probable disperser species. The likeliest suspects would be the animal dispersers of IN plants with seed or fruit morphologies and/or nutritional values similar to those produced by the NI plants. The dispersal repertoire of these animals may simply expand to include similar seeds of generalist NI plants.

2. Invasive plants tend to be ecological generalists with the potential to be dispersed by a variety of animals rather than by a single animal specialist. From a management perspective, NI plants with multiple dispersers would be expected to be especially problematic because of the greater degree of ecological complexity favoring their population growth and spread. Therefore, when the seeds of a particular NI plant are dispersed by animals, it is important to identify all possible disperser species (both native and introduced) and to evaluate each in terms of specific dispersal function. Only when these mechanisms are understood can effective dispersal controls be developed and implemented.

3. Naturalization of NI plants is frequently linked to changes in natural disturbance regimens. Although the changes in disturbance type, intensity, and frequency that promote NI plants are typically associated with human activities, IN and NI animals are often the more proximate causes that promote NI plants. By enabling animals to disturb soil and/or foliage in ways that are evolutionarily and ecologically unusual, humans indirectly promote ruderal and early successional NI plants at the expense of IN plants less capable of accommodating new forms of disturbance. Because of this, managers of natural areas affected by NI plants should focus on reversing human impacts. Such efforts might include eradicating feral domesticated animals and eliminating activities such as livestock grazing and forest clearing.

4. Nonindigenous plants would be expected to become particularly problematic when strong interdependencies develop with animals that either disperse the seeds or maintain the habitats through periodic disturbance. This is because efforts to control one species could also negatively affect the species upon which it depends. When a strongly interdependent animal/NI plant relationship is evident, information gleaned from descriptive studies and manipulative experiments is of critical importance if the NI plant is to be successfully controlled and if unforeseen, undesirable management side-effects are to be avoided.

5. By the time it is apparent that a particular animal/NI plant relationship exists, the NI species may be fully integrated into the ecosystem; niche shifts and habitat changes may be difficult to modify. In such cases, comparisons of invaded communities with similar communities where invaders are absent or are not problematic should provide valuable information relevant to habitat management. Not only can such comparisons facilitate identification of specific animal/NI plant interactions, they are the only way to assess the net effects of NI plant naturalization on community composition and function.

These generalizations should serve as useful starting points in the development of guidelines for management and control of NI plants. They are, however, broad and perhaps overly simplistic given the enormous diversity of NI plants and the complexity and novelty of each NI plant/animal interaction. Greater emphasis on scientific research that explicitly addresses interspecific processes involving NI species is the obvious remedy to the many situations where information gaps exist. As more ecological analyses become available, our understanding of the principles underlying the community ecology of invasive NI plants will become more refined and the efficacy of management strategies will improve.

8
Outlook for Plant Invasions: Interactions with Other Agents of Global Change

Laura Foster Huenneke

Japanese honeysuckle
(*Lonicera japonica*)

It is the daunting task of this chapter to peer into the future and speculate about the prospects for plant invasions. We know that the globe is changing in its physical, biological, and cultural features. What impact might these changes have on the rate at which plant invaders arrive in our natural communities and on the effects they have there? One can certainly list some major changes underway in the environment and review the ways in which the arrival or establishment of invaders might be affected; that is how I will begin. However, it is clearly unwise to think of these factors in isolation from one another, and so I go on to discuss possible synergistic relationships among the factors. I will also try to put into a larger management context the entire issue of plant invasions as affected by global change.

I began initially by trying to review the thoughts of others who have worked on these issues. I was surprised to discover the extent to which ecologists studying invasions, and those working on issues of global change, have failed to discuss connections between the two themes. Checking several recent books reviewing global change and its biological consequences, I found surprisingly few cases where invasions, weeds, or nonindigenous (NI) or exotic species were even listed in the index. Those books or chapters that do speak of plant invasions do not treat them as a feature of global change in themselves, or as a process that might be affected by other aspects of global change. Conversely, there has been virtually no mention of global environmental change in major recent reviews of invasions (e.g., di Castri et al. 1990; Drake et al. 1989; Groves and di Castri 1991; Mooney and Drake 1986). Thus I have few models to follow in this discussion.

Primary Agents of Global Change

Vitousek (1994) summarized many of the important changes taking place today on the globe; he emphasized that less attention has been devoted to some well-documented major changes (e.g., changing nitrogen cycling, carbon dioxide increases, or land-use changes) than to the less-certain prospects for global climate change. His article also pointed out that biological invasions are themselves a major agent of global change. With credit to Vitousek, I will go through a list of abiotic and biological changes occurring today, discussing the prospects for them to influence the rate of arrival and establishment of new invaders.

Atmospheric Changes: Increasing Carbon Dioxide Concentrations

The best-documented aspect of global change is the increase of atmospheric concentration of carbon dioxide. Not surprisingly, this is also the area of global change where the most work has been done on potential impacts to weedy invasive plants. Because plants differ in photosynthetic pathway, and because these biochemical pathways differ in their efficiency at low concentrations of carbon dioxide, it seems reasonable that a change in atmospheric concentration of carbon dioxide should alter the competitive relationships among plants with different pathways of carbon fixation. There have been numerous experimental studies comparing the response of C_3 and C_4 pathway species to increased carbon dioxide; most demonstrate that C_3 species benefit more than do C_4 species, growing either alone or in competition (e.g., Bazzaz and Carlson 1984; Carter and Peterson 1983; Patterson 1986; Patterson et al. 1984). This has led some to speculate that C_3 weeds should become more competitive and more troublesome in C_4 crops (or in natural communities dominated by C_4 species), while C_4 weeds should become less problematic with C_3 crops (Patterson and Flint 1980). Similarly, Johnson et al. (1993) speculated that increasing carbon dioxide has affected the balance between C_4 semidesert grasses and C_3 desert shrubs, helping to explain the conversion of many desert grasslands to shrub-dominated systems. One might conclude, then, that C_4-dominated communities may be increasingly vulnerable to invasion by C_3 species. On the other hand, not all C_3 plants respond to increased carbon dioxide with increased biomass or reproduction; there are definite differences among species in the magnitude and even the direction of response (Garbutt and Bazzaz 1984), making generalizations dangerous.

Altered carbon dioxide concentrations may have differential effects, even within a genus: for example, Sasek and Strain (1991) found that the NI Japanese honeysuckle (*Lonicera japonica*) responded much more vigorously in vertical growth and new leaf production than did the North American indigenous (IN) coral honeysuckle (*L. sempervirens*) when both were grown at elevated carbon dioxide. The authors reported that both vines responded more vigorously than has been reported for other growth forms of woody plants, suggesting that vines might be expected to become even more serious invaders. Similarly, kudzu (*Pueraria montana* [*P. montana*]) responded to increased carbon dioxide with more rapid leaf expansion rates, greater height growth, and increased branching (Sasek and Strain 1988, 1989). These studies attributed the tremendous response of vines to the fact that they do not divert much of their increased carbon gain into support structures, relying instead on other plants for support. Because woody vines are already regarded as among the most undesirable of invaders, this is a sobering thought.

Another perspective on the interaction of carbon dioxide and resource allocation was offered by Hunt et al. (1995), who compared a number of grasses grown under ambient and enriched carbon dioxide and found that those species with the greatest ability to shift allocation to root systems had the greatest ability to increase total growth with an increase in carbon dioxide. Morphological plasticity, then, might be a predictor of a species' responsiveness to carbon dioxide increase (and of its relative success in a world of increased carbon dioxide).

Release of some species from water stress may be expected, as the potential for greater stomatal closure may mean less complete exhaustion of soil moisture (e.g., Eamus and Jarvis 1989; Patterson 1986), altering or ameliorating the harshness of dry sites. Many species also respond to increasing carbon dioxide by reducing stomatal density (Woodward and Bazzaz 1988). While there is no generalization that can be made about the relative ability of invading NI plants versus established IN plants to take advantage of this, one can be certain that competition among species will be altered as moisture availability changes.

Because an increase in biomass without corresponding increase in nitrogen uptake means higher C:N ratios and increased nitrogen use efficiency, the suggestion has also been made that highly responsive species (such as the cocklebur *Xanthium occidentale*) will increase in vigor in marginal or low-fertility sites (Hocking and Meyer 1985). Many weed species are currently restricted to high-nutrient sites; carbon dioxide enrichment may loosen these current constraints. One would predict then that a large number of naturalized species could become more invasive in natural or relatively undisturbed settings, and also that low-nutrient ecosystems may become increasingly vulnerable to invasion.

Increases in Ultraviolet Radiation

Ultraviolet-B (UV-B) radiation is predicted to increase as stratospheric ozone depletion continues and becomes more widespread geographically, especially over temperate regions. This solar UV-B increase may have overall negative effects on growth and productivity of some plants but is also predicted to have important indirect effects on competition and ecological processes in natural communities (Caldwell et al. 1989). In particular, increased UV-B levels stimulate changes in morphology and leaf display of plants, particularly monocotyledons, which might result in especially strong effects on the competition in grass-dicot mixtures (Barnes et al. 1990).

Pollution Issues

Sulfates and other components of acid deposition from the atmosphere represent one concern about the potential impact of atmospheric pollution. In one experimental study, the normal detrimental effect of sulfates on C_3 species was ameliorated by high carbon dioxide levels, while C_4 plants were more severely damaged than at ambient carbon dioxide (Carlson and Bazzaz 1982); hence sulfate deposition might exaggerate the advantage of C_3 species over C_4 plants in a world characterized by elevated carbon dioxide. Another possibility, of course, is that acid deposition will alter soil or soil solution pH sufficiently to change species composition. In the case of acidic soils becoming even more acidic, it is difficult to imagine plant colonization; but where the buffering capacity of alkaline soils is overwhelmed by acid deposition, IN plants might find greater competition from more broadly adapted, robust species.

While we worry about stratospheric ozone depletion and subsequent increases in ultraviolet radiation, abnormally high ozone concentrations here in the troposphere are substantial concerns downwind from urban and industrialized regions. As for most of the factors discussed here, ozone represents a severe stress to which different species respond individualistically (EPA 1986); the only general argument one might advance is that an invasive NI plant species, should it prove to be relatively tolerant to ozone's effects, will become increasingly successful relative to nontolerant IN species. We do not have broad-based surveys, however, to judge how many potential invaders might qualify. Westman and Malanson (1992) summarized evidence that ozone concentrations have depressed the performance of IN Californian shrubs relative to that of several NI species.

The last form of atmospheric pollution I discuss is the increased flux of nitrogen. Under natural conditions atmospheric nitrogen is fixed into plant-available forms by lightning and by microbial fixation. In industrial times, however, anthropogenic nitrogen fixation now takes place at a rate roughly equivalent to annual natural fixation (Schlesinger 1991), meaning that twice as much nitrogen is made available each year. While most of this nitrogen is applied directly to agricultural systems, much ends up in the atmosphere and reaches terrestrial systems in various forms in both precipitation and dry deposition. The rate of deposition of nitrogen has increased across the globe, most dramatically downwind of major industrial regions (Galloway et al. 1995) but significantly even in remote areas (e.g., Mayewski and Legrand 1990).

Nitrogen is frequently limiting to plant growth in natural ecosystems, and increases in nitrogen availability can alter the competition among species (Tilman 1988). In particular, a pattern of community response has been observed repeatedly in many natural systems: higher fertility leads to competitive exclusion of slow-growing plants by robust, fast-growing species (especially grasses). This phenomenon of decreased species richness, following the nitrogen-enhanced performance of robust grasses and other weeds, has been observed in several areas of conservation concern worldwide, for example, heathlands (Aerts and Berendse 1988; Heil and Diemont 1983), chalk grassland (Bobbink 1991; Bobbink and Willems 1987), and serpentine grassland (Huenneke et al. 1990). Even in desert systems, nitrogen additions have enhanced the performance of NI plants (Mun and Whitford 1989). This process poses the greatest threat to very low-fertility ecosystems, for these would contain IN species with the lowest potential growth response to nitrogen (Chapin et al. 1986). Marrs (1993) provided an excellent summary of the many problems that increased nitrogen availability can pose for nature reserves, including decreased plant species richness due to increased competitive pressure from grasses.

Land-Use Changes and Fragmentation

Conversion of natural ecosystems to human-dominated systems, or the abandonment of land after human use, is the most conspicuous of human effects on the biosphere (Turner 1994). There are many links between land-use patterns and the likelihood of plant invasions. For one thing, fragmentation of previously intact habitat creates "edges"; edge habitat may itself provide suitable sites for establishment of invasive NI species (e.g., Luken and Goessling 1995) or, at a minimum, may increase the perimeter across which invasions of the interior can occur. Roadways and other transportation corridors associated with fragmentation may serve as avenues of introduction of seeds of weedy plants. For example, tourist autos and buses were found to transport weed seeds into and within an Australian national park (Lonsdale and Lane 1994). Historical analyses suggest that roads and railroads are major pathways for the migration of invasive species (Frenkel 1970; Luken and Thieret 1993) and that ports and major agricultural centers have served as the point of entry for many problem plants (Forcella and Harvey 1988). Increased trade can only enhance this tendency; at least this suggests certain areas that should be more intensively monitored for the appearance of novel species. It also suggests the importance of limiting roadways or other migration corridors in natural areas. Sadly, the appearance of new roads and corridors of disturbance in previously intact natural habitat is continuing; this fragmentation and increasing accessibility are important features of change in land cover today. Armesto et al. (1991) suggested that greater heterogeneity of the landscape (equivalent to greater fragmentation) would facilitate invasion; certainly this describes the decreasing patch size of natural remnants and preserves. Even our well-intentioned desire to provide corridors of natural habitat for the dispersal of IN organisms—for example, as connections among preserves—may backfire by providing avenues for movement of NI invaders (Simberloff and Cox 1987).

Alteration of Disturbance Regimens

Disturbance is an integral feature of many (or perhaps most) natural systems (Pickett and White 1985); the alteration of natural disturbance regimens can create opportunities for invasions (Hobbs and Huenneke 1992). Increases or decreases in typical disturbance frequency and intensity, as well as the introduction of new types of disturbance, can stimulate colonization and spread of NI plants. It is unclear how disturbance regimens are likely to change in the future.

Fire remains the most conspicuous example of human effects on natural disturbance patterns. While we may have become wiser about the ecological costs of fire suppression in forested and chaparral systems, fire prevention remains a primary objective of natural-area management in most areas of North America with significant human population density. On the other hand, human action has increased fire frequency in several regions. Also, alteration of vegetation (including the invasion of fire-carrying grasses [D'Antonio and Vitousek 1992]) has led to greatly increased fire frequency and intensity in some ecosystems. Keeley (1995) found that few species indigenous to fire-dependent communities in California had been threatened by fire suppression, but he suggested that increases in fire frequency had been far more devastating. In particular he identified the deliberate seeding of NI plants after fire as a severe threat to IN species.

Riparian systems also contain IN plants adapted to the disturbance of floods, scouring of sediments, and variable water levels. I found no documentation that alterations of hydrologic regimens (e.g., the ubiquitous impoundment and control of North American rivers) have enhanced invasions of either aquatic or riparian plants, but I would predict that at least some invaders have benefitted from this; field studies of this phenomenon are needed. Certainly the prevention of natural flood disturbance has had a negative impact on the regeneration of IN riparian woody plants of the western United States (e.g., Fenner et al. 1985; Ripple 1990). Woody invaders, such as tamarisk (*Tamarix*), and herbaceous invaders may alter the hydrology of invaded streams (Blackburn et al. 1982; Dudley and Grimm 1994), which in turn might further favor NI species over IN species.

We should keep in mind that a simple cessation of disturbance will not be sufficient to keep the balance tipped in favor of IN species over NI ones. It is the alteration of a natural disturbance regimen (in intensity, frequency, or quality), not necessarily an increase, that favors invaders; any disturbance regimen favors some species over others, and there is no overall generalization (Hobbs and Huenneke 1992). The literature is replete with examples where invasions have occurred or invaders have persisted even in the absence of human activity and disturbance (e.g., Brandt and Rickard 1994).

Alterations of Biotic Interactions

Plant species are often involved in important ecological relationships with other organisms in the community: animal herbivores, pollinators, and dispersal agents; symbiotic mycorrhizae and nitrogen fixers; and plant competitors and facilitators (nurse plants). Because human activity is affecting global patterns of biodiversity (UNEP 1995), one may reasonably expect disruption of many such relationships and subsequent alteration of rates and effects of invasion. For example, if populations of frugivorous birds decline, one might conclude that the rate of spread of a bird-dispersed invader would actually be slowed. On the other hand, introduction of some animals may provide dispersal vectors for invading plants. For example, the rapid population increase of several invading plants in Hawaii is apparently enhanced by activities of introduced birds and feral pigs (Huenneke and Vitousek 1990). A dramatic example comes from Florida, where several species of banyon and strangler figs (*Ficus*) were used as ornamentals for decades without problem; after the accidental introduction of specialist wasps, these trees are now being pollinated and have started to reproduce successfully (McKey and Kaufmann 1991). Finally, the introduction and invasion of plant pathogens can create openings in plant communities that could certainly enhance the likelihood of invasion. I know of no documented cases where invasion of forests was facilitated by disturbances associated with chestnut blight or Dutch elm disease; sadly, I am certain there will be future opportunities to study this phenomenon. The conclusion is that entire "suites" of invaders may enter a system, and the presence of some may serve to facilitate later arrivals.

Trade and Transportation Patterns

With growing human populations and ever more open communication, there is freer economic trade and activity among nations. This economic flow is exemplified by increasing rates of movement of both people and goods across borders, with decreasing emphasis on inspections, quarantines, or other barriers to transport. This is in no way a negative opinion of the desirability of such openness, but simply a recognition that such movements will lead inevitably to further introductions of organisms. Ongoing wars and the resulting displacements of refugees across borders are a tragic form of this same flow. Whether plant seeds arrive in uninspected container shipments at a port, or in the digestive systems of imported livestock, or simply on the muddy shoes of people crossing borders, we can predict a continuation of accidental plant (and pathogen) introductions.

Not all introductions are accidental, of course. The introduction and development of plant varieties for many purposes (improved forage or range plants, ornamentals and landscape plants, etc.) continue and are economically very important. Lonsdale (1994) and Richardson et al. (1994) pointed out that the very features that make desirable plant introductions (for forestry or forage) can contribute to the species' invasive nature as they "escape" from introduction sites and move into natural ecosystems. Despite that, economic interests continue to drive the rapid introduction of plants (and animals) from one portion of the world to another. The growing trend of evaluating landscape plants for "homeoclimatic matching" in order to create water-conserving landscapes in dry regions, for example, merely exacerbates the likelihood that new introductions will become naturalized. Ironically, as ecologists recognize that some purposeful introductions have had negative ecological consequences (e.g., the introduction of invasive dryland grasses into southwestern rangelands [Anable et al. 1992]), resources are still being spent to ensure that an introduced grass (known to be invasive) can survive herbicide treatments directed against an IN plant (Rasmussen et al. 1985).

Climate Change

Projections of global circulation models under a scenario of continuing increases in atmospheric carbon dioxide suggest that climatic patterns may change rapidly in the next century (e.g., Schneider 1993). While these changes are referred to simplistically as "global warming," changes are projected to be complex and highly variable from region to region. For example, polar and temperate regions are likely to experience more dramatic increases in temperature than the tropics, and a general acceleration of the hydrologic cycle will probably not result in uniform increases in precipitation across the globe. In the literature there has been much more concern about the likelihood of climatic change eliminating IN species from reserves (e.g., Hengeveld 1989; Peters and Darling 1985) than about the altered probability of invasions.

Peters (1992) summarized predictions about the ways in which species' geographic ranges might change with altered climate. He pointed out that climate change and subsequent migration imply that many species will face "exotics" or at least novel assemblages of neighbors. His chapter also mentions that drought and fire (two processes that may increase in frequency with changing climate in certain regions) can enhance the likelihood of invasion.

While it is not clear how the current ranges of many species are established by climatic suitability, there are at least some species for which temperature is an important constraint. For example, the ability of some perennial weeds to overwinter is definitely related to minimum winter temperatures (Schimming and Messersmith 1988); if climatic warming does indeed lead to milder winters in northern regions (Schneider 1993), then one would expect these species to become even greater problems in rangeland and natural systems. The likelihood of noxious semitropical and tropical invaders moving north also increases; in that case, those of us living in more temperate climates will no longer feel comfortably distant from the intense invasion pressures facing Florida and California. Soil-warming experiments have altered species dominance in montane meadows (Harte and Shaw 1995); it may be that high elevation and cool boreal environments will be the most susceptible to long-term temperature changes.

An analysis of pine invasions (escapes from plantations in the southern hemisphere) suggests that more open habitats—bare ground, dunes, grasslands—are more vulnerable to invasion than are shrublands and forests (Richardson et al. 1994); thus ecosystems where some aspect of global change removes woody cover are likely to be most vulnerable to continued invasions. At least some forest types in drier areas are predicted by individual-based forest gap models to respond to climatic warming by a decline in tree cover and a general opening of the vegetation (e.g., Smith et al. 1992); these might be susceptible to invasion by new species. On the other hand, nature reserves in arid regions have been most heavily invaded along riparian or wetland sites (Loope, Hamann, and Stone 1988), suggesting that these relatively high-cover systems are also vulnerable to species invasions.

Paleoecological studies have been used as exemplars of the ways in which plant species respond to changing climates (Davis 1989b; Webb 1992). MacDonald (1993) reviewed paleoecological literature of plant migrations and invasions in Europe and North America. The evidence suggests that plant migrations and invasions into regions of suitable climate can be quite rapid, and invasion appears to have been faster for species moving into forests dominated by short-lived or pioneer species than for those migrating into closed forests of long-lived dominants. Another message from paleoecological studies is that many temperate plants are somewhat constrained in their dispersal and migration rates (Adams and Woodward 1992), suggesting that those species with good dispersal ability will be disproportionately favored as climate changes and communities are in disequilibrium. That is, the sort of plant that today is an effective weed may in the future be even more successful, relative to many IN plants of undisturbed communities.

Interactions Among Factors

The list of global changes above is by no means an exhaustive one; many features of the global environment are changing rapidly. The formidable task of predicting ecological responses is made even more difficult by the fact that interactions among agents of change can be strong. For example, it is often speculated that nutrient limitations in natural communities will constrain plants from responding to increased carbon dioxide concentrations, but others have found that plants can increase production even under extreme nutrient stress (e.g., Patterson and Flint 1982). Certainly there are significant interactions among carbon dioxide, nutrient availability, and individual species' characteristics in determining relative performance of plants in competitive mixtures (Zangerl and Bazzaz 1984). It has been suggested by Franklin et al. (1992) that the future of northwestern North American forests depends on interaction between disturbance (i.e., fire regimen) and climatic change; they speculated that this interaction may increase the likelihood of outbreaks of insect pests and pathogens, but they did not mention the equally increased chance of plant invasions.

Changes in nitrogen flux may interact with other factors; increased nitrogen deposition in the form of ammonium can enhance acidification and is implicated in forest decline (i.e., disturbance or opening of the forest canopy) (Sutton et al. 1993). The paleoecological record also suggests that plant invasions will be facilitated by the increase in forest disturbance rates and forest tree turnover caused by anthropogenic activities (MacDonald 1993). Schlesinger (1994) pointed out that many industrial effects lead to an impoverishment of species richness of natural systems, and that this decline in diversity itself can trigger invasions and an enhancement of species loss. Finally, the various agents of global change interact with the biological properties of plant species themselves. Geber and Dawson (1993) emphasized the potential for plants to respond to various features of global change through microevolution or adaptive genetic changes; it appears that species with high evolutionary potential (rapid reproduction, high levels of diversity, and so on) might have the greatest advantage as invaders in a future changing world.

Mooney (1991) summarized the complexity of some of these interactions: rising carbon dioxide itself can alter competition between C_3 and C_4 plants, but this is affected by temperature and by water availability, both of which would be changing with climate. Meanwhile, carbon dioxide enrichment will alter the quality (C:N ratio) of foliage and thus alter herbivore feeding rates, litter decomposition, and nutrient availability; these can further alter the balance among species. Nutrient availability is also affected directly by human influence on cycling of nitrogen and sulfur. And all these changes are acting at different temporal and spatial scales. So a simple scaling-up from studies of plant parts or single ecosystem components will not be very satisfactory for predictive purposes. Hence the difficulty of this chapter's assignment: there are no simple forecasts to be made about the response of invaders (or any other group of organisms) to global change.

Conclusion

The reviews of paleoecological data remind us that invasion by NI species is not qualitatively different from the progressive assembly and migrations of IN plants, which are constantly in flux (see Jackson, Chapter 4, this volume). The same idea was emphasized by Lodge (1993b), who pointed out that invasions have occurred constantly through evolutionary and ecological time; it is only the rate that has been altered by human action. Invasion is a natural biological process. But we should acknowledge that the rate of introductions is unprecedented. Communities have not been bombarded with so many newcomers in so short a time before; we must therefore be prepared to act aggressively if we want to confine composition of some ecosystem to a certain list of species.

The bottom line is that a continuing increase in human population and impact (the product of population size and resource consumption) will continue to drive changes in the global environment. Human activity will continue to import more species into natural areas at an ever increasing rate and to alter the environments in which IN species compete with invaders (Westman and Malanson 1992). The ultimate state is a greater homogeneity of the biosphere, which will be countered only by deliberate actions on our part. This will mean more deliberate control of species composition, both of "utilized" landscapes and of so-called natural areas or reserves.

In North America we tend to focus on preservation of "pristine" ecosystems, which encourages a dichotomous view of natural (worth preserving) versus human-influenced (by implication, not worth preserving). However, in Europe, there is explicit recognition that some of the landscapes of greatest conservation interest (e.g., heathlands, chalk grassland) have been shaped by human activities over centuries or millenia. I suspect that the problem of plant invasions will stimulate us to take a more realistic view of the extent to which future plant communities will be products of deliberate management.

The importance of deliberate action will force us to make more difficult choices, rather than to hope that undisturbed areas will somehow escape intact. We will have to dedicate more resources and give greater importance to schemes for limiting introductions (or at least limiting the damage done after some undesirable plant arrives). We will have to choose in some cases between maintaining natural processes (e.g., the activities of endangered giant kangaroo rats in California grasslands) and excluding invasive plants that benefit from that process or disturbance (Schiffman 1994). We will also have to develop the wisdom to realize when control or exclusion tactics have greater costs than benefits, or when NI species can actually be used to restore some features of environmental quality to a degraded natural system (De Pietri 1992). We must combine short-term aggressive actions to limit local invaders with longer-term educational and research efforts (Schierenbeck 1995). And finally, we must not become too discouraged to exert our best efforts in managing natural areas for those qualities we value most.

Acknowledgments. I thank Jim Luken for presenting me with the opportunity to think about these issues in a serious fashion. My understanding of invasions as a form of global change was sharpened by my participation in a summer 1994 workshop at the Aspen Global Change Institute; I thank the organizers, Peter Vitousek and Carla D'Antonio, for inviting me, and am grateful to all the participants for their insights and perspectives. Preparation of this manuscript was supported by grants from the National Biological Service and by the Jornada Long-Term Ecological Research program (NSF grant DEB-94111971).

9
Experimental Design for Plant Removal and Restoration

Michael L. Morrison

White-top
(*Cardaria draba*)

It would be difficult to argue with the notion that reliable information is central to the development of good ideas about natural-resource management. There is, however, considerable debate regarding how one obtains reliable information; this debate is a centerpiece of biology but is by no means unique to this field (just ask an economist). Although there is a process known as the scientific method, it calls primarily for the testing of hypotheses and the collecting and evaluating of data. There are, in fact, many ways of "doing science," each of which contains merits and shortcomings. However, each has its place in the scientific method, and one method is seldom an adequate replacement for another.

Nowhere is this debate more obvious than in ecology. Steeped in a tradition of observational studies of nature (natural history), ecology today contains numerous subdisciplines, all of which are developing theories that seek to explain distribution and abundance of organisms. However, the organisms studied by ecologists are increasingly affected by human use of the world and the concomitant impacts such as fragmentation of habitats, pollutants, extinctions, introduction of nonindigenous (NI) plants and animals, and a host of other factors. Thus, basic ecology, with its ultimate goal of prediction, occurs in a context of complex interactions due to the human element. Such prediction is difficult with organisms growing in culture on a lab bench, let alone in the field with the added complication of planned or unplanned human intervention.

It is easy to argue that adherence to scientific methods provides a known and repeatable level of knowledge that is essential if we are to advance our understanding of the environment and thus be in a better position to predict the consequences of human actions. Furthermore, it is important that we avoid basing management decisions on preconceived notions and built-in biases deriving from either a lack of study or poorly conceived studies. Recently, the notion that there are basic sciences and applied sciences, and that somehow the latter require less rigor than the former, has crept into the management and conservation vocabulary. Rather, all science requires adherence to rigorous scientific methods; only good science should ever be applied. In large part, ecology as a discipline has failed to advance rapidly (in its ability to provide reliable knowledge to managers) not because it is inherently flawed but, rather, because its practitioners have failed to treat it as rigorous science (sensu Peters 1991; Romesburg 1981).

My goal for this chapter is to summarize the basic principles required for the gaining of reliable information in three basic contexts: assessment of community change due to plant invasion, assessment of treatment effects on target plants, and assessment of community response after treatment to remove target plants. This chapter could easily be adapted to the study of any organism, indigenous or otherwise.

Scientific Methods

Knowledge results from the application of scientific methods. There is not, however, any single best method. Rather, several of them are suitable for different purposes (Romesburg 1981). Romesburg (1981) identified three main scientific methods. First, *induction* applies to the finding of associations between classes of facts. For example, making the observation that a plant species usually grows along the edges of roads, and then explaining this phenomenon by statistically correlating the abundance of this plant with soil moisture. The more times this association is made, the more reliable it becomes. Thus induction is reliable if done properly, but it does not provide insight into the process causing the relationship. Management decisions based on induction often fail because users extend the relationship well beyond the observed conditions, for example, by applying a relationship in a geographic region or time period differing from that used in its development.

Second, *retroduction* refers to the statement of hypotheses about associations that are explanations or reasons for observed relationships. In our previous example, we might hypothesize that the association was due to drainage from the road, thus providing extra water for the plant (as compared to direct rainfall). Unfortunately, retroductive explanations often provide unreliable information because there are usually many possible alternatives. In the example, although our initial explanation seems intuitive, the resulting relationship might actually be caused by excess roadway salts, which eliminate competitors of the plant. This problem with retroduction (and actually induction as well) brings us to Romesburg's third category, the *hypothetico-deductive*, or H-D method. The H-D method complements retroduction in that it starts with a hypothesis—usually developed through retroduction—and then makes testable predictions about other classes of facts that should be true if the hypothesis is true. This is fundamentally a method of determining the reliability of our ideas. In our example, application of the H-D method would involve an experiment to test the influence of water, salt, and other factors on plant growth.

Unfortunately, most studies in ecology are based on inductive or retroductive methods. This is one reason why many researchers now stress the importance of studies seeking to determine why relationships occur (Gavin 1989). If the underlying processes are understood, then there is a better chance of applying results obtained in one area or time to another spatio-temporal context. Again, this does not diminish the usefulness of inductive-retroductive studies. Rather, it means only that we should not expect them to do more than they are designed to do; we should avoid the temptation to draw conclusions and develop management policies on potentially biased, limited, and unreliable information.

Monitoring as Research

A common misconception among resource managers is that monitoring is something different than research and, as such, requires less rigor in its design and implementation. This misconception is widespread in some agencies and has lead to many poorly designed studies and to subsequent management decisions based on invalid data (Morrison and Marcot 1994). As defined by Green (1979), a monitoring study has the purpose of detecting a change from the present state (e.g., an increase in population of an invading plant species). The resulting data can provide a baseline against which future changes can be compared. However, the precision at which such changes can be assessed will depend on the quality of the monitoring study. Thus, just "going out there and collecting some data" in the vague hope that the data will be useful in the future is foolish.

In developing a monitoring study, then, researchers must decide a priori what level of change is to be considered an impact. This will determine the sampling intensity (i.e., number of study plots) and frequency (i.e., time between sampling events). For example, suppose a manager is interested in tracking change in abundance of an NI plant thought to be expanding in population size. The amount of change considered to be a problem in biological terms must be determined. Specific steps can then be taken to design the sampling protocol needed to detect this change with a desired degree of certainty. This will be discussed in greater detail later in the chapter.

Principles of Design

In what is certainly a classic work on study design, Green (1979) developed basic principles of sampling design and statistical analysis relevant to any study on the assessment and management of plant invasions. I summarize below the most salient aspects of Green's "Ten Principles."

1. *State concisely to someone else what question you are asking*: *your results will be as coherent and as comprehensible as your initial conception of the problem*. The goals of a study flow from initial statement of the problem (e.g., do NI plants have any impact on indigenous [IN] species?) to statement of the specific null hypothesis (e.g., the abundance of NI species x does not significantly reduce the abundance of IN species y). It is absolutely critical that this step be well developed, or all that follows might be for naught. Asking if a particular chemical reduces the cover of NI species requires a much different design than asking if a reduction in cover has a positive influence on IN organisms. Further, contrary to popular belief, there are "bad" questions. For example, knowing (through observation) that an NI species reduces the density of IN organisms leaves to supposition the effects of this reduction on the IN organisms. A good ecologist can put a desired spin on any outcome (i.e., retroduction). In the previous situation, a reduction in density can be viewed as a positive outcome due to increased growth and vigor of the remaining plants (released from competition), or it can be seen as negative for a host of reasons.

In establishing goals for a study, it is critical that the spatial and temporal applicability of the results be determined. If the goal is to determine the best means of reducing population size of a widely distributed NI species, then establishing all study plots in a restricted space and time will unlikely have broad applicability (because of variations in environmental conditions). Initial goals will determine spatial distribution of the sampling areas and the temporal nature of sampling.

Additionally, the confidence one desires in the results must be stated during planning. Asking if a pathogen reduces an NI species by 50% ± 10% requires much more rigorous sampling than asking if it reduces it by 50% ± 25%. Similarly, even asking if the pathogen changes the cover of this species is different than asking if it reduces the cover. There is an obvious interplay among the questions you ask, the sample size needed to answer them reliably, and thus the generality you can expect from your results (see "Desired Precision: Statistical Significance and Power," this chapter).

2. *Take replicate samples within each combination of time, location, and any other controlled variable; replication is thus essential.* The subject of replication is detailed in "Experimental Design," this chapter, including the often-encountered problem of pseudoreplication.

3. *Take an equal number of randomly allocated replicate samples for each combination of controlled variables.* An absolutely critical aspect of study design is random placement of study plots and, usually, random sampling in the plots. This prevents introduction of a major source of observer bias into any study. The topic of randomization is detailed in "Experimental Design," this chapter.

4. *Test whether an impact has an effect.* Collect samples both where the impact is present and where the impact is absent but where all else is similar; an effect can be shown only by comparison with a control. Lack of proper controls is a major weakness in many studies of environmental impacts. Determining the degree of similarity among replicated controls is a major problem with study design; it is discussed below in "Experimental Design."

5. *Conduct preliminary sampling to provide a basis for evaluation of sampling designs and options for analyses.* As stated by Green (1979:31), "Those who skip this step because they do not have enough time usually end up losing time." Preliminary work allows verification that your sampling procedures are actually sampling the segment of the population or environment that you intended; it allows evaluation of necessary sample sizes (see "Desired Precision: Statistical Significance and Power," below, for details); and it allows modification in sampling techniques before you become locked into a particular methodology. This step, usually skipped, certainly leads to more wasted time and money than any other step in a research study. I usually advise that one should first study the study before charging into plot establishment.

6. *Verify that your sampling device or method is measuring the population as a whole and is sampling it with equal and adequate efficiency over the range of conditions likely to be encountered.* Variation in efficiency of sampling from area to area biases among-area comparisons. In studies that require capturing animals, it is well established that different species, and ages and sexes within species, have differential capturability. Different sampling tools and methods are often necessary to sample adequately the species composition and vegetational profile of an area. Bias may not be a problem, especially if a sampling method is applied consistently across all study locations. However, the appearance of bias in a finished study can greatly diminish the importance that others will apply to your results. Potential biases should be identified in the planning stages, and steps implemented to remove their influence on results.

7. *Divide the area into relatively homogeneous sub-areas if the area to be sampled is environmentally heterogeneous; allocate samples to each sub-area in proportion to sub-area size.* As noted in principle 3, allocation of samples should be based on some type of randomization; steps to achieve this are provided in "Experimental Design," below. The geographic area over which study results are to be applied determines the spatial scale at which sampling should be conducted. For example, all else considered equal (e.g., time or funding), the application of study results to wide geographic areas might allow basing the distribution of sampling plots on gross vegetation types, whereas the need to apply results to a single nature reserve likely will require a finer level of resolution (e.g., based on floristics) of sampling. These are issues that must be deter-

mined during development of study goals (principle 1).

8. *Verify that your sample unit size is appropriate to the sizes, densities, and spatial distributions of the organisms you are sampling. Then estimate the number of replicate samples required to obtain the desired precision.* This principle is closely allied to principle 5 (conducting preliminary sampling). As suggested by Green (1979), the spatial distribution of organisms is important in establishing a sampling unit. As discussed below, most ecologists simply gather as many data as they can, somehow hoping that their efforts will be adequate to convince peer reviewers to recommend publication of their work. Of course, seldom are analyses of sample size requested by journal editors. As such, we are either over- or undersampling; neither scenario is a productive use of time.

9. *Test your data to determine whether the error variation is homogeneous, normally distributed, and independent of the mean.* This principle should be explored during the preliminary phase of the study (principle 5). As noted by Green (1979), environmental studies fall into two general categories: those that ignore the fact that there are assumptions at all, and those that are paranoid about assumptions and rely on nonparametric procedures.

Green argued that (1) the assumptions of the statistical method should be understood at the time it is chosen, (2) the likelihood and consequences of violation should be addressed, and then (3) use of the method should proceed with awareness of risks and possible remedies. Assumptions such as normality and homogeneity of variances are seldom met (especially for multivariate data). However, violation of assumptions does not negate the use of many parametric tests because most are robust enough to tolerate violations of assumptions given adequate sample sizes (and, in appropriate situations, equal sample sizes among groups). There is no reason to believe that biological phenomena will be linear or have equal distributions among groups; such conditions are the abnormal. Green (1979) and most general statistics texts provide detailed discussion of transformations and related topics.

10. *Choose the best statistical methods to test your hypothesis; stick with the result. An unexpected or undesired result is not a valid reason for rejecting the method and hunting for a "better" one.* Data dredging, as it is commonly called, has no place in ecological research. Transforming data, discarding outliers, creating new variables, rearranging study contrasts (e.g., changing pairs of grids), and so on are usually not appropriate unless planned before the study begins. As noted by Green (1979), it has always been possible to arrange things so that you will be told what you want to hear. Analyses should be specified, and P-values established, before the study begins.

Optimal and Suboptimal Designs

There are three prerequisites for an optimal study design: (1) the impact (a general term used for change, e.g., construction or treatment) must not have occurred, so that pre-impact baseline data can provide a temporal control for comparing with post-impact data; (2) the type of impact and the time and place of occurrence must be known; and (3) nonimpacted controls must be available (Fig. 9.1, scenario 1). As noted by Green (1979), an optimal design is thus an areas-by-times factorial study in which evidence for an impact is a significant areas-by-times interaction. For example, in an ideal situation for natural areas, plots would be in place to monitor population trends of IN plants. Ideally, with this monitoring system the initial time of plant invasion would also be known. And there would be areas where no invasion is occurring. Given that the prerequisites for an optimal design are met, the choice of a specific sampling design and statistical analysis should be based on your ability (1) to test the null hypothesis that any change in the impacted area does not differ from the control, (2) to relate to the impact any demonstrated change unique to the impacted area, and (3) to separate effects caused by naturally occurring variation unrelated to the impact (Green 1979).

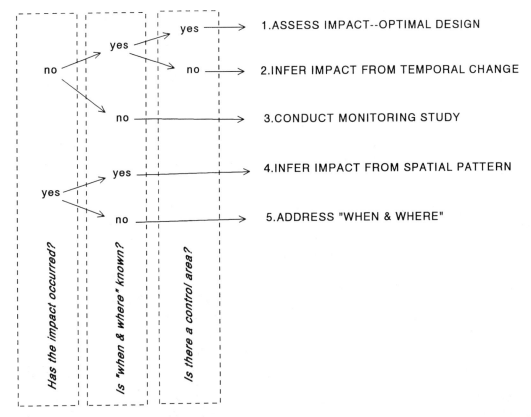

FIGURE 9.1. The decision key for determining the type of design necessary for an environmental study. (From Green [1979]. Figure 3.4, reprinted by permission of John Wiley and Sons, Inc.)

It is often not possible, however, to meet the criteria for development of an optimal design. Impacts often occur unexpectedly. Under these conditions suboptimal designs may be used. If location and timing of the impact are known and no control areas are available (Fig. 9.1, scenario 2), then significance of the impact must be inferred from temporal change alone (e.g., a flood alters severely the conditions in an entire natural area). If location and timing of the impact are not known (e.g., an NI plant is noted in sites surrounding a natural area, and conditions are conducive for entry of the plant into the natural area), the research becomes a baseline or monitoring study (Fig. 9.1, scenario 3). If the study is properly planned spatially, then it is likely that nonimpacted areas will be available to serve as controls if and when the impact occurs. This again indicates why monitoring studies are valid research: they can be subject to rigorous experimental analysis.

Unfortunately, the impact has often already occurred without any preplanning by the land manager (e.g., the first collection of an NI plant is made in a natural area). This common situation (Fig. 9.1, scenario 4) means that impact effects must be inferred from among areas differing in degrees of impact (e.g., an NI plant invades many different natural areas but establishes various population densities); study design for this situation is discussed below. Finally, situations exist (Fig. 9.1, scenario 5) where an impact is known to have occurred, but the time and location are uncertain (e.g., a large population of an NI plant is suddenly found in a natural area).

As outlined above, Green (1979) developed many of the basic principles of environmental sampling design. Most notably, his "before-after-control-impact," or BACI, design (Fig. 9.1, scenario 1) is the standard upon which most current designs are based. It has been noted, however, that any difference between conditions before and after the application of a treatment may not be caused by the treatment of interest (Hurlbert 1984; Underwood 1994). Several authors (Bernstein and Zalinski 1983; Stewart-Oaten et al. 1986) expanded BACI to include replicated times of sampling both before and after the impact. Additionally, Osenberg et al. (1994) developed the BACI design with paired (P) sampling, or BACIP. BACIP requires paired (simultaneous or nearly so) sampling several times before and after the impact at both the control and the impacted sites. The measure of interest is the difference (or delta in statistical terms) in a parameter value between the control and impacted sites as assessed on each sampling date.

Repeated sampling from the same location usually raises concerns of pseudoreplication, a sampling problem popularized by Hurlbert (1984). Ecologists commonly sample from the same location at different times (e.g., weekly or monthly), considering each sampling event an independent sample. Such a design usually artificially inflates the degrees of freedom, leading to erroneous statistical analyses. Even if a case can be made for repeated sampling from the same location, this temporal modification of BACI does not solve the problem of lack of spatial replication. As is well known to any ecologist, organisms may have different temporal trends in different locations (Underwood 1994). This is especially true for plants because of their sensitivity to subtle variation in soil properties. Although replicated treated sites are often not available, or are impractical (e.g., in large-scale experiments), replicated control sites are usually possible. A common misconception is that control sites must be identical to the treated site(s). Rather, controls need only be a random group of sites with the general features of physical characteristics, weather conditions, species, and the like. They should represent the range of conditions in which the treatment has occurred or is expected to occur. A treatment affecting the abundance of the sampled population in a site must cause the temporal pattern of abundance in that location to differ from the temporal pattern in the control locations (Fig. 9.2). The selection of study sites is detailed under "Experimental Design."

Desired Precision: Statistical Significance and Power

Recorded data may take a variety of forms when the subject of study is assessment or management of plant invasions. Whether data are in a qualitative form—"I observed that species A had more coverage than species B"—or of a more quantitative form—"I found a significantly ($P < 0.01$, t-test) greater cover of species A ($\bar{X} = 1.5$, SD = 0.20) than of species B ($\bar{X} = 1.0$, SD = 0.15)"—we have been provided information on an observed phenomenon. Given this information, I suspect that most of us would put more faith in the second, more quantitative example. But why should we? There really is no reason to trust one observation over the other. If I had added "On 75 occasions I observed..." to the first example, and "$df = 6$" to the second, I would certainly put my trust in the former (even though the latter does present a statistically significant difference). I conclude this because, without providing and justifying one's sample size, no data set should be assumed valid regardless of associated P-value (alpha level). Finding a significant P in no way justifies a conclusion. Erroneous and often contradictory conclusions may be reached with variations in sample size because alpha levels vary as sample size increases (Morrison 1984). In addition, results can be quite different depending on whether the data are collected using visual observations (an observer's best guess) or using measurement instruments. The topic of observer and measurement variability is discussed below.

The probability of committing a Type I error (alpha, rejection of the null hypothesis when it is actually true) is inversely related to the probability of committing a Type II error (beta,

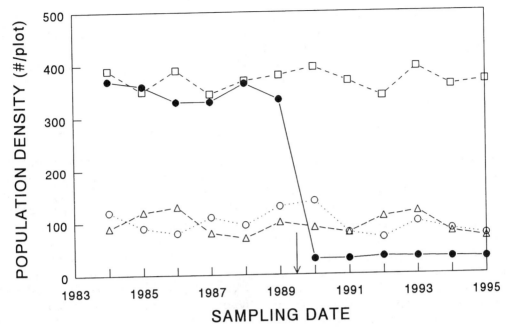

FIGURE 9.2. Depiction of a simulated environmental impact or treatment application (*arrow*) in one location (*closed circles*), with three controls, all sampled six times before and after the impact or treatment. The impact or treatment results in an alteration of temporal variance of the original mean. (From Underwood [1994]. Figure 3D, used by permission of The Ecological Society of America)

failing to reject the null hypothesis when it is in fact false) for a given sample size. Lower probabilities of committing a Type I error are associated with higher probabilities of committing a Type II error; the only way to minimize both errors is to increase sample size. Improving the power of a test has special importance to land managers. Power of a statistic tells us the likelihood that we will falsely reach a conclusion of no effect due to a treatment (or impact). If a test does not reject the null hypothesis of no difference, but has low associated power, we are immediately warned that we just might be wrong. In such a case, caution is advised and further study is indicated.

Many formulas are available for estimating the sample size necessary to achieve a reliable result; virtually all basic statistics books provide them (e.g., Cochran 1977; Sokal and Rohlf 1981; Zar 1984; see also Petit et al. 1990). An advantage of many of these methods is that they force researchers to indicate the difference they want to achieve between their experimental units (to indicate treatment effect) and the alpha and beta desired. This, then, leads to estimation of sample size necessary to reach these goals. One difficulty is that most of these methods require a good estimate of the variance components. If such parameters are unknown, they can either be obtained from the literature or estimated as the study progresses. Such sequential sampling allows refinement of methods and effort as the study progresses, and prevents one from simply gathering large amounts of data, hoping that the effort will be adequate. Sequential sampling methods are discussed by many workers (e.g., Block et al. 1987; Green 1979; Kuno 1972; Morrison 1988).

As indicated above, the number of samples required depends on the precision of the desired answer. Green (1979) showed that, for a wide range of field data, the number of samples required to reach a desired precision is independent of the unit of measurement (e.g., density) and is about equal to the inverse square of the desired precision. As an example, if plant density must be estimated with a precision such that 0.95 confidence limits are ±20% of the mean, then precision approximates 0.10 and the sample size approximates 100. If ±40% is adequate, then the sample size drops to about 25 (see Morrison 1988 for review). The number of replicates necessary can also vary depending on the statistical test planned. For example, in randomized block designs (see "Experimental Design," below), a minimum of six-fold replication is necessary before significant ($P < 0.05$) differences can be demonstrated using the nonparametric Wilcoxon's signed-rank test, whereas only a four-fold replication is necessary using the Mann–Whitney U test (Hurlbert 1984).

Sampling from populations of rare or threatened species presents special problems. In such situations, it is impossible to gather a large, independent sample. Stratified sampling in time and space, when possible, combined with a thorough analysis of the influence of sample sizes on results, will lend at least some confidence to your research. Further, various workers are developing statistical methods that specifically address this special case (e.g., Cochran 1977; Green and Young 1993).

Johnson (1981) offered three general guides for determining sample size. First, examine the stability of the estimates, including both the mean and the variance. Second, investigate sources of variability and how they compare in magnitude. Observer variability, temporal variability, and methodological variability all increase variance in your estimates. Rather than simply increasing sample sizes to try to overcome this variability (which might not work in any case) it is better to reduce variability from the start. And third, in the case of multivariate analyses the necessary sample size increases as variables are added to the equation. For each group in such an analysis, a bare minimum sample might be 20 observations, plus three to five additional samples for each variable used.

Obscure statistical tests and complicated transformations of data cannot overcome a study with an inadequate data set. Although we can all provide reasons (a nice way of saying "make excuses") for failing to gather more data, there is no excuse for failing to provide quantified justification for the sampling design and ramifications of this design on our conclusions.

Experimental Design

Hurlbert (1984) outlined five components of a properly designed experiment: (1) generating the hypothesis, (2) developing the experimental design, (3) implementing the design, (4) analyzing the data, and (5) interpreting the data. Component 1 is obviously the most critical step in an experiment; any weakness here could potentially render component 5 meaningless with regard to your original purpose. Experiments can be broken into two major classes: mensurative and manipulative. Mensurative experiments predict a set of observations under conditions that are usually not controlled; no manipulations are conducted and the hypothesis is tested on the basis of the observations. In contrast to mensurative experiments, manipulative experiments test predictions by altering a component of the system in one area(s) and comparing the outcome with a control area(s) where no manipulation was conducted (Sinclair 1991). Neither type is necessarily more robust than the other; it all depends on the goal of the study. Manipulative experiments have the advantage of being able to test treatments and evaluate the response of something to different levels of treatment and to do so in a manner usually within your control (e.g., various reductions in density of an invading plant). The relationship between monitoring and mensurative and manipulative experiments, as applied to the study of NI plants, is presented in Table 9.1.

Critical in development of any experiment is proper consideration of replicates, use of controls, randomization, and independence. Hurlbert (1984) presented an excellent review of steps necessary for developing a strong experimental design; much of what follows (in this section) is summarized from that article.

9. Experimental Design for Plant Removal and Restoration

TABLE 9.1. Suggested means of answering questions related to nonindigenous (NI) plants.

Study question	Research required
Is an NI plant present?	Reconnaissance
Is the NI plant population increasing?	Long-term monitoring
Is the NI population increase unique to a single area?	Multiple-site, long-term monitoring
Is the NI plant negatively affecting elements of conservation value?	Mensurative or manipulative experiments
What methods will best control the NI plant?	Manipulative experiments
Are the control methods applicable to many areas?	Multiple-site, manipulative experiments
What is the community response to NI plant removal?	Manipulative experiments

Everyone is familiar with the general need for experimental controls, whether the study be mensurative or manipulative. But as noted by Hurlbert (1984), "control" has several important meanings in the context of experimental design. The control may be an untreated treatment (no experimental variable added), a procedural treatment (e.g., using trees injected with water as controls for trees injected with water and an herbicide), or simply as a different treatment. Controls are required because biological systems exhibit temporal variation. If a given system showed stability in its properties over time in the absence of an experimental treatment, then separate controls would be unnecessary (i.e., pretreatment data from the to-be-treated site would be an adequate comparison). Additionally, in many experiments, controls allow separation of the effects of different aspects of the experimental procedure (e.g., in the above example, separating different times of tree injection or using different types of injection methods).

Another definition of controls includes all the design features associated with an experiment. Controls *control* for temporal change and procedural effects. Randomization *controls* for (reduces) experimenter biases in assignment of experimental units to treatments. Replication *controls* for among-replicates variability inherent in any study. Interspersion (of study plots) *controls* for regular spatial variation in experimental units. The term *control* also refers to the homogeneity of experimental units, to the precision of treatment procedures, or to the regulation of the physical environment in which the experiment is conducted. As emphasized by Hurlbert (1984), the adequacy of an experiment is based both on your ability to control physical conditions during the experiment and on the use of an adequate number of treatment controls (e.g., replicated control plots).

Replication reduces the effects of random variation (often referred to as noise) or error, thus increasing precision of an estimate. Randomization reduces potential bias by the experimenter, thus increasing accuracy of an estimate. For example, assume that the null hypothesis is that the cover of an NI plant species does not differ among treated (e.g., herbicide application) and control sites. If the sites are a mosaic of meadow and forest, and the species in question regularly occurs only in the meadow, then random sampling over the site (meadow and forest) would be an inefficient design unnecessarily raising the error variation. This is because the error includes meadow versus forest differences; the ratio of variation among (i.e., between the control and treated areas) to variation within would be reduced, thus reducing the power of the test against the null hypothesis. This example was adapted from Green (1979), who provided other examples of how sampling effort should be allocated.

Independence refers to the probability of occurrence of one event regardless of whether another event has or has not occurred. For statistical analyses, it should be assumed that the error terms are independently distributed. Departures from independence occur from correlations in time and/or space of the experimental samples (Sokal and Rohlf 1981). A somewhat common misconception is that application of nonparametric statistics relieves this assumption; it does not.

Hurlbert (1984) illustrated several ways in which treatments have been interspersed in a two-treatment experiment (Fig. 9.3). The boxes in the figure could represent any experimental unit (e.g., plots in a greenhouse, separate fields, individual animals); the "A" design types are considered acceptable, whereas the "B" types are not. Below I briefly summarize the main features of each design.

Completely randomized (Fig. 9.3, A-1). This is the most basic method of assigning treatments. It is seldom used in field experiments, however, as it is usually difficult to achieve adequate spatial interspersion of the treatments. This is because potential treatment plots are usually clumped in some fashion in nature (e.g., because of soil type); thus one runs the risk (due to chance) of having treatments established in close proximity to each other.

Randomized block (Fig. 9.3, A-2). This commonly used method in field studies overcomes much of the problem associated with completely randomized design. Because treatments and controls are paired, it reduces variance due to different environments among plots.

Systematic (Fig. 9.3, A-3). This is another acceptable design in which one randomly assigns the first treatment and then alters the remaining treatments and controls. A potential problem with field establishment of this design is that experimental plots might be spread along some environmental gradient (e.g., moisture, temperature, and soil type). Blocking (Fig. 9.3, A-2) helps reduce the impacts of such problems on test results.

Simple, clumped, and isolative segregation (Fig. 9.3, B-1, B-2, B-3). These designs are seldom used in the field because they have the obvious disadvantage of segregating treatments

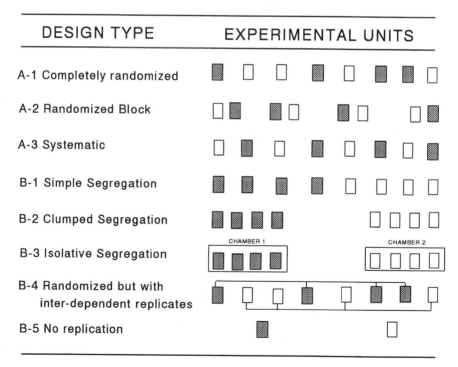

FIGURE 9.3. Schematic representation of various acceptable modes (A-1 to A-3) of interspersing replicates (*boxes*) of two treatments (*shaded and unshaded boxes*), and various ways (B-1 to B-5) in which the principle of interspersion can be violated. (From Hurlbert (1984). Figure 1, used by permission of The Ecological Society of America).

from controls (B-3 is primarily a laboratory design). Thus certain differences existing prior to the experiment might render test results spurious; these problems may never be known to the experimenter. For example, differences in soil micronutrients because of local site conditions might be responsible for differences in plant response to some treatment, regardless of the treatment effect. Additionally, any unplanned changes after the experiment begins might affect treatments or controls only because of their segregation.

Randomized, but with inter-dependent replicates (Fig. 9.3, B-4). This design links experimental units to some common device such as a chemical source, heating device, or water source. Although this approach can be used in any of the above designs, it is subject to generation of spurious treatment effects and, in addition, to failure if the common device fails. Thus for experimental design and practical considerations this design should be avoided. A separate mechanical device should be used for each experimental unit, thus eliminating this design per se (i.e., it becomes one of the "A" designs).

No replication (Fig. 9.3, B-5). Although the weaknesses of this design are intuitive, it is nevertheless employed frequently. This is because many of the ecological systems in which we work are extremely large or by their very nature cannot be replicated at the scale of interest. For example, few managers are willing to replicate treatments such as plant removal both within and among nature preserves. This problem is especially relevant to environmental impact assessment (e.g., the accidental pollution of a stream by a toxic substance). Nonreplicated studies usually suffer from severe pseudoreplication (Hurlbert 1984); even if they do not, their results cannot be generalized outside the immediate study area. However, as long as one does not attempt to generalize results outside the study area, then lack of replication does not negate an otherwise properly designed study. In addition, there are specific statistical procedures, such as repeated measures analysis of variance, that are designed to deal with repeated sampling from the same plots.

Scientists and managers are often confronted with assessing the results of a treatment that cannot be replicated. For example, if the area of interest is a small reserve, then often only one appropriate treatment is possible. This single treatment can be matched with multiple controls; ideally, pretreatment data will be available for all areas. Here, a difference is sought between the time course in the treated location and that in the controls. The treatment will be detected either as a different pattern of interaction among the times of sampling or at the larger time scale of before and after the treatment (see Fig. 9.2). Additional means of dealing with difficult-to-replicate studies were discussed previously (see especially Underwood 1994).

Applications: Managing Study Plots and Data

When initiated, most studies have a relatively short time frame (a few months to a few years). As such, maintaining the sites usually requires only occasional replacing of flags, stakes, or other markers. It is difficult to anticipate, however, what sorts of long-term research may eventually emerge that makes use of the database. In southeastern Arizona, for example, few anticipated the dramatic range increase by NI love grasses (*Eragrostis*) introduced as cattle forage. However, several studies on the impacts of grazing on wildlife included sampling of grass cover and species composition, thereby inadvertently establishing baseline information on love grasses. By permanently marking such plots, researchers can now return, relocate the plots, and continue the originally unplanned monitoring of these species. Unexpected fires, chemical spills, urbanization, and any other planned or unplanned impact will undoubtedly affect all lands sometime in the future. Although it is unlikely that an adequate study design will be available serendipitously, it is likely that some useful comparisons can be made using "old" data (e.g., to determine sample size for a planned experiment). To ensure that all studies can be used in the future it is absolutely necessary that study plots be permanently marked and referenced. All management agencies, universities, and private foundations should establish a central record-keeping protocol. Unfortunately, even dedicated research organizations often fail to do so; no agency or university I am aware of has established such a protocol for all research efforts.

It is difficult to imagine the amount of data that have been collected over time. Only a tiny fraction of these data resides in permanent databases. As such, except for the distillation presented in research papers, these data are essentially lost to future scientists and managers. Here again, it is indeed rare that any organization requires that data collected by its scientists be deposited and maintained in any centralized database. In fact, few maintain any type of catalog referencing even location of the data, contents of data records, or other pertinent information.

Perhaps one of the biggest scientific achievements of our time would be the centralization, or at least the central cataloging, of data previously collected (those that are not lost). It is an achievement unlikely to occur. Each research or management organization can, however, initiate a systematic plan for managing study areas and data in the future.

Conclusions

1. The field of ecology will fail to advance our knowledge of nature until rigorous scientific methods are followed in the design, implementation, and analysis of research. Resource managers are ill-served by poorly designed studies that produce unreliable knowledge.

2. Resource managers must clearly understand the basic methods of conducting science (i.e., induction, retroduction, and the H-D method) so that they can better evaluate the data available to them for making decisions.

3. Monitoring is a form of research that requires rigorous study design and statistical analyses. Thus resource managers must go through all of the steps in designing a monitoring study that one would go through in designing a controlled experiment.

4. Study goals must be clearly elucidated, and the spatial and temporal applicability of results determined, before initiation of sampling. It is critical that managers determine how and where study results will be used so that results match needs. Hoping for the best will not work.

5. The statistical significance and power of a study must be carefully chosen prior to initiation of field sampling. Simple steps can be taken so that the researcher is aware of the likely precision of results before the study begins. By doing this, resource managers will have a known confidence in decisions they make based on study results.

6. All studies must be adequately replicated so that results cannot be assigned to chance alone. Establishing replicates is often difficult in field situations, but replication can usually be achieved with planning.

7. Pseudoreplication must be avoided so that erroneous results do not mislead resource managers into making unwise decisions.

8. Extreme care must be taken to protect study sites from unwanted disturbance. Data must be duplicated, protected, and stored in a manner readily retrievable and usable by others.

10
Response of a Forest Understory Community to Experimental Removal of an Invasive Nonindigenous Plant (*Alliaria petiolata*, Brassicaceae)

Brian C. McCarthy

Garlic mustard
(*Alliaria petiolata*)

The available data on invasion by nonindigenous (NI) plants suggest that most species do not readily enter mature, relatively stable habitats (Mooney and Drake 1986). When NI species do gain access to such mature habitats, the invasion is often correlated with a window of opportunity associated with a disturbance. Observations suggest that forested lands may be somewhat more resistant to invasion compared to vegetation dominated by other life forms. For example, many of California's managed forests are regularly disturbed but are not particularly susceptible to invasion by NI plants (Mooney et al. 1986). Those few plants that can establish do not appear to contribute to problems of tree regeneration.

In contrast, many hardwood forests of eastern and midwestern North America appear to be undergoing a massive invasion by a variety of NI species, for example, garlic mustard (*Alliaria petiolata*), Amur honeysuckle (*Lonicera maackii*), Japanese honeysuckle (*L. japonica*), multiflora rose (*Rosa multiflora*), princesstree (*Paulownia tomentosa*), and tree-of-heaven (*Ailanthus altissima*). The diversity of forest types, disturbance regimens, geological substrates, and life-history patterns of the species involved in these invasions defy simple generalizations. Thus, even relatively stable, mature, largely undisturbed forests appear to be susceptible to invasion by certain species. In particular, garlic mustard is capable of invading both upland and lowland broadleaf forests, with either open or closed canopies. It gains access via small disturbances, but these disturbances are neither required for entrance nor necessary for long-term persistence. The species does not appear to diminish with succession, suggesting it may have major long-term impacts on other species within the system.

Although garlic mustard is uniformly recognized by land managers as a threat to native forests, I am aware of no published study in which experimental data were gathered to evaluate the specific impact of this species on a forest understory community. In fact, experimental studies assessing the community-level impact of invasive plants seem to be rare in general. Moreover, many land managers subjectively conclude that the appearance and perceived dominance of a particular NI species will be disruptive to the native plant community. The purpose of my study was to evaluate quantitatively the effect of garlic mustard on the herbaceous and woody seedling understory layer of a heavily infested flood plain forest community and to assess the efficacy of hand weeding as a low-impact method of control for this species.

Biology of the Study Species

Alliaria petiolata (= *A. officinalis*), Brassicaceae, or garlic mustard (hereafter GM) is a naturalized European biennial that invades forest communities throughout much of eastern North America. The species was first recorded in North America in 1868 on Long Island, New York (Nuzzo 1993). In the next 20 years, it spread throughout the northeastern seaboard and began to migrate westward. By 1991, GM had spread throughout southeastern Canada and was found in 30 northeastern states of the United States (Cavers et al. 1979; Nuzzo 1993).

The biology, life history, and control of GM have all been well studied in recent years and are summarized in Baskin and Baskin (1992), Byers (1988), Cavers et al. (1979), and Nuzzo (1991, 1993). In short, the species possesses many characteristics of what could be termed an "ideal major weed" (sensu Baker 1965, 1974): it germinates under a wide range of environmental conditions, shows rapid seedling growth, spends only a short period of time in vegetative condition, is self-compatible, has high seed output, can produce some seed in a diversity of environmental conditions, and can compete via special means (e.g., rosettes). One of the only life history features this species does not exhibit that would benefit its weedy habit is a persistent seed bank (Baskin and Baskin 1992).

Garlic mustard readily invades wet to dry-mesic deciduous forests, often gaining entrance via disturbances. It regularly invades naturally disturbed flood plain and riverbank forests as well as anthropogenically disturbed areas associated with forest roads, trails, and edges (Nuzzo 1993). Once gaining access, the species quickly assumes dominance in the understory (personal observation; Yost et al. 1991). Control of the species is difficult, particularly after it becomes well established. Suggested methods of control include harvesting the flower stalk prior to fruiting, burning in the dormant season, and/or applying herbicides (Nuzzo 1991, 1994; Nuzzo et al. 1991).

Methods

Study Area

The site selected for this study was the Fort Hill Preserve, which is owned and managed by the Maryland office of The Nature Conservancy. Fort Hill is located at approximately 39°29' N, 78°53' W in extreme southwestern Allegany County, western Maryland. Most of the Fort Hill Preserve lies on uplands underlain by a cherty limestone of the Elliber series. The upland areas are relatively dry-mesic oak forests including a considerable amount of GM in the understory. The specific study site is a somewhat disturbed (via off-road vehicles) lowland area with soils classified as Alluvial (Stone and Matthews 1977) and that

10. Response of a Forest Understory Community to Experimental Removal

FIGURE 10.1. Study site and garlic mustard (*Alliaria petiolata*). (A) Understory of Fort Hill Nature Preserve, Allegany County, Maryland (May 1991). Note the prominent understory dominance by garlic mustard (all mature second-year plants). Range pole at photo center is 2 m. (B) Understory dominance by basal rosettes in June 1992 (note meter stick at center). (C) Overview of a 1 m² circular quadrat where garlic mustard was experimentally removed. Note immediate release of *Impatiens* sp., which colonized from the seed bank (July 1991). Garlic mustard in the surrounding community can be seen toward the upper corners of the photograph.

originate from an adjacent stream feeding the North Branch of the Potomac River (approximately 100 m away). Observations suggest that the soils are well drained and can be quite droughty during drier periods. The site supports overstory vegetation characteristic of many eastern North American flood plains and includes silver maple (*Acer saccharinum*), boxelder (*A. negundo*), green ash (*Fraxinus pennsylvanica*), sycamore (*Platanus occidentalis*), and slippery elm (*Ulmus rubra*). Garlic mustard is abundant in the understory (Fig. 10.1a).

Field Methods

Prior to the spring flush of ephemeral herbs (early April 1991), 18 circular quadrats (1 m^2) were set out in a paired plot design ($n = 9$). Each control quadrat was paired with an adjacent experimental quadrat in which GM was removed continuously for the length of the study. The paired plot design was used to minimize environmental heterogeneity. The quadrats were located on a flat flood plain bench defined by a levee break on one side and a steep slope break leading to the uplands on the other. The entire study area was approximately 25 × 65 m in extent.

The 18 quadrats were sampled for percent cover by species near the middle of May, June, July, and August for a period of 3 years (1991, 1992, and 1993). Visual estimates of percent cover were made separately by the investigator and an assistant; the mean was used for each species. Cover values were determined by layer, and thus the cover of an individual plot could exceed 100%. The removal treatment was maintained throughout the experiment. The cover of GM in experimental removal plots was recorded for each sample period to evaluate new germinants; these were subsequently weeded out after all other data were collected.

Strausbaugh and Core (1982) was used as the primary floristic reference; additional information was obtained from Gleason and Cronquist (1981). Voucher specimens were not collected except in cases of difficult identifications. Some plants could not be identified to species and thus there are a number of "unknowns." This situation arose because I wished to score newly germinated seedlings, but these often suffered heavy mortality due to drought or herbivory prior to full development. Plants were monitored outside the study plots to evaluate phenology and to confirm species identifications. Therefore, I am reasonably confident that most of the unknowns are not just juvenile or cotyledonary forms of the mature species that were already recorded.

Data Analysis

During the first sample period (May 1991) all GM aboveground biomass from each of the nine experimental plots was harvested, dried (80°C), and weighed. To test the hypothesis that there was a relationship between GM biomass and species diversity, biomass estimates were regressed against species richness data for those plots. A significant positive relationship might suggest that high resource microenvironments support both a great amount of GM and a high richness of other understory species. A significant negative relationship might suggest that this species captures most available resources and depresses species richness on a fine scale. No relationship might suggest that the microenvironment is so heterogeneous that there is no direct relationship between GM and species richness.

Data on percent cover for each of the 72 species from each of the 12 sample periods were relativized to a 100% cover scale. Mean ($n = 9$) cover for each species in both the experimental removal and control plots was calculated. The number of species occurring in each plot was used as a measure of richness. To better account for the number of relatively rare or uncommon species found in the study, the Shannon–Weiner diversity index was also calculated (Krebs 1984). Magurran (1988) provided a variance estimate for this measure, which was also determined and used for testing the difference in diversity between experimental and control plots for each sample period. To account for the presence of GM in control plots this species was subtracted prior to all diversity analyses.

Mean cover data for all species in experimental and control plots for all sampling dates were entered into a 72 species by 24 samples (2 treatments × 4 samples per year × 3 years) data matrix for a subsequent ordination analysis. Data were checked for multivariate normal data structure and transformed as needed. Similarities were calculated using Euclidean distance; the resulting hemimatrix was subjected to a detrended correspondence analysis (DCA) using the Multi-Variate Statistical Package (MVSP) (Kovach 1987).

To evaluate what life form groups (guilds) of plants might be most affected by GM, I used both Gleason and Cronquist (1981) and Strausbaugh and Core (1982) to classify each species into one of several functional groups: annuals, graminoids (grasses, sedges), perennial herbs, shrubs, trees, vines, or GM. In the case of the unknowns (all at very low abundance), plant morphology was used as an indicator for the different categories. Most of these latter category assignments were made with a reasonable degree of certainty. Data were evaluated using a log-linear analysis (Hintze 1990) and graphed as 100% bar charts.

Results

The abundance of GM in the control plots varied considerably both within and among years (Fig. 10.2). Greatest abundance of GM was observed in May 1991 when the species dominated the habitat with a mean cover of 70% (Fig. 10.1a). Cover in 1991 declined into June after the species fruited and began to senesce. By July, GM was no longer the dominant. A few seedlings were observed late in the season (less than 5% cover); these overwintered as rosettes to May 1992. During the 1992 field season, I observed no mature fruiting individuals in the habitat but only seedlings and rosettes, which continued to mature

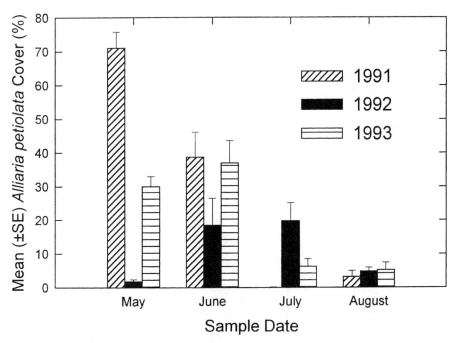

FIGURE 10.2. Population dynamics of garlic mustard (*Alliaria petiolata*), measured in control plots as percent cover, in the flood plain community of Fort Hill Nature Preserve, Allegany County, Maryland, for 1991 to 1993. Data are means ± SE, n = 9.

throughout the year (Fig. 10.1b). Garlic mustard was again dominant in spring 1993 but not to the same extent as in 1991 (Fig. 10.2). I suspect that drought in spring 1993 may have been responsible for the decreased abundance of mature individuals.

Diversity, measured as mean species richness, ranged from a low of about 4 in late summer 1991 (year of greatest GM abundance) to 11 in early spring 1993 (year following lowest abundance of GM, the species seen only as rosettes). Although the mean richness was almost always greater in removal quadrats, only June and July 1992 and May 1993 were significantly different ($P < 0.01$; paired t-test). The Shannon–Weiner index proved to be of greater value because it accounted for the less-abundant species and thus showed more pronounced trends (Fig. 10.3). H′ was significantly ($P < 0.01$) greater in the removal plots for all sample periods in 1992 and early 1993 but oddly reconverged ($P > 0.05$) at the end of 1993 (Fig. 10.3). H′ varied but did not markedly change in control plots across the 3-year study period. H′ actually appeared to be somewhat lower in the control plots in the year in which the population existed only as rosettes (1992).

Dry weight biomass (g/m^2 removed at t_0) of GM was regressed against species richness for each of the nine experimental quadrats for each of the four sample periods of 1991, 1992, and 1993 (data not shown). None of the 12 least-square regressions was significant (all $P > 0.05$), indicating that there was no simple linear relationship between initial biomass and species richness. Removal of one outlier (based on regression diagnostics and influence

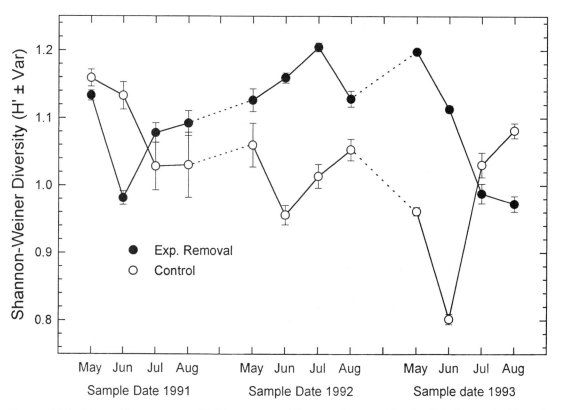

FIGURE 10.3. Mean ($\pm$ var, n = 9) Shannon–Weiner diversity in control versus experimental plots where garlic mustard (*Alliaria petiolata*) was continually removed (Fort Hill Nature Preserve, Allegany County, Maryland). H′ was significantly (adjusted $P < 0.05$, paired t-test) greater in experimental plots from June 1992 through June 1993.

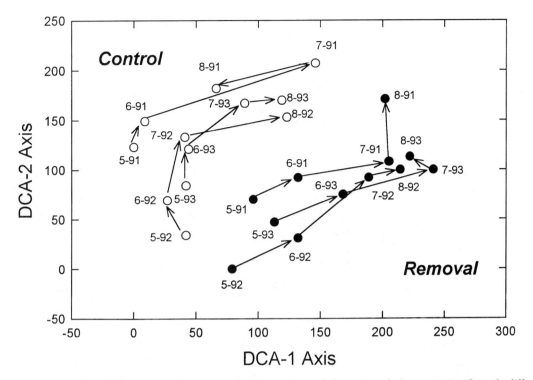

FIGURE 10.4. Detrended correspondence analysis (DCA) of plant communities in a garlic mustard (*Alliaria petiolata*) removal experiment (Fort Hill Nature Preserve, Allegany County, Maryland). Data were obtained from 18 plots sampled 12 times over a 3-year period. Vectors connect subsequent sample periods within a year and treatment. Vegetation of the removal plots was significantly different from the control plots ($P < 0.001$, ANOVA) indicating strong garlic mustard effects. The removal treatment was applied 6 weeks prior to the sample date (May 1991). Open circles are control plots; closed circles are removal plots.

measures) improved the coefficient of determination (R^2) for many of the regressions; however, 11 of 12 regressions were still nonsignificant ($P > 0.05$).

A DCA ordination of the community sample data (Fig. 10.4) was able to explain only 27.50% of the variance in the first axis and an additional 10.75% in the second axis (38.25% cumulative). However, the ordination clearly demonstrated that vegetation in experimental removal plots versus control plots diverged strongly from the beginning and maintained a similar trajectory throughout the experiment. A one-way analysis of variance (ANOVA) of the DCA axis scores confirmed a significant overall multivariate difference ($F = 25.27$, $df = 1$, $P < 0.0001$) between removal and control plots. The detrending of the data prevents a legitimate test of the significance of difference between experimental and control plots for the second DCA axis. These data also highlighted the considerable change in flora from spring to late summer. Within a treatment, the vectoral changes from month to month were generally greater than those from year to year (Fig. 10.4). In particular, the spring flora (May through June) differed dramatically from the summer flora (July through August).

An ordination of the species data (Fig. 10.5) yielded less clear patterns, but two major groups were discernible. Species with high loading scores on the DCA-1 axis and low scores on the DCA-2 axis tended to be more abundant in the experimental removal plots compared to those with low loadings on DCA-1 and high loadings on DCA-2,

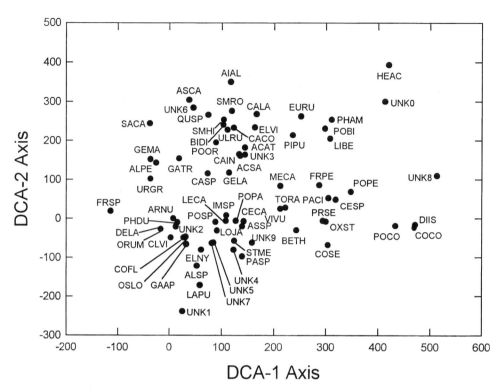

FIGURE 10.5. Detrended correspondence analysis (DCA) of community species data in a garlic mustard (*Alliaria petiolata*) removal experiment (Fort Hill Nature Preserve, Allegany County, Maryland). Data were obtained from 18 plots sampled 12 times over a 3-year period. Acronyms correspond to first two letters of the generic name and first two letters of the specific epithet. Full names of species can be found in Appendix 10.1.

TABLE 10.1. Partial chi-squares and associated probabilities from a log-linear analysis of cover data to determine the relationships among GM removal treatment, year, month, and life form (guild). A step-down model search revealed that the best model to explain the data necessitated the inclusion of month, treatment, and guild.

Effect	df	chi-square	Probability
Year	2	0.00	0.9998
Month	3	0.00	1.0000
Treatment	1	0.00	0.9784
Guild	6	563.10	<0.0001
Year * Month	6	3.18	0.7866
Year * Treatment	2	15.90	0.0004
Year * Guild	12	217.70	<0.0001
Month * Treatment	3	9.99	0.0186
Month * Guild	18	223.31	<0.0001
Treatment * Guild	6	591.68	<0.0001
Year * Month * Treatment	6	15.70	0.0154
Year * Month * Guild	36	152.09	<0.0001
Year * Treatment * Guild	12	64.53	<0.0001
Month * Treatment * Guild	18	38.34	<0.0035

which were generally more abundant in the control plots. Within each of these groups, DCA-1 generally separated out species with respect to seasonality. Species with low loadings were generally early spring species; those with high loadings were late summer species. The most prominent exception to this trend was hepatica (*Hepatica acutiloba*). While this species is generally considered an early spring species, a single plant was discovered in August in one of the control plots in 1991, but did not reappear in 1992 or 1993.

To evaluate the more generalized patterns of species response to removal of GM, species were placed in functional groups based on life form: annuals, perennial herbs, graminoids, shrubs, vines, trees, and GM. Cover data were then evaluated by guild across years, sample periods, and treatment. Log-linear analysis (Table 10.1) indicated that single- and double-term models were generally inadequate to describe patterns of plant coverage and distribution. Life forms responded differently among months and treatments (Table 10.1). The latter indicates that the treatment effect was discernible and transferable to a level above that of species.

To better assess individual life form patterns, the mean relative cover was plotted for all guilds in each month for all 3 years (Fig. 10.6). Although the data are complex, several patterns emerge. First, annuals generally tended to be more prominent in plots where GM was removed. This was observed almost immediately in May 1991 (Fig. 10.1c). The IN annual jewelweed (*Impatiens*) (which never flowered due to herbivory by mammals) was

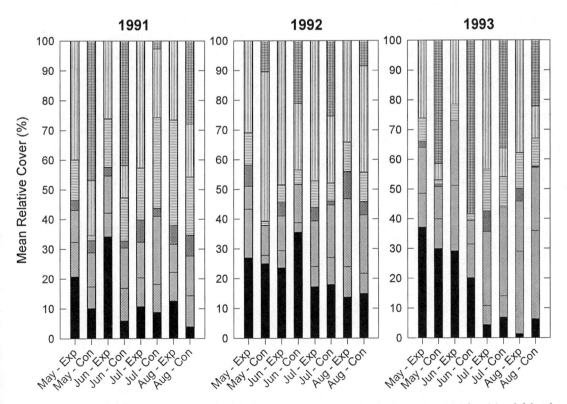

FIGURE 10.6. Mean relative cover (%) by life form type for plant species in garlic mustard (*Alliaria petiolata*) removal experiment (Fort Hill Nature Preserve, Allegany County, Maryland). From bottom to top of 100% bars: *filled bars* = annuals; *rising-right hatch* = herbaceous perennials; *rising-left hatch* = graminoids; *cross-diagonal hatch* = shrubs; *horizontal hatch* = vines; *vertical hatch* = tree seedlings; *cross-hatch* = garlic mustard.

very quick to respond to removal of GM. Perennial herbs, graminoids, and shrubs were constant throughout the study and showed little response to the removal. Suprisingly, tree seedlings were observed to increase considerably in cover with the removal. Likewise, vines generally increased in the absence of GM.

Discussion

The population response of GM is intriguing. The plants studied in western Maryland appear to be on a synchronous biennial production schedule alternating between years of vegetative and reproductive dominance. In and of itself, this result is not particularly surprising given the obligate biennial nature of the species and a likely single founder event. However, this synchrony has been observed to occur at numerous other populations of the species in the region, all of which are many kilometers (50+) apart. In fact, I and other researchers have observed similar phenomena in geographic regions as disparate as Illinois, Indiana, Kentucky, Maryland, Ohio, and Ontario (Baskin J, Cavers P, and Nuzzo V, pers. com.). Although small patches or demes can vary phenologically in these populations, all were generally observed to flower in the same years. These observations and the known high self-compatibility of the species suggest that there still may be strong holdover effects from the initial North American colonization event(s).

In the treatment plots, GM needed to be continually removed throughout the study. Baskin and Baskin (1992) found that GM seeds are dormant at maturity and require a period of cold stratification to come out of dormancy. They found seed germination on moist soil to be high (greater than 90%) in late winter and early spring. Seeds sown on the soil were found to be germinable for 3 to 4 years following dispersal. In the field, plants originating from seeds in the seed bank generally died when allowed to compete with established rosettes, but 3% survived to set seeds when rosettes were removed (Baskin and Baskin 1992). Similar observations were made in my study. Each spring, GM seedling establishment was observed to be greater in experimental removal plots compared to control plots. This might suggest a mild form of autoallelopathy relative to seed germination and/or seedling establishment. Alternatively, it may represent simple space availability for recruitment.

Seedlings continued to appear in the removal plots throughout all sampling periods of each growing season. Their total cover was always low (1% to 2%) but persistent. Most seemed to germinate immediately in late winter or early spring, but apparently some were able to germinate throughout the growing season. Garlic mustard is known to be able to resprout from the base if the roots are not fully harvested by weeding. No doubt, a few new individuals could be attributable to resprouts each sample period; however, a conscientious effort was made to remove the roots when weeding. Thus, germination and establishment appear not to be confined to early spring and may occur as space or other resources become available.

Assuming that variation in GM populations in the field reflects a heterogeneous resource base (i.e., light, moisture, nutrients), I was unable to demonstrate a relationship, positive or negative, between GM biomass and species diversity. Plots with the greatest GM biomass did not have predictably higher or lower species diversity. This suggests that diversity may not be controlled by resource heterogeneity or the quantitative aspects of GM (interspecific competition) but rather the qualitative aspects of whether the species is present or not. Kelley and Anderson (1990) suggested that allelopathy may be an important mechanism in the competitive abilities of this species. Indeed, sinapine-O-β-D-glucopyranoside has been shown to be an important allelotoxin in a variety of mustards and is abundantly present in GM (Larsen et al. 1983). Studies are currently underway to evaluate the possible intraspecific and interspecific effects of GM allelopathy. The former may explain why my study and that of Baskin and Baskin (1992) observed higher seedling establishment in plots with no GM seedlings.

Field observations and community data showed that species with high dispersal abilities (in both time and space) were most likely to regain cover in experimental removal plots. For example, flowering dogwood (*Cornus florida*), fox grape (*Vitis vulpina*), green ash, *Impatiens* sp., Japanese honeysuckle (*Lonicera japonica*), moonseed (*Menispermum canadense*), and motherwort (*Leonorus cardiaca*) all responded quickly, via increased cover, to GM removal. Many of the perennial herbs, grasses, sedges, and small shrubs growing mainly through slow vegetative expansion did not exhibit a clear release following removal of the invasive species. GM may compete largely by co-opting physical space. In the event that space becomes available, only a specific subset of species can readily respond: those species with fast vegetative growth rates (e.g., many vines), species with a persistent seed bank (e.g., many annuals), or species with high dispersability (e.g., wind- or bird-dispersed tree seeds).

Hand removal (weeding) may be an effective method of limiting the entrance of GM into a previously nonoccupied habitat. Observations I made in Maryland, New Jersey, and Ohio suggest that the local strategy of GM is often one of demic spread, that is, various small clusters become established throughout the habitat (as opposed to a moving front or line approach). Annual weeding of these small demes may effectively keep the weed out of a habitat (Cantino PD, pers. com.). On the other hand, weeding in heavily infested areas is almost futile. While the strategy may seem preferable to herbicide and/or fire in certain sensitive natural areas or habitats surrounded by large human populations, weeding is clearly not an efficacious control strategy.

Fire and herbicides have resulted in mixed effects. A single autumn burn may in fact increase the abundance of GM in the community (McCarthy 1994; Nuzzo V, pers. com.). A series of successive burns (where fuel load and environmental conditions permit) or a combination of a surface burn followed by herbicide does a reasonably good job at control but will not eradicate the species (Nuzzo 1994; Nuzzo et al. 1991). The widespread nature of the species, its dispersal abilities, and the vigor of invasiveness argue against attempting eradication from a heavily infested habitat. Within any given period of time, removal of the NI species will likely release only a subset of the flora. However, a strong argument could be made for maintaining control efforts to provide the local flora with occasional release periods. Biocontrol may ultimately provide an option for control and/or eradication; however, it too suffers the drawback of introducing yet another NI species.

Conclusions

1. My study provides experimental evidence that garlic mustard (*Alliaria petiolata*), a biennial weed invading many eastern North American forests, is having demonstrably negative effects on composition and structure of the understory community. The removal of garlic mustard (GM) resulted in immediate release and proliferation of annuals, herbaceous and woody vines, and tree seedlings. The effects on slow-growing perennial herbs, grasses, sedges, and shrubs were less clear; these species either require more time to show an effect or are resistant to the presence of GM.

2. Management of GM, as well as many other invasive exotics, will likely require a diversity of approaches depending on life history characteristics of the target species, level of infestation, type of ambient vegetation (community composition), site quality, and presence of human impacts. The management objective for habitats heavily infested with tenacious weeds such as GM should probably not be eradication but, rather, provision of release periods for native species. Release events may offer windows of opportunity for establishment and growth of herbaceous perennials and seedlings of woody plants, and for regeneration of the seed bank produced by annual species.

3. The first step in any land-management consideration should be a clear demonstration that a particular invasive plant is indeed having a negative effect on the plant community in question. Management for removal of a particular species may not be justified if a community-level effect cannot be demonstrated. The management "cure" may be worse than the invasive "disease" in affecting the surrounding plant community.

4. Simple removal experiments appear to be the easiest and most direct experimental approach to determining a species' impact on a community. To account for the full realm of species diversity in a habitat, a multivariate statistical approach is necessary. Use of species life form groupings may also be appropriate for detecting changes in community structure and may be a preferable approach where human resources are limiting.

5. Because of various life history patterns among nonindigenous and indigenous community constituents, removal experiments may require a minimum of a 3 to 5 years to yield useful data. Careful consideration of space and time are also mandated. Where possible, removal experiments should be conducted across the range of habitats (vegetation types) and site qualities in which the species occurs.

Acknowledgments. Primary funding for this study was provided by the Maryland chapter of The Nature Conservancy; additional support was provided by the Department of Environmental and Plant Biology and the Office of Research and Graduate Studies, Ohio University. Thanks are due to D. Bailey, A. Conklin, M. Droege, and M. McCarthy for assistance in the field; to M. Droege for encouragement at initial stages of the project; and to J.F. Meekins for assistance with library work and comments on the text.

APPENDIX 10.1. Species list

Scientific name	Common name	Family
Acer negundo	Boxelder	Aceraceae
Acer saccharum	Sugar maple	Aceraceae
Actinomeris alternifolia	Wingstem	Asteraceae
Ailanthus altissima	Tree-of-heaven	Simaroubaceae
Alliaria petiolata	Garlic mustard	Brassicaceae
Allium sp.	Wild onion	Liliaceae
Aralia nudicaulis	Wild sarsaparilla	Araliaceae
Asarum canadense	Wild ginger	Aristolochiaceae
Aster sp.	Aster	Asteraceae
Berberis thunbergii	Japanese barberry	Berberidaceae
Bidens discoidea	Small beggarticks	Asteraceae
Carex incomperta	Sedge	Cyperaceae
Carex laxiculmis	Sedge	Cyperaceae
Carex sp.	Sedge	Cyperaceae
Carya cordiformis	Bitternut hickory	Juglandaceae
Celtis occidentalis	Hackberry	Ulmaceae
Cerastium sp.	Mouse-ear chickweed	Caryophyllaceae

APPENDIX 10.1. *Continued*

Scientific name	Common name	Family
Cercis canadensis	Redbud	Caesalpiniaceae
Claytonia virginica	Carolina springbeauty	Portulacaceae
Commelina communis	Asiatic dayflower	Commelinaceae
Convolvulus sepium	Hedge bindweed	Convolvulaceae
Corydalis flavula	Yellow corydalis	Fumariaceae
Dentaria sp.	Toothwort	Brassicaceae
Digitaria ischaemum	Smooth crab grass	Poaceae
Ellisia nyctelea	Waterpod	Hydrophyllaceae
Elymus virginicus	Virginia wild rye	Poaceae
Eupatorium rugosum	White snakeroot	Asteraceae
Fragaria sp.	Strawberry	Rosaceae
Fraxinus pennsylvanica	Green ash	Oleaceae
Galium aparine	Cleavers	Rubiaceae
Galium triflorum	Sweet-scented bedstraw	Rubiaceae
Geranium maculatum	Wild geranium	Geraniaceae
Geum laciniatum	Rough avens	Rosaceae
Hepatica acutiloba	Hepatica	Ranunculaceae
Impatiens sp.	Touch-me-not	Balsaminaceae
Juglans nigra	Black walnut	Juglandaceae
Lamium purpureum	Purple dead-nettle	Lamiaceae
Leonurus cardiaca	Motherwort	Lamiaceae
Lindera benzoin	Spicebush	Lauraceae
Lonicera japonica	Japanese honeysuckle	Caprifoliaceae
Menispermum canadense	Canada moonseed	Menispermaceae
Ornithogalum umbellatum	Star-of-Bethlehem	Liliaceae
Osmorhiza longistylis	Smooth sweet cicely	Apiaceae
Oxalis stricta	Yellow wood-sorrel	Oxalidaceae
Panicum sp.	Panic grass	Poaceae
Parthenocissus quinquefolia	Virginia creeper	Vitaceae
Phacelia dubia	Small-flower phacelia	Hydrophyllaceae
Phytolacca americana	Poke	Phytolaccaceae
Pilea pumila	Clearweed	Urticaceae
Platanus occidentalis	Sycamore	Platanaceae
Poa compressa	Canada bluegrass	Poaceae
Poa palustris	Fowl bluegrass	Poaceae
Polygonatum biflorum	Common Solomon's-seal	Liliaceae
Polygonum orientale	Prince's-feather	Polygonaceae
Polygonum persicaria	Lady's-thumb	Polygonaceae
Potentilla sp.	Cinquefoil	Rosaceae
Prunus serotina	Wild black cherry	Rosaceae
Quercus sp.	Oak	Fagaceae
Robinia pseudoacacia	Black locust	Fabaceae
Sambucus canadensis	Common elder	Caprifoliaceae
Smilax hispida	Hispid greenbrier	Smilacaceae
Smilax rotundifolia	Common greenbrier	Smilacaceae
Stellaria media	Common chickweed	Caryophyllaceae

APPENDIX 10.1. *Continued*

Scientific name	Common name	Family
Toxicodendron radicans	Poison ivy	Anacardiaceae
Ulmus rubra	Slippery elm	Ulmaceae
Urtica gracilis	Stinging nettle	Urticaceae
Vitis vulpina	Fox grape	Vitaceae
Vitis sp.	Grape	Vitaceae

Ten unknowns were species in juvenile form (cotyledons present); they never flowered and were usually present in only one sample period.

Section III
Direct Management

11
Management of Plant Invasions: Implicating Ecological Succession

James O. Luken

Amur honeysuckle
(*Lonicera maackii*)

The modern synthesis of community ecology recognizes that all ecological systems are inherently unstable (Pickett et al. 1992). Emerging from this aphorism is the conclusion that any attempt to manage ecological systems for conservation goals must concentrate efforts on the rates of system change, on the directions of system change, and on the processes contributing to system change. Although the problem of plant invasion is superbly suited to application of these straightforward guidelines, such application has not occurred to any great extent. Instead, much management effort aimed at modifying plant invasion has concentrated on plant control measures and on the success of such measures in achieving population reductions of nonindigenous (NI) plants (see, e.g., Doren and Whiteaker 1990a; Luken and Mattimiro 1991; Nuzzo 1991). This can perhaps be traced to the influence of weed science with its corresponding principles derived from commodity production systems. Hobbs and Humphries (1995) argued that such an approach to the problem of plant invasion is not ecologically sound and will not likely be successful. They stressed the need for multilevel understanding with management efforts directed at the invasive species, at the attributes of the invaded system, and at human activities that may influence both system attributes and species availability (Hobbs and Humphries 1995).

In parks, natural areas, or nature reserves, active management of invasive plants typically occurs for one or both of the following reasons: to stop the influx of new invaders and to reclaim those communities already invaded. Halting the influx of new invaders is cost effective and successful by quarantine efforts, limiting the activities of people, and early detection and eradication of small satellite populations of invasive plants (Hobbs and Humphries 1995; Zamora et al. 1989). However, there remains the problem of plant communities where invaders are well established, where such invaders have been present for many years, and where invading plants are presumably affecting ecological processes. The decision to manage these communities invokes a clear understanding of management goals as well as of historical and extant factors that culminated in the current stage of ecological succession. Furthermore, choice of a management technique should be based on projected long-term responses of the system when managed and in the absence of management action.

This chapter will focus on terrestrial plant communities where invasion has already occurred and where it has presumably caused an undesirable change in system structure or function. First, models of ecological succession will be introduced in an effort to demonstrate how plant invasion occurs and how system attributes contribute to historical and future system trends. Second, the decision to manage a plant community will be examined relative to conservation goals and system attributes. Last, effects of various management techniques (physical, chemical, and biological) commonly used to eliminate invading plants will be critically analyzed relative to their impact on processes influencing community development.

Succession: The Rise and Fall of Populations

Ecological succession is widely viewed as directional change in the species composition of plant communities. Regardless of whether

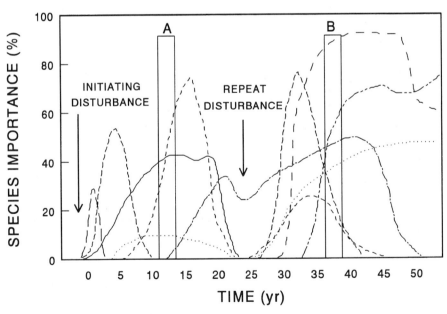

FIGURE 11.1. Changes in importance (*lines*) of plant species through time. Succession is initiated by a disturbance; another disturbance may reinitiate succession leading to changed community development. Note that the community assessed at A, 12 years after the initiating disturbance, differs from the community assessed at B, 12 years after the repeat disturbance.

this process is set in motion on bare substrates that have not previously supported life (i.e., primary succession) or on sites that have previously supported life but are under the influence of an initiating disturbance (i.e., secondary succession), the overall expression of change (Fig. 11.1) is similar: individual species rise and fall in importance. Community development is driven by specific events in the lives of plants (e.g., establishment, growth, and senescence); it is also a record of how plants respond to both internally generated and externally imposed environmental factors.

Several aspects of the successional picture in Figure 11.1 warrant emphasis. First, the number of distinct communities or phases identifiable along the time axis is large; indeed, this number approaches infinity. Second, stability is a relative condition never achieved. There may be periods of time where change is more or less rapid, where one species achieves greater importance than all others, but the system continues to change without reaching what has traditionally been recognized as a climax. Lastly, disturbance operating at various scales, frequencies, and intensities will occasionally reset the state conditions and in turn will initiate new expressions of community development.

Much effort has been devoted to understanding and interpreting the succession process. This effort has generally been directed at two sides of the same coin. What contributes to the process of change? What contributes to the condition of relative stability? The seminal paper of Connell and Slatyer (1977) developed three alternative models of succession: facilitation, tolerance, and inhibition. Although these models were developed from situations where a disturbance initially created an open space with enhanced resource availability, the emphasis was on species replacement following the disturbance. As such, these models have great potential relevance to systems where novel plant species enter and eventually dominate plant communities.

1. Facilitation: Late-successional species establish and grow only after early-successional species have modified (ameliorated) environmental conditions.

2. Tolerance: Late-successional species establish and grow in the presence of early-successional species. This is possible because late-successional species tolerate the low resource availability of the early-successional environment.

3. Inhibition: Late-successional species establish and grow to maturity only after early-successional species become senescent or are killed.

Supporting evidence for facilitation and tolerance was scant (Connell and Slatyer 1977). Facilitation appeared to operate only on severe sites where an early-successional species greatly changed the nutrient status of the substrate. Tolerance was rejected in many studies due to the fact that late-successional species were not able to tolerate low resource availability but were instead able to grow slowly and then respond to a sudden increase in resource availability. Most evidence (descriptive and experimental) supported inhibition as a viable model of succession; however, later work suggested that both tolerance and inhibition may operate simultaneously when a diversity of plant growth forms is found in the same early-successional community (Hils and Vankat 1982; Walker and Chapin 1987).

Although competitive interaction leading to tolerance and inhibition provided the general tone of succession, further work was required to explain the great diversity of successional trajectories observed at different sites. Noble and Slatyer (1980) maintained that this diversity could be explained based on "vital attributes" of individual participating species. They modeled succession based on the method of arrival or persistence of propagules at a site, the conditions that allow these propagules to become active and establish plants, the longevity of these plants, and the time it takes these plants to reach critical life history events. Recognizing the importance of interaction between site conditions and species characteristics, Pickett et al. (1987) developed a conceptual framework for succes-

TABLE 11.1. General causes of ecological succession, contributing processes, and modifying factors.

General causes	Contributing processes	Modifying factors
Site availability	Disturbance	Size, severity, time, dispersion
Species availability	Dispersal	Landscape configuration, dispersal agents
	Propagules	Land use, time since last disturbance
	Resources	Soil, topography, site history
Species performance	Ecophysiology	Germination response, assimilation rates, growth rates, genetic differentiation
	Life history	Allocation, reproductive timing, mode of reproduction
	Stress	Climate, site history, prior occupants
	Competition	Competition, herbivory, resource availability
	Allelopathy	Soil chemistry, microbes, neighboring species
	Herbivory	Climate, predators, plant defenses, patchiness

From Pickett et al. (1987).

sion based on three general causes: site availability, differential species availability, and differential species performance (Table 11.1). Their concept used a hierarchical approach. Various processes contributing to the three general causes of succession were identified. Modifying factors that in turn influence these processes were also listed (Table 11.1). For example, differential species performance can be linked to the differing ecophysiology of individual species; the germination requirement of species is one modifying factor defining the ecophysiology of species. The strength of the approach presented by Pickett et al. (1987) is that succession can be explained at any site if sufficient information is available on the disturbance regimen, site conditions, species availability, and characteristics of these participating species. Furthermore, succession can be explained in the event that a novel species is suddenly added to the competitive mix (i.e., plant invasion).

Plant Invasion: A Successional Interpretation

Ecological succession involves and indeed depends on some degree of plant invasion. Such invasion by indigenous (IN) species may contribute to management goals or may work against them (Luken 1990). Plant invasions of most critical concern to resource managers typically involve NI species capable of dramatically changing rate and direction of succession and also of causing a corresponding decrease in species richness. For example, Lehmann love grass (*Eragrostis lehmanniana*), a species imported from South Africa and introduced to semidesert grasslands of southern Arizona, now dominates 145,000 ha of land where it contributes nearly 90% of the grass biomass (Anable et al. 1992). The following discussion will focus on those situations where NI species are introduced to new geographical ranges and subsequently invade established plant communities.

From a successional perspective, the events of primary importance in problematic invasion by NI species are (1) entry of species into established plant communities, (2) performance of species once established, (3) asymmetric competitive interactions with surrounding IN species, and (4) long-term persistence.

Entry

Entry of NI plant species into established plant communities indicates that inhibition from previous colonists is somehow overcome. It is generally accepted that a change in the prevailing disturbance regimen allows NI plants to circumvent inhibition from competing plants (Hobbs and Huenneke 1992). However, there is considerable debate regarding actual mechanisms. Although the creation of bare ground, one direct result of many disturbances, is cited as an important factor in allowing invading plants to establish (Cronk and Fuller 1995), it is likely that other changes (e.g., increased resource availability, damage to previous occupants) linked to the disturbance regimen may also be of importance (Hobbs and Huenneke 1992). For example, DeFerrari and Naiman (1994) noted that young riparian patches on the Olympic Peninsula, Washington, supported a relatively high percentage of NI species due to frequent flooding that disturbed soil and removed understory plants. In populated landscapes where human impacts are widespread, entry of NI plants into plant communities may not be caused by a single environmental change but may instead be associated with a suite of environmental changes accompanying human presence (Knops et al. 1995). Conversely, IN species and their continued regeneration may not be favored in the face of such environmental changes (McIntyre and Lavorel 1994). In the absence of overt anthropic influence, NI plants may experience varied entry success based on traditional ecological factors such as soil type and light availability (Myers 1983), but even here there may subtle anthropic factors that interact.

Performance and Competition

Once invading plant species have established seedlings in a community, growth to maturity requires that inhibition from surrounding plants must be overcome. Various theories have been proposed regarding the apparent superior abilities of NI species to grow, compete, and achieve dominance (Cronk and Fuller 1995).

One theory proposes that successful NI invaders possess unique traits allowing them to outcompete and overgrow IN species (Cronk and Fuller 1995). This has spawned numerous attempts to identify attributes of successful invaders (Bazzaz 1986; Newsome and Noble 1986). Although in many instances it can be demonstrated that an NI plant does grow faster and compete better than a closely related or coexisting IN plant, always there are conditional factors such as herbivory and resource level complicating this scenario (Harrington et al. 1989; Schierenbeck et al. 1994). Thus, the search for unique plant attributes may be futile unless searching is restricted to specific plant growth forms invading specific habitat types (see, e.g., Forcella 1985a).

Another theory proposes that successful NI invaders need not possess unique traits provided that growth and reproduction are optimal under the extant environmental conditions at any particular point in time and space (i.e., a genotype/environment match as a result of preadaptation). Relatively strong growth and reproduction by the NI species may occur because the environment has changed and thus no longer favors the established IN species (McIntyre and Lavorel 1994), because established IN plants have been damaged, thus releasing resources (Deferrari and Naiman 1994; Mack 1981), or because of unique interactions emerging between the NI invader and either NI or IN animal species (Bossard 1991; D'Antonio 1990b).

Long-Term Persistence

Various degrees of persistence by NI species are theoretically possible when invasion occurs (Fig. 11.2). A changed disturbance regimen may allow an NI plant to become established, but if inhibition by surrounding species is not overcome, then the invasion may fail (i.e., a condition of temporary coexistence [Fig. 11.2a]). The invading plant simply failed to grow well or could not achieve reproductive success. Many such invasions likely occur, but they are seldom noticed nor are they reported in scientific literature. A disturbance or environmental change may allow plant invasion to occur and may also shift differential species performance in favor of the invading plant. In the event that the invader eventually senesces and is replaced by other species, this is a case of temporary dominance (Fig. 11.2b). Senescence of the invader could

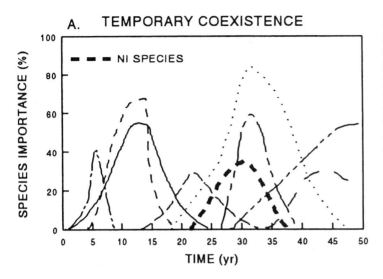
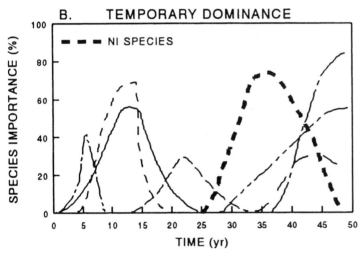
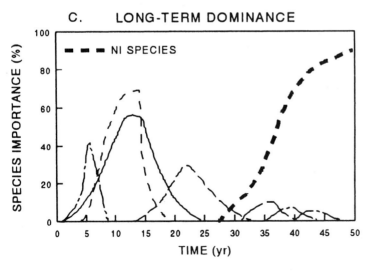

FIGURE 11.2. Various degrees of persistence for a nonindigenous species invading a community following disturbance. (A) Temporary coexistence occurs if the invader does not assume dominance and fails to persist. (B) Temporary dominance occurs if the invader assumes dominance but fails to persist. (C) Long-term dominance occurs if the invader assumes dominance and also maintains this dominance for long periods of time as a result of competitive interaction, regeneration, or changed disturbance.

occur as maximum longevity is reached, as reproduction fails, as growth slows due to changing environmental conditions, or as a result of herbivore or pathogen attack. If an invader is long lived, if it has well-developed adaptations for persistence, if it changes the disturbance regimen to favor its own reproduction, or if there are no species-specific herbivores or pathogens, a condition of long-term dominance may occur (Fig. 11.2c).

There are descriptive data suggesting that invasive NI plants may participate in succession with a variety of roles and with various degrees of persistence. For example, Vankat and Snyder (1991), using a chronosequence of old fields, savannas, and forests in Ohio, showed that NI species of a variety of growth forms participated in succession much like IN species (i.e., importance values increased and then decreased as sites increased in age). However, all stages of succession did support some NI species (Vankat and Snyder 1991). Robertson et al. (1994) also showed that invasive NI species growing in the Pennypack Wilderness of Pennsylvania were sorted in importance along a successional gradient leading to forest communities; low persistence and limited colonization were associated with low light availability and lack of establishment sites (Robertson et al. 1994). Long-term community dominance by invasive NI species has also been demonstrated. Such dominance may occur when invading plants have several mechanisms for population persistence (Luken and Goessling 1995; Luken and Mattimiro 1991), when NI species change the disturbance regimen so that their own regeneration is favored while regeneration of IN plants is not favored (Anable et al. 1992; Myers 1983), or when NI plants expand in coverage with a concomitant decrease in coverage of community associates (Weiss and Noble 1984).

Identification and description of NI plant invasion leading to long-term dominance must be made with caution due to the youth of most well-studied invasions. Still, resource managers may be concerned if such a condition exists for only a decade. It must be stressed that no plant community, even one dominated by a monospecific stand of a long-lived invader, will remain balanced and stable (see, e.g., Johnson and Mayeux 1992). In those situations where an invasive NI plant has slowed the pace of community development, the important questions are: What long-term changes can be expected? and How might management influence this condition of relative stability?

The Decision to Manage

Considering the diverse roles that NI plants may play in structure, function, and development of preserved systems and considering the potential for management activities to change functional roles of both NI and IN species, it is important to assess reasons for management and projected effects of management. The reasons for management of NI plants are related to management goals. A first step in managing plant invasions is the formal statement of these goals. After it is concluded that management is indeed warranted, a concerted effort should be made to predict community-level change both with and without active management. This final exercise may reveal supervening effects of management that in turn make necessary either new management approaches or may force a change in conservation goals.

Goals for Preserved Nature

Historically, management goals for North American plant communities in parks, natural areas, or preserves centered around the presumed equilibrium condition existing prior to European settlement (i.e., the pre-Columbian period). Presettlement plant communities assumed historical significance and served as standards for success or failure of modern conservation efforts (Luken 1994). This concept of successfully preserved nature emerged despite abundant historical evidence suggesting that native Americans modified many plant communities and also extended plant ranges as a result of their quest for food, fiber, or fuel (Williams M 1993). Invoking the standard that presettlement ecological systems were balanced and undisturbed by humans, invasion of natural areas by NI species is almost universally viewed as undesirable for several reasons: the plants are not indigenous and they do not belong there; the plants change the structure and function of ecological systems from that existing in the presettlement times; the plants are present due to the activities of people; and the plants threaten to decrease diversity by outcompeting IN species.

In an elegant treatise outlining the potential biological virtues of newly formed national parks, Grinnell and Storer (1916) stated a philosophy toward NI species that has persisted in various forms to this day: "Equal vigilance should be used to exclude all non-native species from the parks, even though they may be non-predaceous. In the finely adjusted balance already established between native animal life and the food supply, there is no room for the interpolation of an additional species."

More recently Cronk and Fuller (1995) reiterated a similar idea with their definition of an invasive plant: "an alien plant spreading naturally (without the direct assistance of people) in natural or seminatural habitats, to produce a significant change in terms of composition, structure or ecosystem processes."

In an effort to apply a modern synthesis of ecological theory to the practice of conservation, Pickett et al. (1992) suggested that goals for preserved systems should center around the idea of change rather than stability. Furthermore, they suggested that conservation should proceed with knowledge of (1) the

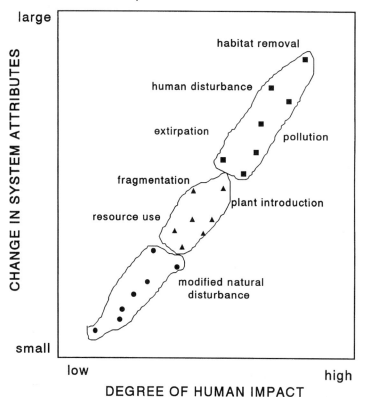

FIGURE 11.3. Relationship between degree of human impact in ecological systems and change in attributes of these systems. Pristine systems are minimally changed due to human impact; urban/cultural systems are maximally changed due to human impact. Various human impacts are identified.

processes governing a system, (2) the context of a system, (3) the historical change in a system, (4) the historical and extant impacts of humans in a system, and (5) the limits of organisms now comprising the system (Pickett et al. 1992). These guidelines parallel the management recommendations for invasive NI plants developed by Hobbs and Humphries (1995) where system attributes and human impacts on these attributes are assessed and considered.

Organizations involved in acquisition and management of many nature reserves might view the problem of plant invasion within the context of system attributes and how system attributes change along urban to rural gradients (Fig. 11.3). McDonnell et al. (1993) predicted that ecological systems at the urban end of such a gradient would be characterized by greater numbers of NI species, increased biotic limitation due to these NI species, decreased probability of influence by natural disturbance, decreased diversity in feasible management options, and increased pressure to provide services to human populations. System attributes at the urban end of the gradient pose unique and challenging problems for those resource managers desiring to move these systems toward presettlement conditions primarily because the available pool of species is different than in presettlement times and because management options are relatively limited. As such, resource managers are faced with either mitigating human impact or changing conservation goals relative to human impact, or some hybrid combination of these two approaches.

Reassessment of conservation goals relative to system attributes would likely produce a range of conservation values for NI plants (Williams, Chapter 3, this volume). An undesirable NI species would be one associated in a causal fashion with decline in specialized IN species. On the other hand, NI plants could be viewed as desirable if the goal was to quickly move succession to forest in the aftermath of severe anthropic disturbance. Nonindigenous plants could be viewed as benign if they simply participate in ecological processes without causing any measurable change in these processes. The emergence of novel ecological systems (Bruce et al. 1995) and interactions (Schiffman, Chapter 7, this volume) as a result of plant invasion offers unique opportunities to assess and reformulate conservation goals so that NI plants assume greater ecological value.

The Response to Management

Management of plant communities where NI species are well established generally proceeds under two different philosophical influences. The traditional path is that of weed science. It involves directing chemical, physical, or biological control methods at the problem species; success is typically measured in terms of kill. In contrast, management activities can be directed at the full range of processes contributing to system change through time (Luken 1990) with success measured as both negative and positive change in importance of participating species. Clearly, the later approach is preferred, but it is a more complex action and it likely requires more effort to assess success or failure.

Species response to management practice will be influenced by (1) availability of propagules at the time of management, (2) propagule interaction with management, and (3) IN species performance under conditions of changed resource availability caused by management.

Propagules and the Seed Bank

Propagule availability is commonly assessed as seeds in the soil seed bank or as presence of plant parts capable of regenerating whole plants. The make-up of the seed bank is influenced by local vegetation sources, distant vegetation sources, seeding efforts, seed germination, seed decay, seed predation, and physical destruction of seeds (Luken 1990). Furthermore, germination of these seeds is controlled by various dormancy mechanisms interacting with site conditions.

The seed bank variously records plant species that have been present on the site as well as those that could be present if environmental conditions become conducive to germination. The role of the seed bank as a determinant of plant persistence during disturbance and community development is well studied. Managers of plant invasions should be particularly sensitive to the fact that make-up of seed banks and thus community response to management will depend on site history and proximity to sources of NI species. Sites with long and complex histories of human impact and sites adjacent to areas with histories of human impact will likely have seed banks with high representation of NI species (Brothers and Spingarn 1992; Hester and Hobbs 1992; Rapoport 1993; Roberts and Vankat 1991). Non-seed propagules (i.e., plant fragments, stem bases, root systems) also record previous and potential participants in succession. These propagules assume increasing importance when management practice stimulates the activity of vegetative reproduction or when long periods of site occupation allow the accumulation of large amounts of belowground biomass.

The increasing NI component of seed banks toward the urban end of urban-to-rural gradients suggests that it is unrealistic to hold all sites within a region to a similar standard of system development. Furthermore, it is difficult to predict system response to management in the absence of site-specific research. Hobbs and Huenneke (1992) presented a paradox of management where designed disturbances could have both negative and positive effects on NI species. This occurs because management activities commonly change resource levels; in turn, seed germination of NI and IN species may be stimulated. Gibson et al. (1993) demonstrated this with a study of tallgrass prairie in Kansas. Burning and mowing produced different community development pathways and also had various effects on the establishment of NI species (Gibson et al. 1993). Likewise, Huffaker and Kennett (1959) noted that NI species increased in importance in some plots after successful biological control of Klamath weed (*Hypericum perforatum*). Finally, Luken and Mattimiro (1991) found that removal of Amur honeysuckle (*Lonicera maackii*) shrubs from forests and open sites increased honeysuckle seedling establishment and also stimulated resprouting from stem bases. The relative contribution of seeds and non-seed propagules to community development after management was site specific.

The preceding discussion reinforces the idea that system attributes (e.g., the seed bank) may be changed as a result of invasion. This change will likely be greater the longer a site has been invaded. Any management practice should be assessed not just by virtue of effects on NI species but on the overall environmental change that may stimulate seeds or propagules in the soil.

Species Performance

It is often assumed that once both NI plants and propagules of these plants are eliminated, then IN species will reassert dominance due to relaxation of competitive pressure and subsequent increased performance. Unfortunately, there are few studies where long-term effects of management practice, specifically plant removal, have been documented in terms of community response. The research of Marrs and Lowday (1992) is an exception in that community response to bracken (*Pteridium aquilinum*) removal from grass heath and from heather (*Calluna vulgaris*) heath was measured over a 10-year period. The management goal was to re-establish species-rich heath communities that historically dominated the landscape. Bracken was controlled by either cutting or herbicide spraying. Establishment of desirable IN species such as *Calluna* and sheep fescue (*Festuca ovina*) was facilitated by artificial seeding. Although removal of bracken did allow establishment of target IN species, management activities also inadvertently allowed/caused the establishment of nontarget grass species. These grasses in turn formed dense, monospecific stands that directed succession away from the desired path. Marrs and Lowday (1992) hypothesized that high deposition of atmospheric nitrogen may have changed the system attributes so that the species-rich heath community typical of low-fertility sites was no longer a realistic goal.

The above-mentioned research demonstrates the importance of current system attributes as determinants of system development following management. Where systems have been radically changed by past disturbance, chronic anthropic effects, or long-term occupation by NI species, the reestablishment of a typal community may not be possible without complete system replacement. Even when such radical restoration efforts are undertaken, there is no assurance that success will be achieved (Doren et al. 1991).

A Paradigm for Management

There now exists a relatively well-defined body of knowledge on the initiating causes and extent of plant invasions. In contrast, there are relatively few data on the overall conservation impact of these invasions (see, however, Pakeman and Marrs 1992) or on the long-term effects of management practice. When long-term effects of management practice are known, typically one finds that plant invasion has been favored rather than inhibited, for example, effects of large wapiti (*Cervus elaphus*) populations on NI species invasion in Yellowstone National Park (Wagner and Kay 1993). In short, we know why plant invasion occurs but we do not know what should be done or even if something should be done. Historically, plant invasion has been viewed through the eyes of the weed scientist. Here, revision is required so that this issue can be placed in the context of nonequilibrial systems.

The recommendation of Hobbs and Humphries (1995) is cogent: system attributes allowing invasion to occur must be targeted. If such attributes are not addressed, it is possible, even likely in some systems, that management practice aimed at manipulating succession will lead from one invasion of NI plants to the next (Fig. 11.4, scenario 2). The ability of resource managers to modify system attributes will likely vary along urban to rural gradients. As such, new conservation goals

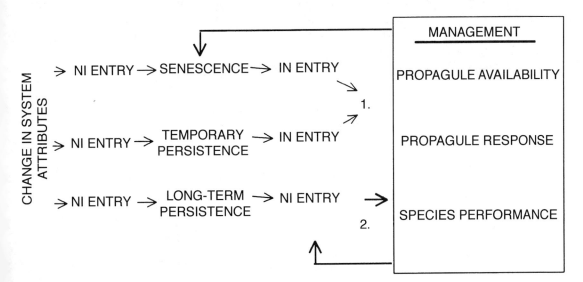

FIGURE 11.4. A management paradigm for communities where nonindigenous plants are established as a result of changing system attributes. Active management and the corresponding factors influencing plant response can lead to either (1) communities with mostly indigenous species or (2) communities where invasion of nonindigenous species occurs sequentially.

need to be explored and developed with greater emphasis placed on the potential ecological functions of NI species, especially where such species may dominate the seed bank. The observation—that as the visitation rate of nature reserves increases, the number of established NI plants also increases (Usher 1988)—poses a unique challenge, especially when conservation organizations are mandated to encourage human visitation of reserves. Lastly, new management approaches for plant communities already invaded need to be developed. The emerging practice of succession management (Luken 1990) consisting of designed disturbance, controlled colonization, and controlled species performance should replace the traditional practice of weed control. Creative design of reserves coupled with management practice might then lead to systems that are inherently better able to reduce or resist invasion by NI plants while at the same time expressing greater representation of IN species through time (Fig. 11.4, scenario 1).

Conclusions

1. Management of plant invasions has historically occurred in natural vegetation with the assumption that plant control will eventually allow establishment of a balanced system dominated by IN species. This is not likely to occur.

2. Targeting individual NI species for control may not be successful in achieving traditional management goals without first addressing historical and extant processes that change system attributes.

3. System attributes determine the rate and direction of community development. Such attributes affect availability, entry, performance, persistence, and management potential of NI species.

4. Nonindigenous species, once established in a biological community, can assume a variety of roles with various degrees of persistence.

5. For plant communities heavily invaded by NI species, system response to management will be determined by the propagule pool, its response to management, and performance of all species after management alters site conditions.

6. Intensity of human impact (extant and historical) will determine the realistic conservation goals for any preserved system.

7. Because system change is inevitable, resource managers should determine the types of system changes desired. Various conservation standards should be developed based on the intensity of human impact.

8. Management practice aimed at eliminating NI species should manipulate both NI and IN species with the goal of gradually producing a dynamic system that satisfies explicit management goals.

12
Methods for Management of Nonindigenous Aquatic Plants

John D. Madsen

Water-hyacinth
(*Eichhornia crassipes*)

Numerous terrestrial plant communities have been invaded by nonindigenous (NI) plant species that have disrupted community dynamics and ecosystem processes. Aquatic communities, too, have been modified by many recent arrivals. Some, such as purple loosestrife (*Lythrum salicaria*), can inhabit both wetland and terrestrial sites; others, such as Brazilian elodea (*Egeria densa*), are strictly aquatic, growing submersed in water. Invaders range in size from tall trees, for example, melaleuca (*Melaleuca quinquenervia*), to the duckweed-sized mosquito fern (*Azolla caroliniana*). Many are nationwide problems (e.g., Eurasian watermilfoil [*Myriophyllum spicatum*]); some are localized geographically (e.g., water-chestnut [*Trapa natans*]). However, all these aquatic species follow similar principles in their impacts on aquatic sites and invoke similar approaches to their management.

Three NI aquatic plants are currently considered nationally prominent problem species in the U.S. Army Corps of Engineer's Aquatic Plant Control Program: Eurasian watermilfoil (Fig. 12.1), hydrilla (*Hydrilla verticillata*) (Fig. 12.2), and water-hyacinth (*Eichhornia crassipes*) (Fig. 12.3). Most aquatic plant management efforts nationally are related to NI species; less commonly, indigenous (IN) species create weed problems.

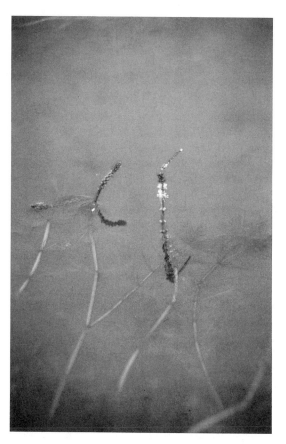

FIGURE 12.1. Eurasian watermilfoil (*Myriophyllum spicatum*): closeup of inflorescence. Damselfly is approximately 5 cm long.

Why Manage Nonindigenous Aquatic Plants?

The value of littoral vegetation in ecological processes has been documented by several studies. Such vegetation assists in stabilizing sediments and in binding nutrients (Maceina et al. 1992). Aquatic invertebrates, which are the essential food of many fishes and semiaquatic animals, are often more diverse (Butler et al. 1992) or abundant (Cyr and Downing 1988; Schramm and Jirka 1989) in aquatic macrophyte beds as opposed to unvegetated areas. Littoral vegetation is often a significant food source to semiaquatic animals such as waterfowl and some mammals (Hocutt and Dimmick 1971; Hunt and Lutz 1959; Jones and Drobney 1986; Perry and Uhler 1982). Aquatic vegetation is an important habitat for fishes, both for young-of-the-year (Dewey and Jennings 1992) and desirable sportfishes (Crowder and Cooper 1982; Savino and Stein 1989). For these reasons, IN aquatic plants are an important component of lake ecosystems (Grimm and Backx 1990; Ozimek et al. 1990). However, the introduction and spread of NI species may significantly alter community structure of aquatic ecosystems. Rarely are these changes beneficial.

The initial rationale for controlling these nuisance-forming NI species was, and remains, their impact on human safety and activity. Nuisance growths of aquatic plants interfere with commercial and recreational navigation, increase risk and height of floods, and, in some instances, contribute to increased populations of insects that may be vectors for human diseases (Gallagher and Haller 1990). However, ecological degradation is also an important rationale for controlling NI aquatic plants. Invasions by these plants alter structure of plant communities, resulting in changes in water quality, IN plant populations, and interactions among animal populations (Fig. 12.4).

Typically, NI species create dense canopies, which result in decreased oxygen exchange, increased nutrient loadings, and increased water temperatures (Honnell et al. 1993; Seki et al. 1979). For instance, ponds at the Lewisville research facility in Texas planted with Eurasian watermilfoil, hydrilla, and water-hyacinth all had nighttime dissolved oxygen levels below 5 ppm, which may cause physiological stress to fishes (Fig. 12.5) (Honnell et al. 1993). An IN pondweed (*Potamogeton*) with floating leaves also produced conditions of low dissolved oxygen. However, a pond containing a mix of submersed IN species had acceptable dissolved oxygen concentrations.

The NI plants that spread rapidly are, by their very nature, "weedy" exploitative species (Smart and Doyle 1995). The three major nuisance species in the United States combine

FIGURE 12.2. Hydrilla (*Hydrilla verticillata*): (A) Habit. This dense bed is 500 m wide growing in >4 m water depth; (B) Closeup. Leaves are approximately 1 cm long.

the weedy, rapid colonizing characteristics of early-successional species with a competitive, canopy-forming morphology that greatly reduces importance of other species once a site is colonized. Because of these characteristics, they tend to form dense monospecific stands. Invasions of Eurasian watermilfoil (Aiken et al. 1979; Grace and Wetzel 1978; Smith

FIGURE 12.3. Water-hyacinth (*Eichhornia crassipes*): (A) Habit. This pond is 40 m wide and 100 m long; (B) Floral closeup. The flower is approximately 10 cm across.

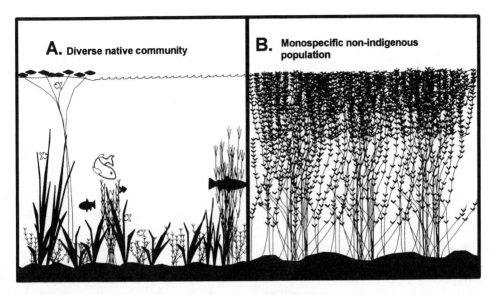

FIGURE 12.4. Schematic representation of (A) a diverse native plant community structure versus (B) community structure of monospecific nonindigenous plant population

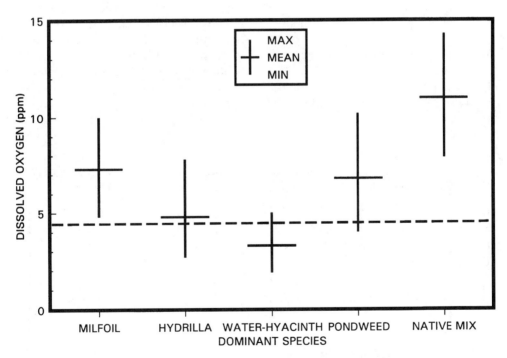

FIGURE 12.5. Maximum, minimum, and mean daily dissolved oxygen concentrations (ppm) over a 48-day period in replicate ponds planted with Eurasian watermilfoil (*Myriophyllum spicatum*) (MILFOIL), hydrilla (*Hydrilla verticillata*) (HYDRILIA), water-hyacinth (*Eichhornia crassipes*) (WATER-HYACINTH), a floating-leaved pondweed (*Potamogeton* sp.) (PONDWEED), and a mix of submersed indigenous plants (NATIVE MIX). The line at 5 ppm represents the level at which fishes experience oxygen stress. Data were collected at the Lewisville Aquatic Ecosystem Research Facility, Lewisville, TX (Honnell et al. 1993).

and Barko 1990), hydrilla (Swarbrick et al. 1981), and water-hyacinth (Gowanloch 1944; Penfound and Earle 1948) have all been accompanied by a decline of IN plants, resulting in reduced diversity and abundance. However, community change was not quantitatively documented in tandem with invasion, giving rise to fallacious speculation that these species invaded only habitats with degraded or disturbed plant communities. Study of a newly invading population of Eurasian watermilfoil presented quantitative evidence that this species can invade a dense, healthy stand of IN plants and, over several years, develop a dense canopy (Madsen et al. 1991b). A permanent 6-by-6 m grid placed over a newly developed bed found that, over a 3-year period, cover of Eurasian watermilfoil increased from 25% to 97% (Fig. 12.6). During this time, the total number of species (species richness) in the grid dropped from 21 to 9. In addition, the cover of IN plants dropped precipitously. This study documented a decline in diversity and abundance of IN species in tandem with invasion of an NI plant species.

Several traits have been suggested for the ability of these NI species to outcompete IN plants, including more efficient photosynthetic pathways, superior nutrient uptake, and lower light compensation point (Grace and Wetzel 1978). Although many factors may contribute to success of these species, formation of a dense canopy over the IN species appears the most important (Madsen et al. 1991a).

In addition to affecting water quality and reducing diversity and abundance of IN plants, invasion by NI plants resulting in dense monospecific stands alters animal communities utilizing littoral zones and wetlands. Aquatic predator/prey relationships are involved and express numerous complex, weak interactions; but basic predictions can be made from theoretical models and empirical observations (Savino et al. 1992; Wiley et al. 1984). As a dense, monospecific stand is created, forage species will be better able to hide from predators, resulting in large numbers of small forage fishes and reduced growth and vigor of predatory species (Lillie and Budd 1992). Therefore, these NI species create problems not only for the human users of ecosystems, but they also are deleterious to water quality and to the indigenous biotic communities they infest.

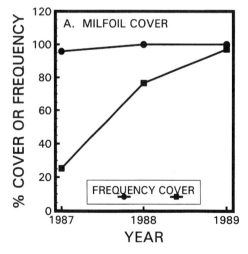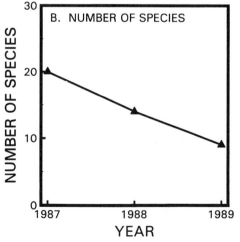

FIGURE 12.6. Change in (A) Eurasian watermilfoil (*Myriophyllum spicatum*) percent frequency and percent cover, and (B) total species richness within a 6 m × 6 m experimental frame in Lake George, New York, over a 3-year period. (Redrawn from Madsen et al. [1991b])

Given the strong impetus to solve these problems, are management measures effective in not only eliminating NI species but also in restoring the original plant community? Although management techniques vary in their environmental and ecological consequences, research indicates that management of target species can lead to community restoration. Treatment with the Experimental Use Permit (EUP) herbicide triclopyr not only successfully eliminated Eurasian watermilfoil in study plots on the Pend Oreille River, Washington State, but diversity of IN plants was restored for up to 2 years after one treatment (Madsen et al. 1994b). Similar restoration of IN plant communities resulted after treatment with the herbicide 2,4-D (Getsinger et al. 1982) and fluridone (Getsinger 1993).

Other management techniques can successfully reduce the importance of NI species and increase diversity of IN species. Community restoration has been documented after management with benthic barriers (Eichler et al. 1995) and diver-operated suction harvesting (or dredging) (Eichler et al. 1993). Once the dense canopy of NI species is removed, IN species often will regenerate from new introductions or from the propagule bank.

Management Techniques

Developing an appreciation for management techniques takes an investment of time and effort, but in the long run it is well worth it. One important rule to remember is that no management technique is intrinsically superior to another, nor will one management technique (e.g., a single chemical, or herbicides as a group) be sufficient for all situations in a management program. Rather, all techniques should be considered tools in the manager's toolbox. Some are more expensive but will better control dense populations in larger areas. For small infestations (<0.1 acre) or new colonies, hand picking may actually be the best approach. Each site should be evaluated and management techniques selected based on the desired level of control as well as on environmental and economic constraints.

A large volume could be written on management techniques alone. The following is an overview of those techniques currently used to control aquatic plants. I separate these into the areas of biological, chemical, mechanical, physical, and institutional management techniques.

Biological Management Techniques

Development of biological control technology has been based largely on the premise that, because target NI plants have been brought to a new habitat without their accompanying herbivores and pathogens, the best approach to manage NI plants is to find their IN pests and introduce these naturally associated control agents from their native land. Although this approach sounds attractive in its simplicity, it ignores both the effort and potential perils of the approach as well as the many contradictions to the model of using highly selective agents to manage a target species. For these reasons, biological control has typically targeted either insects or pathogens as control agents. These offer the greatest hope for highly selective management provided that growth rates of the agents are sufficient to maintain populations in check.

Because biological control efforts are research intensive and require long periods of discovery and quarantine with little hope of economic return, most introduction, rearing, and release programs have been done by federal and state government agencies (see also DeLoach, Chapter 13, this volume). Without government involvement, it is unlikely that biological control efforts in their present form will continue. It may take up to 20 years to discover even one biocontrol agent in the native land, evaluate its selectivity and effectiveness, bring it through the overseas and continental quarantine examination, and finally successfully rear and release the agent to have self-sustaining populations in the wild. Considering that current biocontrol projects attempt to find three or four agents to release per target species, this presents a substantial long-term research and development effort. And despite intense scrutiny, many control agents simply are not effective in operational situations. However, a significant minority of agents may be extremely successful.

The greatest success with insect-based selective biological control of an aquatic plant remains the control of alligatorweed (*Alternanthera philoxeroides*) by several South American insects, primarily the alligatorweed flea beetle (*Agasicles hygrophila*) (Gallagher and Haller 1990). Before introduction of these control agents, alligatorweed was a significant nuisance in most southeastern states in North America. After introduction of the beetle, alligatorweed is essentially under control in Florida and the Gulf Coast states (Gallagher and Haller 1990; Quimby and Kay 1983). However, in the northern fringe of its range and in other isolated areas, it is not controlled by the beetle. This demonstrates that biological control will not necessarily be effective throughout the entire new range of the invading species.

Many NI and IN organisms have been used for biological control programs (Gallagher and Haller 1990); however, current operational or research and development efforts

TABLE 12.1. Summary of biological management methods for aquatic plants.

Management method	Description	Advantages	Disadvantages	Systems where used effectively	Plant species response
Grass carp (white amur)	Herbivorous fish	Long-term (decades), relatively inexpensive	Cannot control feeding sites, difficult to contain in water body, tendency for "all or none" community response	Isolated water bodies, effective against hydrilla and other preferred species. Operational	Fish have strong preference for hydrilla, avoid Eurasian watermilfoil, generally do not prefer floating plants
Neochetina spp.	Water-hyacinth weevils	Species selective	Not effective in reducing areal coverage in many situations	Released in Florida, Gulf Coast states. Developmental	Leaf scars, some reduction in growth
Hydrellia spp., *Bagous* spp.	Hydrilla fly, hydrilla stem weevil	Species selective	Has not yet been established	Released in Florida, Alabama, Texas (research)	Limited
Euhrychiopsis lecontei	Weevil—native or naturalized	Already established in the U.S.	Less selective, currently under R & D	Currently under study in Vermont, Minnesota (research)	Plants lose buoyancy, weevil interferes with transfer of carbohydrates
Mycoleptodiscus terrestris (Mt)	Fungal pathogen; acts as a contact bioherbicide	Low dispersion, fairly broad spectrum	Expense, cross-contamination, inconsistent viability and virulence of formulation	Under R & D for both Eurasian watermilfoil and hydrilla	"Contact bioherbicide"; plants rapidly fall apart, but regrow from roots
Native plant community restoration	Planting of desirable native plant species or community	Provides habitat, may slow reinvasion or initial invasion	Expensive, techniques still under development	Under R & D around the country	Native plants provide ecosystem benefits, slow invasion

center on a few: grass carp or white amur (*Ctenopharyngodon idella*), introduced insects for water-hyacinth and hydrilla, naturalized pathogens for Eurasian watermilfoil and hydrilla, and naturalized insects for Eurasian watermilfoil (Table 12.1).

Grass carp are a popular control agent for aquatic plants, especially in small ponds or isolated bodies of water, and are particularly effective in controlling hydrilla. These fish have strong feeding preferences (Pine et al. 1990; Pine and Anderson 1991; Sutton and Van Diver 1986) and will selectively feed on plants in a mixed community from the most to the least preferred. If hydrilla is the target plant, this may be beneficial—at least until the hydrilla is eaten (Van Dyke et al. 1984). If Eurasian watermilfoil is the target, all other plants may be eaten first, and grass carp may in fact never completely remove Eurasian watermilfoil (Fowler and Robson 1978). In addition, there are many concerns about using grass carp, including the length of time they remain in the system, the difficulty of controlling where and what they eat, the highly variable results for large systems (>500 acres), the escape of carp from the managed system, the impact of their feeding on nontarget plant and animal species, and the difficulty of removing them when control is no longer needed (Bonar et al. 1993; de Kozlowski 1994; Durocher 1994; Hammond 1994). An initial concern regarding reproduction of grass carp (Stanley et al. 1978; Webb et al. 1994) has been addressed largely through the use of sterile triploids (Durocher 1994).

The effectiveness of grass carp is strongly influenced by water temperature and seasonality, with northern ecosystems typically requiring substantially higher stocking rates than southern ones (Stewart and Boyd 1994). In addition, stocking rates can vary by an order of magnitude, depending on whether adequate results are required in 3 years as opposed to the need for more immediate results (Stewart and Boyd 1994). The problem of lag time can be moderated by combining stocking of grass carp with herbicide treatments in the first year (Eggeman 1994; Haller 1994; Jaggers 1994). However, a strong tendency for obtaining either no perceived control with understocking or complete plant elimination with overstocking remains—this has been termed the "all-or-none" dilemma. If achieving an intermediate density of plants is possible using grass carp, it is certainly very difficult and must be based on a more sophisticated understanding of interacting factors than have been considered in the past.

To clarify the following discussions on biological control techniques, it should be clearly stated that the only operational successes with biological control have been with grass carp and the alligatorweed flea beetle, and even these successes are not without caveats. The remainder of the discussion deals with research and development directions that are not operational. Likewise, the only biocontrol approach currently usable by individuals or private groups is grass carp. Other approaches are either in the earliest stages of development or are a function of governmental agency action.

Biological control of water-hyacinth has been primarily through rearing and establishment of water-hyacinth weevils in the genus *Neochetina*: *N. bruchii* and *N. eichhorniae* (Center and Van 1989; Gallagher and Haller 1990). Despite extensive establishment efforts, biological control agents for water-hyacinth have not been found to be nearly so successful as those for alligatorweed (Gallagher and Haller 1990). Water-hyacinth is managed throughout most of its range in the United States by maintenance applications of herbicides, which cost almost $3 million for 1992 in Florida alone (Schardt 1993).

Like those for water-hyacinth, the insect biocontrol agents currently under research and development for hydrilla were discovered from overseas investigations of native habitats and brought in through the biocontrol "pipeline" (Cofrancesco 1994, 1995). Hydrilla biocontrol agents include the flies *Hydrellia pakistanae* and *H. balciunasi* (Buckingham and Okrah 1993) and the weevils *Bagous hydrillae* and *B. affinis* (Grodowitz et al. 1994, 1995). Although several introduced biocontrol agents feed in a complementary fashion to stress hydrilla populations, it is too early in the research and development process to predict operational-scale success. For instance, mathematical models of *H. pakistanae* growth rates suggest that even if the fly were successful in central Florida, its development rate may be too slow in the colder climate of northern Alabama to be effective (Boyd and Stewart 1994).

Although foreign surveys for biocontrol agents for Eurasian watermilfoil have been recently initiated (Buckingham 1995), most effort has been spent looking at naturalized or IN insects that feed on this species (Kangasniemi 1983; Oliver 1984). In particular, laboratory, mesocosm, and field research have been vigorously pursued on the pyralid moth *Acentria nivea* (Creed and Sheldon 1992, 1994) and on the weevil *Euhrychiopsis lecontei* (Creed and Sheldon 1993, 1994; Newman and Maher 1995). *Euhrychiopsis lecontei* looks promising in that it is capable of cutting off the flow of carbohydrates to root crowns, reducing the plant's ability to store carbohydrates for overwintering (Newman et al. n.d.) and reducing the buoyancy of the canopy (Creed et al. 1992). However, no strategy for using these naturalized insects at an operational level has yet been tested.

Pathogens, like insects, are usually discovered in the "neoclassical" method of searching overseas for pathogens in the IN range of the target plant. Despite overseas searches (Harvey et al. 1995; Theriot et al. 1993), no foreign pathogen agents are currently "in the pipeline." Actually, the best pathogen control agent for submersed aquatic plants appears to be an IN fungus species, *Mycoleptodiscus terrestris* (Mt) (Shearer 1995). At one time, it was thought that Mt was a selective agent for *Myriophyllum spicatum* (Gunner et al. 1990), and another supposedly selective endemic pathogen, *Macrophomina phaseolina*, was being examined as a candidate for pathogen biocontrol of hydrilla (Joye 1990). However, subsequent research discovered not only that neither pathogen was selective but, in fact, that *Macrophomina phaseolina* was actually *Mycoleptodiscus terrestris* (Shearer 1993, 1994). Subsequent field tests indicated that Mt was an effective mycoherbicide and acted like a contact herbicide with little spread or drift (Shearer 1995). In addition, Mt has shown promise as part of an integrated management strategy in which applications of Mt combined with low dosage rates of the herbicide fluridone act synergistically (Netherland and Shearer 1995). However, more research and development effort must be accomplished before an effective marketable mycoherbicide is available for use.

The last type of biological management technique, IN plant restoration, is an ecological approach aimed at managing for a desired plant community. The basic idea is that restoring a community of IN plants should be the end goal of most aquatic plant management programs (Nichols 1991; Smart and Doyle 1995). Lakes currently lacking communities of IN plants should have such communities actively established (Smart et al. 1996). Extant communities of IN plants should be protected from invasion by NI species through mechanisms detailed later. In communities only recently invaded by NI species, a propagule bank probably exists that will restore the community after management of the NI plant (Madsen et al. 1994b). However, in communities that have had monospecific NI plant dominance for a long period of time (e.g., greater than 10 years), IN plants may have to be reintroduced after a successful maintenance management program has been instituted. A healthy community of IN plants might slow invasion or reinvasion by NI species and will provide the environmental and habitat needs of an aquatic littoral zone. However, even well-developed communities of IN plants may eventually be invaded and dominated by NI species.

Chemical Management Techniques

Chemical management techniques have changed dramatically in the past 20 years. Increased concern about the safety of pesticide use in the 1950s and 1960s changed the review process for all pesticides, particularly for products used in water. Currently, no product can be labeled for aquatic use if it poses more than a one in a million chance of causing significant damage to human health, the environment, or wildlife resources. In addition, it may not show evidence of biomagnification, bioavailability, or persistence in the environment (Joyce 1991). The greatest change for herbicides came with the Federal Insecticide, Fungicide, and Rodenticide Act (FIFRA) passed in 1972 and amended in 1988 (Getsinger 1991; Nesheim 1993).

Due to more stringent standards for testing, fewer compounds are now available for aquatic use. In 1976, 20 active ingredients were available; as of 1995, only six are available (Table 12.2), with one additional compound (triclopyr) undergoing the registration process. However, the compounds no longer registered for aquatic use are not necessarily too dangerous; rather, in most cases, the companies marketing them opted not to pursue registration due to economic reasons. Their reluctance to invest in registration is understandable—it can take $20 to $40 million and 8 to 12 years to navigate successfully the registration process and its accompanying series of laboratory and field testing (Getsinger 1991). What remains are six active ingredients that not only are ensured safe for aquatic use

TABLE 12.2. Characteristics of U.S. Environmental Protection Agency–approved aquatic herbicides.

Compound	Trade name	Company	Formulation contact vs. systemic	Mode of action	Bluegill 96 hr. LC_{50} (mg/L)
Complexed copper[1]	Cutrine-Plus, Komeen, Koplex, K-Tea	Applied Biochemists (Cutrine), Griffin Corporation (All Ks)	Various complexing agents with copper, superior to $CuSO_4$; systemic	Plant cell toxicant	1250
2,4-D[1]	Aqua-Kleen, Weedar-64, Wee-Rhap, A-6D, several others	Applied Biochemists, Rhone-Poulenc, Inter-Ag	BEE salt, DMA liquid, IEE liquid, systemic	Selective plant-growth regulator	1.1–1.3 123–230
Diquat[1]	Reward	Zeneca	Liquid, contact	Disrupts plant cell membrane integrity	10–140
Endothall[1]	Aquathol K, Hydrothal 191, Aquathol granular	Elf Atochem (all formulations)	Liquid or granular, contact	Inactivates plant protein synthesis	125 0.06–0.2
Fluridone[1]	Sonar, Sonar SP	SePro	Liquid or granular, systemic	Disrupts carotenoid synthesis, causing bleaching of chlorophyll	9–12.5
Glyphosate[1]	Rodeo	Monsanto	Liquid, systemic	Disrupts synthesis of phenylalanine	4.2–14
Triclopyr[2] (EUP only)	Garlon 3A (EUP)	Dow Elanco	Liquid, systemic	Selective plant growth regulator	148

[1] Westerdahl and Getsinger (1988a, 1988b).
[2] Triclopyr is currently not approved for aquatic use; however, Dow Elanco is pursuing an aquatic label.

(when used according to the label) but have manufacturers that are committed to the aquatic market.

The important caveat to remember is that these products are safe when used according to the label. The U.S. Environmental Protection Agency (EPA)–approved label gives guidelines protecting the health of the environment, the humans using that environment, and the applicators of the herbicide. In most states, there are additional permitting or regulatory restrictions on the use of these herbicides. Additional state laws may require that these herbicides be applied only by licensed personnel.

Herbicides labeled for aquatic use can be classified as either contact or systemic. Contact herbicides act immediately on the tissues contacted, typically causing extensive cellular damage at the point of uptake but not affecting areas untouched by the herbicide. Typically, these herbicides are faster acting, but they do not have a sustained effect, in many cases not killing root crowns, roots, or rhizomes. In contrast, systemic herbicides are translocated throughout the plant. They are slower acting but often result in mortality of the entire plant.

An important consideration for treatment of emergent plants is that these plants all have thick cuticles, requiring the use of a surfactant or adjuvant (typically a strong detergent) to penetrate this waxy layer (Westerdahl and Getsinger 1988a). In addition, it is far more effective to treat emergents such as waterhyacinth when they are at low density and biomass than waiting for dense stands of large plants to develop (Lopez 1993). This explains, in part, the success of Florida's maintenance management program (Joyce 1991).

Although additives are generally not a consideration for submersed plants, a more difficult problem emerges for these species. In this case, we are actually treating the water with a herbicide and allowing the plant to take up herbicide from the water. This creates a situation in which the person applying the herbicide needs to know the exchange rate of water

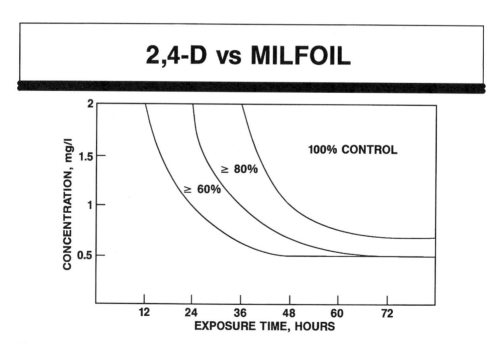

FIGURE 12.7. Concentration (mg/l) versus exposure time (hr) relationship for 2,4-D on Eurasian watermilfoil (*Myriophyllum spicatum*), indicating zones of <75% control, 75% to 95% control, and >95% control. (From Netherland [1991])

TABLE 12.3. Application restrictions of U.S. Environmental Protection Agency–approved aquatic herbicides.

Compound	Persistence (half-life, in days)	Maximum application rate	Maximum water concentration	Safety factor	Application notes	WES Recommended for
Complexed copper[1]	3	1.5 gal/ft/acre	1.0 mg/L	>50	Algicide/herbicide	Hydrilla, other submersed spp.
2,4-D[1]	7.5	0.5 gal/acre	0.1 mg/L	>25	Some formulations for special permits only	Eurasian watermilfoil, water–hyacinth, and others
Diquat[1]	1–7	2 gal/acre	2 mg/L	5	Binds with particles (suspended solids) in water	All
Endothall[1]	4–7	13 gal/acre	5.0 mg/L (Aquathol) <1.0 (Hydrothal)	>10	Fish are sensitive to Hydrothal 191, over 1 mg/L may cause fish kill	All submersed spp.
Fluridone[1]	21	1.1 qt/acre	0.15 mg/L (150 ppb)	>20	Applications have been successful below 10 ppb	Most submersed spp.
Glyphosate[1]	14	2 gal/acre	0.2 mg/L	>20	Aerial portions only, not for submersed plants	Most emergent and floating spp.
Triclopyr[2] (EUP only)	na	na	2.5 mg/L	>50	EUP/special needs only, U.S. EPA label expected in 1997	Eurasian watermilfoil, water-hyacinth, others

[1] Westerdahl and Getsinger (1988a, 1988b).
[2] Triclopyr is currently not approved for aquatic use; however, Dow Elanco is pursuing an aquatic label.

to have a successful application (Getsinger et al. 1991). The exposure time of the plant to the herbicide is determined predominantly by the water exchange rate. The response of different plant species to different herbicides is a function of the properties of both the plant and the herbicide. One needs to match an herbicide with an appropriate concentration and exposure-time relationship for the target species (Fig. 12.7) (Netherland 1991). The concentration and exposure-time relationship for a given compound has been determined from laboratory experiments. For instance, if it is known from water exchange studies that the exposure time will ensure only 24 hours of contact with 1 mg/liter of 2,4-D if applied at full label rate, then a 75% control rate for Eurasian watermilfoil can be expected (Fig. 12.7). If longer exposure times are expected, then lower concentrations can be applied. One goal of this area of research is to allow for lower application rates, both to save money on herbicides and to introduce a lower total amount of herbicide into the aquatic environment. For faster exchange rates, one will need to use higher concentrations of the contact herbicides such as diquat or endothall; slower exchange rates may allow use of systemic herbicides (Tables 12.3 and 12.4).

Some herbicides (e.g., 2,4-D and triclopyr) are intrinsically selective, being very effective for controlling water-hyacinth and Eurasian

TABLE 12.4. Use suggestions for U.S. Environmental Protection Agency–approved aquatic herbicides.

Compound	Exposure time (water)	Advantages	Disadvantages	Systems where used effectively	Plant species response
Complexed copper[1]	Intermediate (18–72 hours) (ref. not avail.)	Inexpensive, rapid action, approved for drinking water	Does not biodegrade, but biologically inactive in sediments	Lakes as algicide, herbicide in higher exchange areas	Broad-spectrum, acts in 7–10 days or up to 4–6 weeks
2,4-D[1]	Intermediate (18–72 hours) (Green and Westerdahl 1990)	Inexpensive, systemic	Public perception	Water-hyacinth and Eurasian watermilfoil control	Selective to broad-leaves, acts in 5–7 days up to 2 weeks
Diquat[1]	Short (12–36 hours) (Westerdahl 1987)	Rapid action, limited drift	Does not affect underground portions	Shoreline, localized treatments, higher exchange rate areas	Broad-spectrum, acts in 7 days
Endothall[1]	Short (12–36 hours) (Netherland et al. 1991)	Rapid action, limited drift	Does not affect underground portions	Shoreline, localized treatments, higher exchange rate areas	Broad spectrum, acts in 7–14 days
Fluridone[1]	Very long (30–60 days) (Netherland 1992)	Very low dosage required, few label restrictions, systemic	Very long contact period	Small lakes, slow flowing systems	Broad spectrum, acts in 30–90 days
Glyphosate[1]	Not applicable	Widely used, few label restrictions, systemic control	Very slow action, no submersed control	Nature preserves and refuges	Broad spectrum, acts in 7–10 days, up to 4 weeks
Triclopyr[2] (EUP only)	Intermediate (12–60 hours) (Netherland and Getsinger 1992)	Selective, systemic	Not currently labeled for general aquatic use	Lakes and slow-flow areas, purple loosestrife	Selective to broad-leaves, acts in 5–7 days, up to 2 weeks

[1] Westerdahl and Getsinger (1988a, 1988b).
[2] Triclopyr is currently not approved for aquatic use; however, Dow Elanco is pursuing an aquatic label.

watermilfoil but not hydrilla (Table 12.4). Other herbicides may be used selectively but only through application based on the target and nontarget plant's biology.

The future of herbicide use may include applying plant growth regulators, such as flurprimidol and paclobutrazol, which reduce plant elongation rather than cause plant death (Lembi and Chand 1992; Van 1988). The future of this approach dimmed considerably in the United States when Du Pont Corporation did not pursue the registration of bensulfuron methyl, which showed great promise in restricting tuber formation in hydrilla (Haller et al. 1992; Langeland 1994; Van and Vandiver 1992) and growth in Eurasian watermilfoil (Getsinger et al. 1994).

A second area in the future of herbicide use is integrated control, where herbicides are used in conjunction with other management techniques to improve their effectiveness. Herbicides have been used with grass carp (Eggeman 1994; Jaggers 1994), insect biocontrol agents (Haag et al. 1988; Haag and Habeck 1991; Van 1988), and pathogens (Netherland and Shearer 1995; Sorsa et al.

1988) to increase their effectiveness. Combining herbicides with mechanical and physical control techniques is also possible. However, integrated management approaches are still largely in the research phase with few published operational successes.

Mechanical Management Techniques

Mechanical management methods have been widespread in attempts to control aquatic plants (Table 12.5). Possibly nothing else quite excites the imagination and ardor of potential inventors as the hope of inventing the ultimate mechanical answer to aquatic plant problems. Yet all too often the approach to a solution is strictly engineering rather than applying engineering to a knowledge of biology and ecology of the target organism. Likewise, the erstwhile inventor often neglects a concern for environmental implications of use of the mechanical control, confirmed in the belief that it must be better than "using poisons."

The most common form of mechanical control is actually the use of hand cutters, rakes, or bare hands (no tools) to remove vegetation. Not only is this the most common method worldwide, but it is the method most widely used by most lakeshore owners in the United States. In a recent do-it-yourself guide, McComas (1993) listed a large number of hand implements and other small-scale devices for mechanical control.

Larger-scale control efforts require more mechanization (Table 12.5). The first uses a

TABLE 12.5. Characteristics of mechanical management techniques.

Management method	Description	Advantages	Disadvantages	Systems where used effectively	Plant species response
Hand-cutting/pulling	Direct hand pulling or use of hand tools	Low-technology, affordable, can be selective	Labor-intensive, cost is labor-based	Most of the undeveloped world, volunteer labor pools	Very effective in very localized areas
Cutting	Cut weeds with mechanical device (typically boat-mounted sickle bar) without collection	More rapid than harvesting	Large mats of cut weeds may become a health and environmental problem, may spread infestation	Heavily infested systems	Nonselective, short-term
Harvesting	Mechanical cutting with plant removal	Removes plant biomass, reduces spread	Slower and more expensive than cutting	Widespread use with chronic plant problems	Like cutting, it is cosmetic, nonselective, short-term
Diver-operated suction harvester	Vacuum lift used to remove plant stems, roots, leaves; sediment left in place	Moderately selective (based on visibility and operator), longer-term	Slow and cost-intensive	Useful for smaller infestations in which plant density is moderate	Typically have minimal regrowth for Eurasian watermilfoil; not effective for tuber-setting hydrilla
Rotovating	Cultivator on long arm for tilling aquatic sediments	Disrupts Eurasian watermilfoil stembases, intermediate-term results	May spread large numbers of fragments	Used extensively in the Pacific Northwest and British Columbia with mixed results	Effective in disrupting Eurasian watermilfoil dense stands; not selective and only intermediate-term

mechanical cutter, which is typically a boat with a sickle-bar cutting blade. Although cutting alone is relatively rapid, it leaves large mats of plants that can not only spread the infestation but create a floating obstacle, wash up on shorelines, and cause water-quality problems through decomposition. Because of these problems, cutting operations are typically combined with plant removal. However, in some applications, removal is not necessary, in which case cutting alone is sufficient.

In mechanical harvesting, cutting operations are combined with plant removal. Occasionally, there are separate cutting and harvesting boats. More often, the harvesters have both a sickle-bar cutting blade with a conveyor belt that loads the cut material on a boat. The plant material is carried away by disposal vehicles.

One neglected aspect of harvesting operations is disposal of plant material. The plant material is generally more than 90% water, is not suitable as a feed, and cannot be sold or made into anything truly useful. The common response is to use it as mulch.

Several studies have indicated that one harvest per year provides only brief control, whereas two to three harvests of the same plot in a given year are required to provide adequate annual control. However, cutting three times in a year may also reduce growth the following year (Madsen et al. 1988; Nichols and Cottam 1972). Most researchers directly ascribed successful control to reductions in total stored carbohydrates (Kimbel and Carpenter 1981; Perkins and Sytsma 1987), although one researcher could not correlate harvest success to this factor alone (Painter 1988). The likeliest explanation for the requirement of multiple harvests is in the reduction of stored carbohydrates. Although many claim that harvesting is environmentally superior to herbicide use, most neglect to consider that harvesting removes large numbers of macroinvertebrates, semiaquatic vertebrates, forage fishes, young-of-the-year fishes, and even adult gamefishes (Engel 1990; Mikol 1984). The harvester acts as a large, nonselective predator "grazing" in the littoral zone.

Not all secondary effects of harvesting are negative, however. Removal of large amounts of plants can improve the diel oxygen balance of littoral zones and rivers, particularly in shallower water (Carpenter and Gasith 1978; Madsen et al. 1988). At this point, no studies have indicated whether IN plants respond preferentially to harvesting.

In the past, harvesting was widely touted as a mechanism to remove nutrients from lake systems. However, ecosystem studies indicated that harvesting was not likely to significantly improve the trophic status of a lake. For instance, harvesting all available plants in Lake Wingra, Wisconsin, removed only 16% of the nitrogen and 37% of the phosphorus net influxes into the lake; these removals were insignificant compared to the lake's internal pools of those nutrients (Carpenter and Adams 1976, 1978). Plant harvesting in Southern Chemung Lake, Ontario, removed 20% of the annual net phosphorus input (Wile 1975). In a more eutrophic system (Sallie Lake, Minnesota), continuous harvesting of aquatic plants in the littoral zone during summer removed only 1.4% of the total phosphorus input (Peterson et al. 1974). In a less eutrophic system (East Twin Lake, Ohio), harvesting the entire littoral zone would have removed from 26% to 44% of the phosphorus and from 92% to 100% of the nitrogen net loadings to the lake over a 5-year study period (Conyers and Cooke 1983). Harvesting aquatic plants is not an effective tool for reducing nutrient loads in a lake; in none of the above scenarios was the internal nutrient pool reduced. In the best case scenario, removing all the plants in the lake only kept pace with the amount of external nitrogen loading and with not quite half of the external phosphorus loading. Because no operational control program is going to remove all plants in the littoral zone, it is unlikely that any operational harvesting program will significantly modify the internal nutrient balance of the system.

The use of diver-operated suction harvesting (or dredging, as it is often called), is a fairly recent technique. I call this harvesting rather than dredging because, although a specialized small-scale dredge is used, sediments are not removed from the system. Sediments are resuspended during the operation, but these effects are mitigated by using a sediment curtain. Divers use this device to remove plants from the sediment (NYSDEC 1990). The technique can be very selective; divers can literally choose the plants to be removed. Removal is efficient and regrowth is limited. The system is very slow (100 m^2 per person-day) (Eichler et al. 1993), and disposal of plant material must also be resolved. However, it is an excellent method for small beds of plants or areas of scattered clumps of plants too large for hand harvesting.

The last major mechanical management technique is rotovating, which is widely used in the Pacific Northwest and, formerly, in British Columbia for management of Eurasian watermilfoil. This method uses rotovator heads on submersible arms to till up the bottom sediments and to destroy the root crowns. Rotovating is relatively rapid and can effectively control dense beds of Eurasian watermilfoil for up to 2 years (Gibbons and Gibbons 1988). However, it spreads Eurasian watermilfoil fragments, resuspends large amounts of sediments and nutrients, causes high levels of turbidity, disrupts benthic communities, and is nonselective.

Physical Management Techniques

Physical management methods may or may not utilize large equipment but are distinguished from mechanical techniques in the following manner: in mechanical techniques the machines act directly upon the plants; in physical techniques the environment of the plants is manipulated, which in turn acts upon the plants. Several physical techniques are commonly used: dredging, drawdown, benthic barriers, shading or light attenuation, and nutrient inactivation (Table 12.6).

Dredging is usually not performed solely for aquatic plant management but to restore lakes that have been filled in with sediments, have excess nutrients, have inadequate pelagic and hypolimnetic zones, need deepening, or require removal of toxic substances (Peterson 1982). However, lakes that are very shallow due to sedimentation typically do have excess plant growths. This method is effective in that dredging typically forms an area of the lake too deep for plants to grow, thus opening an area for riparian use (Nichols 1984). By opening more diverse habitats and creating depth gradients, dredging may also create more diversity in the plant community (Nichols 1984). Results of dredging can be very long term. Biomass of curly pondweed (*Potamogeton crispus*) in Collins Lake, New York, remained significantly lower than predredging levels 10 years after dredging (Tobiessen et al. 1992). Due to the cost, the environmental impacts, and the problem of disposal of dredge soil, dredging should not be done for aquatic plant management alone. It is best thought of as a lake remediation technique.

Drawdown is another effective aquatic plant management technique that alters the plant's environment. Essentially, the water body has all of the water removed to a given depth. It is best if this depth includes the entire depth range of the target species. Drawdowns, to be effective, need to be at least 1 month long to ensure thorough drying (Cooke 1980a). In northern areas, a drawdown in the winter that will ensure freezing of sediments is also effective. Although drawdown may be effective for control of water-hyacinth and hydrilla for 1 to 2 years (Ludlow 1995), it is most commonly applied to Eurasian watermilfoil (Geiger 1983; Siver et al. 1986) and other milfoils or submersed evergreen perennials (Tarver 1980). Drawdown requires that there be a mechanism to lower water levels. Although it is inexpensive and has long-term effects (2 or more years), it also has significant environmental effects and may interfere with use and intended function (e.g., power generation or drinking water supply) of the water body during the drawdown period. Lastly, species respond in very different manners to drawdown and often not in a consistent fashion (Cooke 1980a). Drawdowns may provide an opportunity for the spread of highly weedy or adventive species, particularly annuals.

TABLE 12.6. Characteristics of physical management techniques.

Management method	Description	Advantages	Disadvantages	Systems where used effectively	Plant species response
Dredging, sediment removal	Use mechanical sediment dredge to remove sediments, deepen water	Creates deeper water, very long-term results	Very expensive, must deal with dredge spoils	Shallow ponds and lakes, particularly those filled in by sedimentation	Often creates large usable areas of lake, not selective
Drawdown	"De-water" a lake or river for an extended period of time	Inexpensive, very effective, moderate-term	Can have severe environmental impacts, severe recreational and riparian user effects	Only useful for man-made lakes or regulated rivers with a dam or water control structure	Selective based on perennation strategy; effective on evergreen perennials, less effective on herbaceous perennials
Benthic barrier	Use natural or synthetic materials to cover plants	Direct and effective, may last several seasons	Expensive and small-scale, nonselective	Around docks, boat launches, swimming areas, and other small, intensive use areas	Nonselective, plant mortality within 1 mo underneath barrier
Shading, light attenuation	Reduce light levels by one of several means: dyes, shade cloth, planting trees (rivers)	Generally inexpensive, effective	Nonselective, controls all plants, may not be aesthetically pleasing	Smaller ponds, man-made water bodies, small streams	Nonselective, but may be long-term
Nutrient inactivation	Inactivate phosphorus (in particular) using alum	Theoretically possible	Impractical for rooted plants limited by nitrogen	Most useful for controlling phytoplankton by inactivating water column P	Variable

Benthic barriers or other bottom-covering approaches are another physical management technique that has been in use for a substantial period of time. The basic idea is that plants are covered with a layer of growth-inhibiting material. Many materials have been used, including sheets or screens of organic, inorganic, and synthetic materials, sediments such as dredge spoils, sand, silt or clay, fly ash, and combinations of the above (Cooke 1980b; Nichols 1974; Perkins 1984; Truelson 1984). The problem with using sediments is that new plants establish on top of the added layer (Engel and Nichols 1984). The problem with synthetic sheeting is that the gasses evolved from decomposition of plants and normal decomposition activity of the sediments underneath the barrier collect under the barrier and lift it (Gunnison and Barko 1992). Benthic barriers will typically kill the plants beneath them within 1 to 2 months, after which they may be removed (Engel 1984). Sheet color is relatively unimportant; opaque (particularly black) barriers work best, but even clear plastic barriers will work effectively (Carter et al. 1994). Sites from which barriers are removed will be rapidly recolonized (Eichler et al. 1995). In addition, synthetic barriers may be left in place for multi-year control but will eventually become sediment-covered and will allow colonization by plants. Benthic barriers, effective and fairly low-cost control techniques for limited areas (e.g., <1 acre), may be best suited to high-intensity use areas such as docks, boat launch areas, and swimming areas. However, they are too expensive to use over large areas and they have a strong influence on benthic communities.

A basic environmental manipulation for plant control is light reduction or attenuation. This, in fact, may have been the first physical control technique. Shading has been achieved by fertilization to produce algal growth, by application of natural or synthetic dyes, shading fabric, or covers, and by establishing shade trees (Dawson 1981, 1986; Dawson and Hallows 1983; Dawson and Kern-Hansen 1978; Jorga et al. 1982; Martin and Martin 1992; Nichols 1974). During natural or cultural eutrophication, phytoplankton growth alone can shade macrophtyes (Jones et al. 1983). Although light manipulation techniques may be useful for narrow streams or small ponds, in general these techniques are of only limited applicability.

Nutrient inactivation, the final physical management method often discussed, is commonly done for algal or phytoplankton control by adding alum to the water column, which binds phosphorus and thus limits the growth of algae (McComas 1993). However, larger vascular aquatic plants are typically limited by nitrogen rather than phosphorus and derive most of their nutrients from the sediment rather than from the water column. No chemical is available that binds nitrogen as readily as alum binds phosphorus. Additionally, the difficulties of adding a binding agent to the sediment rather than to the water column are obvious. Despite these limitations, nutrient inactivation has been attempted, but with limited success (Mesner and Narf 1987; Nichols 1974). At this point, nutrient inactivation for control of aquatic vascular plants is still in the research and development phase.

Institutional Management Techniques

"Institutional" was selected for this title, in part because these techniques are less direct and rely on some form of government involvement. Although these approaches alone will not control plant infestations, they are a critical part of a long-term management plan. For lack of a better characterization, I have selected four types of institutional approaches to aquatic plant management: quarantine, regulation, prevention, and watershed management (Table 12.7).

Quarantine seeks to control the entry or exit of an NI species from a given area. Of course, the most effective form of quarantine would be to prevent the entry or exit of species from our continent or country. In fact, this is currently attempted by the U.S. Department of Agriculture Animal and Plant Health Inspection Service (APHIS). Although such quarantine could be quite effective for some species, it alone cannot totally prevent entry of new species into our ecosystems. Moreover, many potentially problematic species are not even on quarantine lists. Typically, quarantine efforts are proposed and maintained by state governments. Examples of quarantine efforts are the boat-launch patrols used at infested water bodies and the somewhat less effective approach of marking Eurasian watermilfoil beds with yellow buoys (to keep out boat traffic) by the state of Minnesota (De Steno 1992). At boat launches, volunteer attendants ensure that boaters remove all plant fragments from boats and trailers.

Regulations prohibit the possession and distribution of the target NI species. For instance, several states including Washington have fines for possession of Eurasian watermilfoil or hydrilla. Many other states maintain lists of prohibited NI species. However, a major problem is that these regulations are rarely enforced. There is little monitoring. Even though regulations are currently in place for federal and state governments to control the spread of NI aquatic plants by the aquaculture and aquatic nursery trades, little is done to stop the influx of new, potentially invasive NI species.

Prevention is best accomplished through combining a volunteer or citizens "weed watcher" program with a quick response strategy for new invasions. Likewise, this type of program needs to be organized at the state level, with volunteers from virtually every region or lake. Volunteers, trained to identify the NI species of interest, routinely patrol given water bodies. They are usually trained to key in on potential hot spots such as boat ramps and tributaries. They are also instructed in how to map out or locate new infestations and who to contact when they find one. States with active weed watcher programs include Minnesota (De Steno 1992), New Hampshire (NHDES 1989), and Vermont (Crosson 1987). Other states have special strategies such as preferential funding to control new infestations (Gibbons et al. 1994). California has even declared hydrilla a threat to agriculture and has a state-wide eradication program. Further, states often assist in prevention of new spread by active media campaigns, signs at boat launches, presentations to outdoor groups, and other educational programs (Newroth 1987).

Watershed management is really a holistic approach to improving water quality and reducing nonpoint runoff into lakes and other water bodies by zoning and by encouraging the best land management practices to reduce overall nutrient and sediment loading to lakes. Although rooted aquatic plants do not respond directly to nutrients in the water, nonpoint runoff also adds sediments to the lake, which can encourage the growth of these plant species.

Developing an Integrated Management Plan

An integrated aquatic plant management plan has six basic components (Table 12.8): prevention, assessment, site-specific management, evaluation, monitoring, and education. An aquatic plant management plan should be

TABLE 12.7. Characteristics of institutional management approaches.

Management method	Description	Advantages	Disadvantages	Systems where used effectively	Plant species response
Quarantine	Prevent removal of plants from infested lake; or prevent introduction of plants to uninfested lake	Inexpensive compared to control	Difficult to monitor	Relatively successful at national level, less successful the more localized it becomes	One mechanism to prevent plants from entering a water body
Regulation	Legislation controlling spread or possession of nonindigenous species	Preventive, saves money in long run	Difficult to monitor and enforce	Most effective at national or state level; more difficult to enforce locally	Mechanism for preventing or slowing plant spread from one region to next
Prevention	Monitoring and patrolling activity to control new infestation	Involves volunteers, relatively inexpensive compared to control	Difficult to enlist interest, local effects may be overwhelmed by regional events	Most effectively organized as a lake-group, user-group; several successful state programs	Monitoring and maintenance management geared to prevent establishment and spread within lake
Watershed management	Manage for nonpoint source pollution reduction	Reduce sources of nutrients	Difficult to enforce	Nationally, wide success in nutrient loadings, less impact on plants	Plants may or may not respond to influx of nonpoint nutrients; but does clean up lake

TABLE 12.8. The six components of an integrated aquatic plant management plan.

Component	Subcomponent	Description	Examples
Prevention	• Prevent eutrophication • Prevent nonindigenous introductions	• Quarantine plant introduction • Monitor for plant presence, remove small colonies by hand	• Lake-watchers
Assessment	• Examine problem • Get group involvement, approval of problem definition • Study the problems • Set management goals in concentric layers • Set management goals in project management framework	• State the problem without assuming an answer • User groups, regulatory agencies, funding agencies • Site specific, lakewide, watershed • Master plan including personnel, budget, time line	• Hydrilla interferes with lake use
Site-specific management	• Select management practices tailored to site-specific needs and site priority • Evaluate all management techniques based on technical effectiveness and environmental and economic impacts	• Low-tech approaches for small or scattered colonies • More expensive, higher-tech mechanisms for larger, denser infestations	• Drinking water intakes • Endangered species • High-use areas
Evaluation	• Evaluate management practices quantitatively based on effectiveness and economic and environmental impacts • Manage sites to economic and environmental thresholds	• On-site quantitative assessment of effectiveness of management techniques • Economic cost-benefit analysis	• See Table 12.10
Monitoring	• Monitor the ecosystem for change • Monitor for nonindigenous species and basic conditions of system	• Limnological parameters • Measure target plant spread, nontarget plant impacts, other species	• Volunteer • Utilize available experts • Public meetings
Education	• Public education and awareness • Educate team members • Use opinion leaders • Target needed audience (lake users, local government leaders, etc.)	• Public involvement to build consensus • Group education for decision-making	• Use available media • Workshops • Lectures

an "integrated aquatic plant management plan" in that the managing group should not only use more than one technique to manage the problem plant both spatially and temporally but should also include some level of problem assessment, prevention, evaluation, monitoring, and education with operational treatments. Planning before beginning operational treatments will spell the difference between success and failure (Nichols et al. 1988). Several guides are available for developing lake management plans (Moore 1989; Nichols et al. 1988; NYSDEC 1990) and for developing integrated aquatic plant management plans (Gibbons et al. 1994). An excellent example of integrating all aspects of prevention, assessment, evaluation, and monitoring for British Columbia is presented by Newroth (1993).

Prevention

The best time to develop an integrated aquatic plant management plan is before an infestation occurs. However, few management groups have this level of foresight. A small commitment of volunteer time in a lake- or weed-watch program could save hundreds of thousands of dollars in plant management costs later on. Likewise, once the infestation of one plant species occurs, it is obvious that other species of NI plants and animals could follow.

A great deal can be accomplished through proper training of professional resource management personnel currently monitoring the environment. In addition, steps should be taken to protect the resource from NI species through monitoring boat launches or other public access sites and by educating users of the dangers of NI species.

When new colonies of NI plants are found, immediate action should be taken to eradicate them. These procedures will save large amounts of time and money later. Hand harvesting, spot treatment with contact herbicides, or diver-operated suction harvesting are all useful methods for initial treatments. Under no circumstances should cutting or other methods be used that will create fragments or spread the plant. Likewise, slow-acting systemic herbicides should be avoided. The goal is to remove any source of propagules in the lake. After treatment of these sites, rigorous monitoring of adjacent areas and the treatment area should be conducted to ensure that the plant has not spread.

Assessment

When developing a management plan for public or "common" areas, include all significant user groups such as lakeshore owners, fisherpersons (even different fishing groups), boaters, and local businesses dependent on the lake. In a group context, the problem should be identified and defined. The group as a whole should agree with the problem definition(s). The problem should be more detailed than "plants get in the way" because problem identification is the first important step to finding acceptable management solutions.

Then the problem should be studied and as much information gathered as possible to understand the problem. Management goals should be set in concentric layers, for example, prevent introduction of new species to the lake, set allowable levels of plants at swimming areas and boat launches, and establish monitoring at low-intensity use areas. As part of the problem identification and study process, relevant resource management agencies, regulatory agencies, and various layers of government agencies having overlapping jurisdictions must be included (see also Frandsen, Chapter 17, this volume). Early on, discover who does what and what they will allow. Although some regulatory agencies have a core of unbending rules, they may have another layer of rules or guidelines that can be negotiated. Once the regulatory agency develops some confidence in the management group, it may become more willing to allow a given level of activity. For instance, a core of unbending rules would be the label restrictions on herbicides placed by both the EPA and state pesticide regulatory agencies. However, both levels of government have some discretion in granting either special use or special needs permits. Every locality has different requirements; some states require permits for herbicide applications or even any aquatic management activity, while others do not. Many groups wear themselves out fighting these layers of bureaucracy rather than figuring out how to work with them; in the long run, the latter usually is quicker and more efficient.

As part of the assessment process, information on the system needs to be gathered. This may require hiring persons with special expertise. At a minimum, a species list for the lake should be made and the areas with the problem plant identified. Areal extent can be estimated on maps or through remote sensing (Table 12.9). Even if maps are made, some locations should be surveyed using semiquantitative techniques (Madsen and Bloomfield 1992, 1993).

An important component of making a species list is to identify correctly the species collected and to maintain an herbarium (pressed plant) record of what was collected (Hellquist 1993). Although many projects may never require legal verification of what was collected, such a collection will, in the long run, reduce confusion and aid in proper management. Identification of some species is difficult enough so that verification by specialists may be required. Sometimes even these specialists will resort to molecular techniques to separate species or races of plants that may have management implications. For instance, flavonoid pigment analysis (Ceska and Ceska 1986; Newroth 1993) has been used to separate the often similar species of *Myriophyllum* in British Columbia. Likewise, isozyme analysis (Langeland 1989; Ryan 1989; Ryan et al. 1991) and random amplified polymorphic DNA (RAPD) analysis (Ryan and Holmberg 1994) have been used not only to differentiate between monoecious and dioecious hydrilla but to track different genetic families of these plants around the country.

Assessment methods can also be used to evaluate potential levels of plants that might be acceptable within the ecosystem. These may be based on environmental characteristics (Wong and Clark 1979), acceptable risk to life and property (Madsen et al. 1988; Madsen 1993), or user-group preference (Wilde et al. 1992; Henderson 1994). This provides a quantitative target for aquatic plant management.

Finally, management goals need to be set within a project management framework. Identify available resources (funds, personnel) and a reasonable time line to accomplish management goals. Developing both realistic resources and time lines will aid in scheduling.

Site-Specific Management

Management should be tailored to the priority and goals of each site. All areas within the lake should be categorized as to use, restrictions, and priority. Based on these categories, management techniques can be selected. For instance, swimming beaches and boat launches are high-use areas and should have a high priority. Wildlife areas (e.g., refuges) probably have lower intensity use and also likely some restrictions attached. The high-priority, high-intensity use sites might justify high-cost management techniques such as benthic barriers or diver-operated suction harvesting. Low-intensity use areas might either remain untreated if resources are low or would be categorized for less expensive techniques such as herbicides. Likewise, areas with higher concentrations of plants should receive more resources than areas with no plants or with acceptable levels of infestation. As dense colonies are brought under control, maintenance management approaches can be used (Deschenes and Ludlow 1993). Under no circumstances should management be discontinued once plant densities are low. After an NI species has entered a system, continuous management will be required. If management techniques are very successful, management may entail only monitoring the system and hand-removing individuals that are occasionally found. Scale the control tech-

TABLE 12.9. Summary of aquatic plant quantification methods.

Technique	Effort required	Level of sensitivity	Area of usefulness
Species list	Low	Low	Identification of nonindigenous or rare species
Subjective estimates	Low	Low	Misleading, should not be used
Semiquantitative estimates	Low	Low	Initial surveys, small budgets
Transect methods (cover)	Moderate	Moderate	Distribution of species
Biomass methods	High	High	Abundance of species, localized areas
Remote sensing	Low	Low	Whole-lake distributions, may not be species-specific
Geographic information systems	High	Variable	Visualization of data, planning, correlation to riparian uses
Computer modeling	High	Variable	Potential predictive ability, evaluation of alternatives, economic impacts

Revised from Madsen and Bloomfield (1993).

nique to the level of infestation, the priority of the site, the use, and the availability of resources.

Evaluation

All management techniques should be evaluated on a routine schedule to make sure that they are in fact effective in managing plants, cost-effective economically, and, in some cases, effective in managing plants in acceptable environmental thresholds. Evaluations should be quantitative. This may require that expertise in plant-community quantification be hired. Expertise can be sought either from environmental consulting firms or from universities.

Several quantification methods are useful for evaluation; each has strengths and weaknesses (Table 12.9). Transect methods, biomass methods, and remote-sensing techniques are the most appropriate for evaluations. Transect methods are most appropriate for examining the distribution of species within an area as well as for evaluating species richness or diversity (Madsen and Bloomfield 1992, 1993; Madsen et al. 1994a; Titus 1993). Biomass techniques are the most sensitive to changes in plant abundance; as such, they are best for evaluating the effectiveness of control techniques in given plots (Madsen 1993). For best results, these evaluation studies should be based on rigorous experimental principles (see Morrison, Chapter 9, this volume; Spencer and Whitehand 1993). Remote sensing may provide a large-scale image of the distribution (near the water's surface, at least) of problem aquatic plants. As such, it may help to evaluate the overall effectiveness of management (Anderson B 1990a; Andrews et al. 1984; Frazier and Moore 1993). Remote sensing can also be used to evaluate effects on nontarget plants (Farone and McNabb 1993). Suggested practices for evaluating management techniques are presented in Table 12.10.

Economic analyses have been woefully neglected in aquatic plant management planning. For most, the extent of any cost/benefit analysis has been only an estimate of direct treatment costs of different management techniques (Conyers and Cooke 1983). Although these comparisons are important, the economic effects of aquatic plant management are far broader. Detailed economic models and analyses are important for large management programs (Henderson 1993, 1994, 1995). For instance, the 20-year-old British Columbia program for managing aquatic plants was evaluated in 1991 (Newroth and Maxnuk 1993). Results indicated a benefit:cost ratio of 11.3:1. A direct provincial contribution of $265,000 in 1991 resulted in $3 million in provincial revenues. A contribution of $360,000 by the province in 1990 also resulted in $85 million in tourist revenues, 1700 tourist industry jobs, and $360 million in real estate values. Typically, an extensive cost/benefit analysis of aquatic plant management programs yields a high benefit:cost ratio and an excellent return on the investment.

Other techniques may help in evaluating management techniques and in developing management plans. Geographic information systems (GIS) are computer systems that utilize digital base maps and overlay databases for categorizing information on these maps. In other words, each point on the digital map can be labeled with various attributes such as depth, abundance of a given plant species, or use intensity. Although these systems require a large input of time, personnel, and money, they are well worth the investment for larger management efforts. In addition, the GIS can be combined with simple models to predict expected outcomes such as distribution of plants within a lake (Remillard and Welch 1993). Data acquisition becomes important when a GIS is being used. One piece of equipment helpful here is the Global Positioning System (GPS), which uses U.S. Department of Defense satellites to pinpoint (at various levels of accuracy) the latitude, longitude, and elevation of the receiver. GPS units can be used to map plant distributions, make bathymetric maps, record where herbicides or other management techniques were applied, and find sampling locations (Kress and Morgan 1995). Since these units have become quite inexpensive, simple GPS receivers are within reach of almost any volunteer group.

12. Methods for Management of Nonindigenous Aquatic Plants

TABLE 12.10. Suggested practices for evaluation associated with management activities.

Program or tactic	Minimal practice	Best practice

Preinvasion monitoring	• Citizen monitoring keying on:
	  ▪ Boat launches
	  ▪ Embayments
		• Professional surveys (3–5 yrs)
		• Boat launch surveillance
		• Citizen lake-watcher program

Postinvasion monitoring	• Professional survey (3–5 yrs) • Citizen lake-watcher program	• Professional survey (annual) • Citizen plant-watcher program • Transect survey (3–5 yrs) • Aerial/remote sensing
Management tactic assessment	• Professional site survey • Transect survey (3 yrs) • Assess treated vs. reference	• Professional site survey • Biomass quantification • Assess treated sites (2+, 2 levels) vs. reference (2+)
Specific management tactics		
Biological control	• Transect quantification at treated and reference site before and after treatment • Assess for bioagent damage to plants and agent population	• Biomass quantification at treated and reference sites (2+ of each) before and periodically after release • Assess population levels and feeding activities and effects at treated and reference sites as above
Chemical control	• Transect quantification at treated and reference site before and after treatment	• Biomass quantification at treated and reference sites (2+ of each) before and after treatment and during regrowth
Harvesting	• Transect quantification at treated and reference site before and after treatment	• Biomass quantification at treated and reference sites (2+ of each) before and after treatment and during regrowth
Drawdown	• Transect quantification before and after treatment	• Biomass quantification at 2+ sites before and after treatment and during regrowth • Monitor revegetation at permanent transects or grids
Benthic barrier	• Transect quantification at treated and reference site before and after treatment	• Biomass quantification at treated and reference sites (2+ of each) before and after treatment and during regrowth • Monitor revegetation at permanent transects or grids

Revised from Madsen and Bloomfield (1992).

Computer models in the past have been too complex for the average citizen or resource manager to use for evaluating management practices. However, several user-friendly models in Windows format were developed for resource managers through the U.S. Army Corps of Engineers Aquatic Plant Control Research Program (APCRP) (Table 12.11).

Another excellent source of information on target and nontarget plants and their management is the Aquatic Plant Information Retrieval System, operated by the University

TABLE 12.11. Simulation models for aquatic plant management developed by the Waterways Experiment Station for the Aquatic Plant Control Research Program.

MODEL	Description	Status	Reference
INSECT	Insect biocontrol model for water-hyacinth	DOS version available	Stewart and Boyd 1992
AMUR STOCK	Grass carp stocking model	DOS and WINDOWS versions available	Stewart and Boyd 1994
HERBICIDE	Herbicide dissipation model	WINDOWS version available	Stewart 1994
HARVEST	Mechanical harvester model	DOS and WINDOWS versions available	Stewart 1995

of Florida's Center for Aquatic Plants (904-392-1797). In addition to free bibliographic searches, the center has a variety of educational materials available.

Monitoring

Monitoring is distinct from evaluating plant-management success in that its goal should be to watch for ecosystem changes in target and nontarget plant species, physical and environmental parameters, and other nontarget species such as fishes and macroinvertebrates. Much of the monitoring efforts can be done by volunteer citizens; several guides are available to assist (Michaud 1991). In addition, several states such as New York (NYSDEC n.d.) and Michigan (MDNR 1990) have statewide citizen-based monitoring programs in which the state will assist with data analysis and the minimal costs involved. As with assessment of the plant problem and evaluation of plant management techniques, occasional assistance from experts will probably be required.

In some instances, monitoring for the environmental impact of herbicides may be desired. Herbicide residue levels may be measured directly in the water column or hydrosoil by standard analytical techniques. In addition, immunoassay approaches such as enzyme linked immunosorbent assay (ELISA) techniques are being developed for selected herbicides. These techniques cost much less and can sensitively detect presence or absence of a given herbicide, but, at present, accuracy is highly dependent on the skill of the analyst. In some instances, the effects of herbicides on nontarget plants may be of interest. Several techniques, including peroxidase analysis, have been used to measure stress levels in aquatic plants (Byl et al. 1994; Sprecher et al. 1993; Sprecher and Netherland 1995). The effectiveness of a herbicide on the target plant may also be desired, particularly in the case of a long-acting herbicide such as glyphosate or fluridone, which may take 30 to 90 days to be effective. During this time, symptoms may even be present, but mortality may not occur. Phytoene analyses are being prepared, in the case of fluridone treatments, to be able to predict if plants will in fact die after the required waiting period (Sprecher and Netherland 1995).

Education

Educational programs are not simply a frill in an aquatic plant management plan but are a long-term requirement for success. Education involves creating public awareness of the problem and the potential for resolution, soliciting assistance both for labor and resources, persuading key individuals and the public at large concerning the management solutions decided on, and educating members of the management planning group on the nature of the problems and their solutions. Many activities can be used for education: workshops, public meetings, press conferences and releases, courses, invited speakers, and so on. However, it is important to use all available educational methods, including the media, to get the message out, to build public consensus, and to educate group members and get their assistance in decision making.

Conclusions

1. Many management techniques are available for control of NI aquatic plants. No one technique is superior to others or applicable to all situations. Rather, each is a valued tool with economic, environmental, and technical strengths and weaknesses. A thorough knowledge of these will improve management success.

2. Management should be done in the context of an integrated aquatic plant management plan. The successful integrated aquatic plant management plan has six components: prevention, assessment, site specific management, evaluation, monitoring, and education. Site-specific management should consider all of the five management techniques: biological, chemical, mechanical, physical, and institutional. The integrated aquatic plant management plan may use a mixture of these techniques.

Acknowledgments. Support for the preparation of this article was provided by the U.S. Army Corps of Engineers Aquatic Plant Control Research Program of the U.S. Army Engineer Waterways Experiment Station, Environmental Laboratory, Vicksburg, MS. Permission to publish this document was granted by the Chief of Engineers, Headquarters, U.S. Army Corps of Engineers, Washington, DC. Technical reviews of this chapter were provided by Kurt Getsinger and Mike Stewart.

13
Biological Control of Weeds in the United States and Canada

C. Jack DeLoach

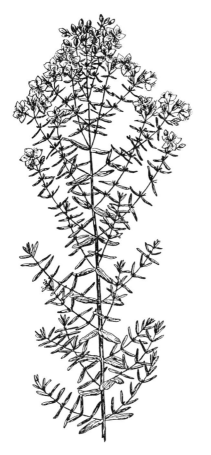

St.-John's-wort
(*Hypericum perforatum*)

The worldwide problem of nonindigenous (NI) pest invasions is as old as human migrations and trade between ecologically isolated areas. This problem began with the discovery and settlement of, and trade among, the Americas, Australia, southern Africa, the Pacific islands, and other areas by Europeans. These settlers brought with them alfalfa, cabbage, wheat, cattle, and chickens as well as seeds that escaped, grew, and later became weeds that devastated indigenous (IN) plants; rats that ate stored food; cats that destroyed birds; cabbage worms and Hessian flies; and smallpox and measles. Some pests traveled back to Europe, for example, ragweeds, potato blight, and syphilis. The former British colonies can be identified by plagues of prickly-pear cacti brought from the Americas to grow cochineal insects used to dye the British Army's red coats. Homesick settlers brought many garden ornamentals; several of these became serious weeds. We cannot blame settlers for bringing these plants, which had never been weeds in their IN range; no one could imagine that these plants would become problems elsewhere.

This avalanche of plant introduction has continued since colonial times. Such purposefully introduced plants include bermuda grass (*Cynodon dactylon*), Johnson grass (*Sorghum halepense*), Kentucky blue grass (*Poa pratensis*), kudzu (*Pueraria montana*), love grass

(*Eragrostis*), Russian-olive (*Elaeagnus angustifolia*), and salt-cedar (*Tamarix*). Of course, not all of these are unmitigated weeds.

Early Beginnings of Biological Control—Worldwide Development

Prickly-pear cactus (*Opuntia vulgaris*) was introduced from southern South America to India where it eventually naturalized. In 1795 a scale insect (*Dactylopius ceylonicus*) was introduced to India from Brazil in the mistaken belief that it was the cochineal insect (*D. coccus*) from which the British hoped to produce red dye. The insect increased rapidly and soon controlled the cactus. In 1836 *D. ceylonicus* was introduced to southern India and in 1865 to Ceylon (Sri Lanka) as the first deliberate attempt at biological control of a weed. It provided complete control of the cactus over vast areas (Julien 1992).

When up to 11 species of cacti (from North and South America) became serious pests in Australia and several other countries, biological control efforts were organized. *Dactylopius ceylonicus*, released in South Africa in 1913 (Moran and Zimmerman 1991), gave complete control of prickly-pear as it did the next year in Australia. Several other insects from the United States released in Australia from 1921 to 1933 controlled *Opuntia inermis*. The Argentine phycitid moth (*Cactoblastis cactorum*), released in Australia in 1926, completely controlled *O. inermis* and three other species of *Opuntia*. Most of the 13 million acres of infested land returned to production within 5 years (Dodd 1940; Mann 1969). Releases have continued to the present; the 22 species of insects introduced to Australia have provided complete or substantial control of 11 cacti. Some of these insects have now been released in 15 countries and islands, with excellent control in most areas (Julien 1992).

A similar program was developed in Hawaii beginning in 1902 to control the exotic lantana (*Lantana camara*), an ornamental shrub from Mexico, which had become a serious weed of rangelands. Although lantana is difficult to control—its many cultivars exhibit different susceptibility to insect attack—excellent to acceptable control was achieved after introduction of 22 species of insects from the IN range of lantana. The same insects have controlled lantana in several other countries, though not always with the same success (Julien 1992).

The first biological control project in mainland North America was in 1945 for control of the poisonous St.-John's-wort or Klamath weed (*Hypericum perforatum*), an introduction from Europe. Almost complete control was achieved in California within 5 years of initial releases of a leaf beetle (*Chrysolina quadrigemina*) from Europe (Holloway and Huffaker 1951) via Australia. Additional insects were later introduced to provide better control in other areas of the Northwest (McCaffrey et al. 1995; Rees et al. 1996).

These initial successes stimulated great interest in biological control of weeds in several countries after World War II. DeBach (1974) calculated that of the 41 "projects" (weeds per country) attempted worldwide, 75% had achieved measurable success: 8 projects were completely successful; 9 achieved substantial control; 14 partial control; and 10, no control. Through 1992, control had been attempted for 94 weed species in 53 countries, for a total of 729 weed-insect per country projects. Foreign insects have been released to control 32 weed species in the conterminous United States, 32 in Australia, 21 in Hawaii, 18 in South Africa, 15 in Canada, 11 in New Zealand, 11 in India, 7 in Fiji, 3 to 6 in 11 countries, 2 in 12 countries, and 1 in 22 countries (Julien 1992). The percentage of weeds successfully controlled is now lower because many new projects have been initiated but have not had sufficient time to achieve success.

A few organizations did most of the original research, including overseas exploration and testing. The U.S. Department of Agriculture's Agricultural Research Service (USDA-ARS) has conducted research since 1945 and has had permanent laboratories in Rome, Italy, since 1958 (moved to Montpellier, France, in

1991), Buenos Aires since 1962, and Japan or Korea since 1975. Laboratories were opened in 1988 in Beijing, China, and in 1989 in Townsville (now Brisbane) Australia. The Canadian Department of Agriculture has long worked on control of rangeland weeds, most of joint interest with the United States. Other laboratories are operated by the Hawaiian Department of Agriculture (HDA); International Institute of Biological Control (IIBC, formerly known as Commonwealth Institute of Biological Control [CIBC]), with permanent laboratories in India (now closed), Kenya, Malaysia, Pakistan, Switzerland, and Trinidad; Australian Commonwealth Scientific and Industrial Research Organization (CSIRO); Queensland Department of Lands near Brisbane, Australia; Plant Protection Research Institute, Department of Agriculture, South Africa; and Department of Scientific and Industrial Research, New Zealand (DSIR, now Manaaki Whenua Landscape Research New Zealand, Ltd.). Various universities and other government agencies have played important roles in several projects in all these areas. Commonwealth countries often obtain much of their overseas research through IIBC. Control agents developed by one country are often exchanged or shared with other countries, which has been of great value to all countries but especially to those that lack funds.

Several reviews describe theory and current progress of projects: in the United States (DeLoach 1991; Goeden 1978; Holloway 1964; Huffaker 1957, 1959, 1964); for aquatic weeds (Andres and Bennett 1975; Brezonik and Fox 1975; Buckingham 1994; Canter et al. 1995); for IN weeds (DeLoach 1995); for crop weeds (Charudattan and DeLoach 1988); in the western United States (Nechols et al. 1995; Rees et al. 1996); in Canada (Kelleher and Hulme 1984); in South Africa (Hoffman 1991); the Ethiopian area of Africa (Greathead 1971); in southern and eastern Asia (Hirose 1992); in New Zealand (Cameron et al. 1989); internationally (Andres et al. 1976; Harley and Forno 1992; Julien 1992; Laing and Hamai 1976; Schroeder 1983); in a series of international symposia on biological control of weeds (Delfosse 1981, 1985, 1990; Delfosse and Scott 1995; Freeman 1978; Moran and Hoffmann 1996); and by the use of pathogens (Charudattan and Walker 1982; Templeton et al. 1979).

Biological Control Defined

Just what is this method of weed control that requires no chemical herbicides, involves little or no effort by the farmer, rancher, or land manager, does not pollute the environment, is very specific, does not harm nontarget plants, and provides permanent control? Why is it not used to control all weeds? Are there dangers in or restrictions to its use? Are there only certain weeds or certain ecosystems in which it can be used? How does it work? Will it control both NI and IN weeds? Will it control weeds in crops, natural areas, and waterways as well as in rangelands? How is a biological control program conducted? Where can control agents be found?

For the purposes of this discussion, I define biological control as the planned use of undomesticated organisms (usually insects or plant pathogens) to reduce vigor, reproductive capacity, or density of target plants (see also Harris 1988). It excludes cultural controls (grazing management, crop rotation, transgenic manipulation of crops, etc.) and natural control (the action of organisms without human direction). It is economical for use in natural areas (Andres 1977; Harris 1979). Although it involves introduction of yet another NI organism, the control agent itself, this is acceptable in natural areas because biological control acts to tip the competitive interaction back in favor of IN species, even though the control agent is still present. Biological control seeks not to eradicate the target species but to reduce its density below the economic threshold or, in natural areas, below some biological threshold, so that other controls are not needed or, at least, so that damage and the use of other controls are reduced. Biological control of weeds has neither eradicated a target species from an ecosystem nor made rare plants more rare (Harris 1988).

Objectives

The objectives of this chapter are the following: (1) to explain the philosophy and methodology of biological control; (2) to describe the great extent to which biological control has been used worldwide, with emphasis on Canada, Hawaii, and the conterminous United States, the number and type of control agents used, and the success obtained; (3) to describe safety procedures protecting nontarget plants; (4) to demonstrate that biological control is environmentally compatible and protects IN plants (including rare and endangered species); and (5) to describe the potential for control of other NI plant species and the research needed to do this.

Comparison of Control Methods

Herbicides

Herbicides generally are broad spectrum and effective against many species. One application will control all or most weeds in an agricultural system; application can be limited to a given area. Cost depends on the area treated and on the duration of the treatment; it increases as more area is involved. Herbicides are most useful where the objective is to control all species except one (or a few), that is, the crop(s) being produced. Herbicides are often too expensive for areas of low economic return such as rangelands since the entire affected area must be treated periodically. Resistance to herbicides is now developing in many weeds. Herbicides also pollute streams, lakes, and groundwater and drinking water supplies. Mechanical control is expensive and allows soil erosion. In natural areas, where the objective is to kill only one (or a few) weeds and leave unharmed the rest of the plants, both herbicides and mechanical controls kill many desirable plants. In Australia, over 50 plant species are considered endangered because NI weeds out-compete them (Bell 1983). In Germany, 89 of 589 rare plants are declining as a result of herbicide applications (Sukopp and Trautmann 1981).

Biological Control

Biological control by augmentation and introduction has the advantage of being species-specific and of not harming nontarget species. It is most useful in systems where the objective is to control only one plant species (the weed) and leave all other species unharmed, which is the opposite of the control objectives of agricultural herbicides. It causes no chemical pollution and, sometimes, can control species difficult to control otherwise. Because of its high specificity, it is most useful where only one weed species causes most of the damage. However, natural enemies for controlling every weed may not exist. Developing a new control requires 5 to 10 years, and the research is costly, though substantially less than for developing a new herbicide.

Two major approaches are available in biological control. First, the introduction of foreign control agents, often called the "introductory approach" or the "classical approach," has been the most used and the most successful. The approach is to find insects or sometimes plant pathogens near the origin or within the natural range of the target weed that can reduce its population. These organisms are tested to insure that they do not harm other plants and then are released in the field (Harris 1993; Huffaker 1957). The agents need to be released at only one or a few sites; they reproduce and spread on their own and actively seek out the weed. The only cost is that of the research. Introduction is permanent and self-sustaining, and it controls the weed in all areas, including those hard to reach by ground equipment. Because of low cost, the method is suitable for controlling weeds in natural areas or in areas of low economic return such as rangelands. A disadvantage is that the released agent is likely to control the weed in all areas, thus damaging target species in areas where it is valued (e.g., landscape plantings). The area treated or the duration of treatment can be limited only with difficulty, if at all; rarely does a control agent spread throughout the range of the weed.

The second approach, including "augmentation" and "conservation," was developed recently (Frick and Chandler 1978; Harris 1993; Templeton et al. 1979). It seeks to increase effectiveness of phytophagous organisms already in an area, whether native or previously introduced. Frick (1974) reviewed various methods of augmentation. These include increasing the number of biotic suppressants attacking the weed by distributing insects from areas of surplus or from laboratory colonies grown on host plants or artificial diets, producing an inundative effect by using plant pathogens as "bio-herbicides," and using insecticides to reduce parasites or predators attacking the biocontrol agent. Among conservation methods are the adjusting of cultural practices to preserve natural enemies of weeds and the preserving of alternate host plants or refuges for these enemies. Augmentation has the advantage of having few or no conflicts of interest (the effectiveness of modifications made or the bio-herbicides applied will spread only a very short distance from where applied), so limited areas can be treated and control can be discontinued at will. A disadvantage is that the method is expensive because all infested areas must be treated periodically.

Mass production of insect herbivores in sufficient numbers to control weeds is expensive and probably not practical except in high-value crops. However, plant pathogens often can be produced at costs competitive with herbicides. These have been used in a few cases such as control of northern joint-vetch (*Aeschynomene virginica*) in Arkansas rice fields (Templeton et al. 1979) and of stranglervine (*Morrenia odorata*) in Florida citrus groves (Ridings et al. 1978).

Throughout the history of biological control, the "introductory" or "classical" approach has been used only in relatively stable ecosystems, for example, rangelands, pastures, natural areas, and aquatic sites, and almost entirely to control NI weeds. It has been used mostly to control herbaceous broadleaf plants or shrubs or sometimes annuals and biennials. Control of grasses and weeds in crops has been little attempted. Grasses present the probability of high conflicts of interest because of the extremely valuable cereal crops. However, some insects and many pathogens are sufficiently host specific so as not to harm cereal grains. Several grasses have severe conflicts of interest within the species, as with Bermuda grass, a severe pest in crops but a major, beneficial turf and pasture grass.

Control of weeds in crops has been attempted only rarely with introduced control agents. Three factors make such control in crops difficult: (1) disturbance from agricultural activities is likely to kill the biological control agents; (2) several species of weeds in a crop usually need control, and the biological control of only one would not reduce conventional controls needed for others; and (3) rapid control is needed to prevent crop damage (Andres et al. 1976). However, methods can be envisioned to circumvent these difficulties. Given that the greatest losses occur in crops (Chandler et al. 1984), biological control in this system deserves further investigation. The opportunities and problems associated with biological control of crop weeds were reviewed by Charudattan and DeLoach (1988).

The causes of an IN plant becoming weedy were wholly or partly different from the causes of NI plants becoming weedy. Indigenous plants become weedy because of environmental changes brought about by humans, for example, overgrazing, creation of disturbed areas, increased landscape fragmentation, increased atmospheric CO_2, and possibly (though not demonstrated) reduction in populations of insects or other organisms that formerly controlled the plant. Non-indigenous plants become weedy primarily because of the absence of natural enemies (DeLoach 1995).

Control of IN weeds historically has been considered more difficult than control of NI weeds. However, a few IN rangeland weeds

weeds. However, a few IN rangeland weeds have been successfully controlled: prickly-pear cactus (*Opuntia tricantha*) in the Caribbean (Simmonds and Bennett 1966), other prickly-pear species (*O. littoralis* and *O. oricola*) on Santa Cruz Island off California (Goeden and Ricker 1981), and manuka weed (*Leptospermum scoparium*) in New Zealand (Hoy 1961).

Some have argued against the concept of biological control of IN weeds as a general principle (Miller and Aplet 1993; Pemberton 1985). However, some IN weeds are very invasive and damaging, often in agricultural and sometimes in natural ecosystems, and today are hundreds of times more abundant than before European settlers arrived. Examples are mesquite (*Prosopis glandulosa*), junipers (*Juniperus*), snakeweeds (*Gutierrezia*) and poison ivy (*Toxicodendron*).

A very sucessful biocontrol program would be unlikely to reduce these even to their pre-Columbian abundance. Johnson (1985) argued that biological control would not result in damage to ecosystems, although individual species could be harmed, and that the "in principle" objections are mostly based on outdated Clementsian theory rather than on the accepted Gleasonian theory of plant ecology. Any proposed candidate IN weed should be carefully evaluated to assure that other species do not critically depend on it and would not be harmed by its control. DeLoach (1995) and Lawton (1985) reasoned that control of carefully selected IN targets could be both feasible and safe.

Most of the more important rangeland weeds of the southwestern United States, many crop weeds, and several weeds of southeastern pastures are IN plants. A long-term project has been underway in Texas to control IN weeds that have invaded rangelands in the southwest (DeLoach 1981; DeLoach et al. 1986) (discussed below). Lawton (1985) began a project on biological control of native bracken fern (*Pteridium aquilinum*) in Britain, using insects from South Africa. At present, no further research is underway on these or other IN weeds.

Community-Level Effects of Invasion and Response to Biological Control

Dense stands of NI weeds decrease plant diversity in a natural plant community. These NI invaders have a competitive advantage, presumably because they have few enemies that feed on them, while the IN plants generally have a large number of herbivore enemies (Harris 1988). This displacement of IN species can drastically affect wildlife (including endangered species), which may not be able to utilize the NI invaders.

Ecologically complex natural communities supposedly are less susceptible to invasion than species-poor ones (Ewel 1986). However, Center et al. (1995) cited cases of highly complex communities, such as the fynbos of South Africa, that also are seriously threatened by invasive NI plants. They proposed that the proximity and availability of a pool of invasive species are preeminent factors little considered by ecologists who study biological invasions. The main source of potential invaders in the United States is the commercial importation of plants.

The danger that an invading NI plant can pose to plant communities and wildlife is illustrated by the conquest of southwestern United States riparian ecosystems by salt-cedar since the 1920s (DeLoach 1990a; Horton and Campbell 1974; Robinson 1965). These riparian areas are the only locations where IN cottonwoods (*Populus*), screwbean mesquite (*Prosopis pubescens*), seepwillow baccharis (*Baccharis salicifolia*), willows (*Salix*), and many herbaceous plants can reach the water table. These plants, especially cottonwoods and willows, are prime breeding habitat for most of the birds and mammals in this region. Riparian areas also are vital green belts through the desert along which these and many other species migrate and where northern species overwinter.

In large areas where salt-cedar has invaded, most of the IN cottonwoods and willows, and to a lesser extent screwbean mesquite and other plants, that formerly occupied these areas have been replaced with virtually pure stands of salt-cedar. The birds and mammals are not adapted to salt-cedar; it produces almost no fruits, seeds, or phytophagous insects that they can eat, and most species do not use it for nesting habitat.

In studies conducted along the lower Colorado River (Anderson and Ohmart 1977, 1984; Anderson et al. 1977), the average density of all bird species in salt-cedar during the entire year was only 59% of that in the average of cottonwood, screwbean, velvet mesquite (*Prosopis velutina*), and willow habitats, and only 39% of that in those habitats in the critical winter period. Along the middle Rio Grande, the preference index for all bird species in winter averaged 50% for cottonwood-willow but only 9% for salt-cedar (Engle-Wilson and Ohmart 1978).

Hunter (1984) reported that eight IN bird species were in danger of extirpation from the lower Colorado River area, mainly as a result of salt-cedar invasion. Recently, the southwestern subspecies of the willow flycatcher (*Empidonax trailii extimus*) was placed on the endangered species list, primarily because its critical cottonwood-willow habitat has been replaced by salt-cedar (Federal Register 1995).

Mammals and reptiles also are affected by salt-cedar. Along the Rio Grande in Big Bend National Park, Texas, Boeer and Schmidly (1977) found that the population of beavers (*Castor canadensis*) had greatly decreased because giant cane (*Arundo donax*), cottonwood, seepwillow baccharis, and willow were being replaced by salt-cedar. On the middle Rio Grande, Hink and Ohmart (1984) reported that salt-cedar communities supported low populations of small mammals among the various community/structural types sampled. On the Rio Grande of western Texas, Engle-Wilson and Ohmart (1978) trapped 33 and sighted 30 reptiles in thornscrub; they trapped none and sighted only 3 in salt-cedar.

Salt-cedar infestations have lowered the water table in many areas, causing springs and ponds to dry up, limiting drinking water for wildlife, and further harming rare fish species. This has even further restricted the range of the endangered peninsula bighorn sheep (*Ovis canadensis cremnobates*) in southern California (Comrack 1986). In Death Valley National Monument, athel (*Tamarix aphylla*) planted around springs in the 1930s caused them to dry up, forcing wildlife out of the area. When the athel was cut recently, the springs flowed again and the wildlife returned (Rowlands 1989). A similar situation recently occurred following salt-cedar clearing near Artesia, New Mexico (Duncan 1994).

Similar cases occurred after invasion by other NI plants, although few documenting studies have been made. In Canada, the spread of leafy spurge (*Euphorbia esula/virgata*) is partly responsible for the recommendation to change the status of the only Canadian population of the northern prairie skink (*Eumeces septentrionalis septentrionalis*) to threatened. The skink requires short grasses and is not found in spurge stands invading this habitat. There is also a concern that the western prairie fringed orchid (*Platanthera praeclara*), which occurs in the Sheyenne National Grasslands, North Dakota, is threatened by leafy spurge and particularly by the program to control spurge with chemical herbicides (Harris 1988). Purple loosestrife (*Lythrum salicaria*) is partly credited with decline of a threatened bulrush (*Scirpus longii*) in Massachusetts; the rare dwarf spike-rush (*Eleocharis parvula*) is threatened in New York. Purple loosestrife is also aggravating the tenuous status of the endangered bog turtle (*Clemmys muhlenbergii*) in the northern part of its range. It makes habitats unsuitable for breeding by the black tern (*Chilonius niger*), a species declining rapidly (Thompson et al. 1987). Austin (1978) reported that the melaleuca (*Melaleuca quinquenervia*) reduces biodiversity 60% to 80% where it invades wet prairie or marsh communities in southern Florida. Harris (1990) concluded that biological control poses little risk to rare species, even if within the host range of the introduced control agent. This is because stenophagous insects (those with moderate host ranges) feed on the most abundant host, as the very scarcity of rare plants makes them a poor investment of adult searching time.

Few studies have been made to document the return or increase in abundance of IN plants and animals following biological control of NI weeds. The best documented study is that of Huffaker and Kennett (1959), who measured vegetation changes following the 1945 releases of the *Chrysolina* leaf beetles in California to control St.-John's-wort. Ten years later, St.-John's-wort had decreased in their four study sites from an average occurrence of 42.9% to only 1.2%, a 97.3% decrease. However, plant species richness (number of species present) of other species had increased by 37%. The number of other weed species increased from an average 3.3 species to 6.3 species (a 92% increase), legumes from 2.0 to 2.8 species (40%), forage grasses from 4.2 to 5.0 (19%), and forbs from 6.2 to 7.5 (21%). However, percent frequency of other weeds increased only from 10.0 to 11.5 (a 14% increase), but legumes increased from 3.9 to 14.2 (264%), forage grasses from 20.8 to 50.0 (136%), forbs from 14.0 to 18.2 (30%), and total forage plants from 38.8 to 82.4 (112%) (Huffaker and Kennett 1959).

Classical biological control of a dominant NI weed usually leads to a mixed community of plants that includes the weed at low density (Harris 1990) and other IN species (CAB 1994). In Saskatchewan, pure stands of musk thistle (*Carduus*) reverted to prairie grassland after biological control (Harris 1990). Vogt et al. (1992) reported that, after alligatorweed (*Alternanthera philoxeroides*) control in Louisiana and Mississippi, 31 species of IN plants had returned or increased in abundance.

Concern has been expressed that released biological control agents might attack nontarget plants or insects causing them to become rare, endangered, or even extinct (Miller and Aplet 1993). Howarth (1985) cited several putative cases of affected IN insects in Hawaii. However, Funasaki et al. (1988) found that only 33 of the 243 established biological control agents (for both insect pests and weeds) were reported to have attacked nontarget species. Of the 71 species of phytophagous insects and one pathogen introduced to control 21 species of weeds, only six fed or reproduced on two IN and nine beneficial plant species; significant damage did not result. During the past 21 years, no purposely introduced control agent has been recorded attacking any IN or otherwise desirable species in Hawaii.

Turner et al. (1987) measured the effects of feeding by the seed-head weevil (*Rhinocyllus conicus*) (introduced to control musk thistle [*Carduus nutans*]), on nontarget, IN species of thistles (*Cirsium*). Over a 3-year period in California, the number of adults per seed head was measured on 12 IN species of *Cirsium*, compared with the naturalized thistles. An average of 11.6 adult weevils per head was obtained from the target weed, musk thistle, 1.6 from the secondary target, milk thistle (*Silybum marianum*), 3.3 from the IN *Cirsium cymosum*, and no more than 0.5 from any other species. The reduction in seeds of *C. cymosum* probably will have little effect on its ability to reseed, and the damage to all the other IN species of *Cirsium* surely is insignificant. Turner and Herr (1996) reported similar attack on the rare IN *Cirsium fontinale* in California; *R. conicus* reduced seed production by 27% but this caused no reduction in stands.

The bug *Teleonema scrupulosa*, native to Mexico, was released at the Serere Research Station in Uganda in June 1963. By April 1995 it had increased enormously and controlled lantana within a radius of 10 miles. As the lantana died and the adult bugs were everywhere, they damaged sesame (*Sesamum indicum*), an experimental crop at the station. However, the bug could not complete its life cycle on sesame, the adults were not attracted to it, and, after the peak wave of bugs from the lantana control passed, further damage did not occur (Greathead 1971, 1973). This is an occasional phenomenon in biocontrol of weeds; no important damage has ever been reported.

Research Protocol

The basic protocol for conducting a research program in biological control of weeds has been established and refined by many workers over the years (Huffaker 1957; Wapshere 1975). It was discussed in detail by Buckingham (1994), Goeden (1977), and Harley and Forno (1992). The basic steps are (1) selecting the most appropriate target weed, (2) carrying out overseas exploration and testing, (3) conducting domestic field studies, (4) completing quarantine studies, (5) releasing and establishing the control agent, and (6) evaluating and integrating biological control with other control methods.

Selecting the Most Appropriate Target Weed

Target weeds for biological control research generally are selected either by development of public awareness of a serious problem and consequent political pressure to solve the problem or by scientific examination of the weeds in a given ecosystem and selection of the most promising for control. In practice, both methods are used. As a target, the weed (1) should have known or potential natural enemies, (2) should not have overriding beneficial values, either from the weed itself or from its close relatives, and (3) should cause enough damage to be worth the research cost.

The following factors should be examined in selecting the target weed for research (DeLoach 1981; Huffaker 1959, 1964): (1) damage caused by the weed, area infested, density in different areas, amount of loss caused by competition with beneficial plants, toxicity, usage of water, degradation of wildlife habitat or recreational value; (2) beneficial values of the weed, for example, aesthetics, crafts, firewood, grazing, honey production, medicines; (3) ecological values of the weed for wildlife food and shelter, erosion control, nitrogen fixation; (4) taxonomic-geographic relationships of the weed; and (5) potential for successful control.

Nonindigenous weeds generally are easier targets than IN weeds because relatively host-specific agents capable of controlling the weed can usually be found more easily overseas. The number of usable control agents will be greater if the origin of the weed's genus also is foreign. Control will be more difficult if the weed is closely related to highly beneficial plants because agents with a high degree of host specificity will then be required. The number and type of natural enemies known to attack the weed and the amount and type of damage they cause give an estimate of success potential. The weed with the greatest potential is one already successfully controlled elsewhere (Goeden 1983). The final analysis should consider all these factors to determine if biological control is feasible.

Carrying Out Overseas Exploration and Testing

Some of the most difficult and important aspects of a biological control program are those that occur overseas. These include finding and identifying insects or pathogens that attack the weed, deciding which among many are the most promising for control, conducting definitive host-range testing, and determining field ecology, life cycle, and behavior of the control agent.

The optimal location to find host-specific natural enemies is in the natural range of the weed and in an area climatically similar to the infested range (Room 1981; Sands and Harley 1981; Wapshere 1981). Surveys should identify natural enemies attacking the weed, the type of damage they cause, their geographic range and climatic tolerance, the degree to which they are parasitized, and whether they attack other species of plants.

Deciding which among the many insects and pathogens attacking the weed in its IN range is probably the most difficult step in the program. Early workers used their lifelong experience. Harris (1973) attempted to devise a numerical system for evaluating characteristics such as reproductive rate, degree of parasitism of the control insect, and part of the plant attacked; Goeden (1983) made some refinements in the system.

The heart of biological control research is to determine safety of the introduced control agent to be certain that it will not damage beneficial plants when released (Huffaker 1959). Careful host-range testing should be done overseas if possible. The phylogenetic method of testing (Harris and Zwölfer 1968; Wapshere 1974), in which plants more and more distantly related to the target weed are tested until the true host range of the control agent is defined, is now universally accepted. Testing methods have been refined and adapted to accommodate different types of insects and life stages and to represent more realistic conditions (Cullen 1990; Wapshere 1981; Zwölfer and Harris 1971). Especially promising is the use of open-field testing, which avoids cage artifacts (Clement and Cristofaro 1995), and new intensive methods of herbivory assessment (Cordo 1985). These methods provide a high degree of reliability.

In addition, overseas workers should determine the field ecology of the control agent (which cannot be determined in the United States before release), should gather sufficient data about its life cycle and behavior to be able to rear it in the laboratory in the United States after introduction, and should understand how it will behave after being released. Finally, when testing is completed, the overseas laboratory must collect and ship (free of parasites and predators if possible) sufficient numbers of the control agent to allow testing in quarantine and eventual field release in the United States.

Initiating Domestic Field Studies

Domestic studies are desirable to discover IN agents presently attacking the weed and to measure the amount and type of damage done. This will prevent wasting time in studying an organism already present in the infested range. Studies also should identify weak points in the present control system. For example, if stem borers are not present, then an effort could be made to find borers overseas.

Completing Quarantine Studies

Before releasing the control agent in the field, studies must be completed with the candidate agent in quarantine in the United States. Quarantine facilities must be secure to prevent premature escape of the agent. Operating procedures and training of personnel for the quarantine facility must be approved by USDA-APHIS-Plant Protection and Quarantine (PPQ). For biological control of weeds, a quarantine greenhouse is also important. The two primary responsibilities of quarantine testing are (1) to complete any needed host-range testing not done overseas, including testing of North American species and varieties, and (2) to produce cultures of the control agent free from predators, parasites, and pathogens, and in numbers sufficient for field release.

Releasing and Establishing the Control Agent

After release permits are obtained from USDA-APHIS-PPQ, the control agents may be released in the field. The release sites should have protection from disturbance (mowing, herbicides, burning) for 5 to 10 years for the agent to establish and spread. The climate should match requirements of the control agent; sufficient quantities of the target weed should be present. Sufficient numbers should be released at each site to allow population growth. These numbers vary with the type of insect and the life stage released but generally should be at least 100, with 1000 being better. The release of only a few insects may not be detrimental to establishment from a genetic point of view, but the probability of loss of the insects through chance events will vary inversely with the number released (Myers and Sabath 1981).

Historically, some insects (e.g., leaf beetles and weevils) have been easy to establish while others (e.g., the European moth *Tyta luctuosa*) have been difficult. Local predators, parasites, or pathogens may destroy the released insects (Goeden and Louda 1976); cages may protect the agents from predators until establishment. Determining the ease of establishment before release is difficult. Several cases exist where the insects appeared not to have established, only to reappear suddenly 5 to 15 years after release and to begin control.

Evaluating and Integrating

This final phase is often neglected because of a shortage of funds. However, if control has not been satisfactory, this step becomes important for understanding how to improve control, or to know if, and what type of, additional agents should be introduced. The step is also important for informing user groups and the public (who paid for the research) of the degree of control achieved and of the improvements made in the agricultural and/or ecological systems. Evaluation should document the change in area of weed infestation as a result of biological control, which sometimes can be accomplished by remote sensing (Everitt and DeLoach 1990).

Regulations and Safeguards

Biological control of weeds entails potentially serious dangers, particularly attack by the control agent on nontarget plant species and attack on the target weed in areas or situations where control is not wanted (Huffaker 1959). In the United States, the Technical Advisory Group (TAG) on the Introduction of Biological Control Agents of Weeds reviews and makes recommendations to USDA-APHIS-PPQ on research and introduction of control agents. APHIS regulates and issues permits for this research (Coulson and Soper 1989).

At the start of a project to control a weed, TAG provides the researcher an opinion on conflicts of interest between beneficial and harmful values of the weed and whether an introductory biological control program should proceed. After a candidate control agent is found and tested overseas, TAG reviews the test results and recommends whether the agent may be introduced into quarantine in the United States or whether additional testing is needed. Introduction into quarantine requires state approval and an APHIS permit. Further testing and the production of clean colonies for release is then done in a quarantine facility. TAG again reviews these test results before recommending whether the control agent may be released. The final step before release is the preparation of an environmental assessment. Release from quarantine into the field requires permission from the departments of agriculture of the states in which releases will be made and another permit from APHIS.

Miller and Aplet (1993) and others have pointed out the need for revising and improving the legal basis of the regulatory process, possibly along the lines of the recent Australian biological control law, which includes both public input and legal protection once decisions are made (Cullen and Delfosse 1985). Regulatory changes in the United States presently are under discussion (Anonymous 1996).

Control of Major Weeds in the United States and Canada

In the United States and Canada, research has been conducted on classical biological control of at least 45 species of NI weeds since 1945. Foreign control organisms have been released on 33 species (DeLoach 1991; Julien 1992); releases since have been made on two others. Another 10 species are current projects for which foreign insects and pathogens are being tested; many other foreign weeds have potential for biological control if personnel and funding were available to conduct required explorations and testing.

Of the 33 weed species targeted for biological control (for which natural enemies have been released), nine have been completely or substantially controlled, at least in major parts of their ranges. These are St.-John's-wort (1940s); puncturevine (*Tribulus terrestris*) (1960s); tansy ragwort (*Senecio jacobaea*) (1960s to 1970s); alligatorweed (1960s to 1970s); water-hyacinth (1970s to 1980s); musk thistle (1970s); and skeleton weed (*Chondrilla juncea*), Mediterranean sage (*Salvia aethiopis*), and water-lettuce (*Pistia stratiotes*) (1990s).

These represent the near ideal in biological control: the uniting of a weed with susceptible characteristics and the proper insect or pathogen to exploit the weed's weaknesses. In all these cases, only one or two species of control agents provided nearly all the control, although additional species may have been added later to improve control in certain areas. Control was easy and fast. This is a tribute to the skill of early explorers such as Albert Koebele in Hawaii, George Vogt on aquatics in the United States, CIBC and Canadian workers, and ARS explorers in Europe in the 1950s and early 1960s. A few other weeds, for example, Russian-olive (*Elaeagnus angustifolia*) and African-rue (*Peganum harmala*) could, with luck, be easy targets. However, the number of promising insects that have failed to establish on other weeds makes any prediction extremely difficult. Most other target weeds have been more like the lantana projects in Hawaii, Australia, and other countries where a series of insects was required to obtain control. Leafy spurge (*Euphorbia esula*), knapweeds and yellow starthistle (*Centaurea*), hydrilla (*Hydrilla verticillata*), salt-cedar (*Tamarix*), and melaleuca would appear to be in this category.

Research on biological control of weeds in the United States and Canada can be discussed by geographical regions, each of which has, for the most part, a group of weeds of importance mostly in that area.

Hawaiian Rangeland and Forest Weeds

In 1902, 14 species of Mexican insects were released to control lantana in Hawaii. Nine became established and provided control, although not in all areas. Three species were released in 1925 and 1926 on nutsedge (*Cyperus rotundus*) and gorse (*Ulex europaeus*), but these gave no control. Renewed interest occurred in the 1950s after the success of St.-John's-wort control in California. Eight additional insects were released on lantana; five of these gave substantial to complete control in many areas (Funasaki et al. 1988; Goeden 1978; Julien 1992).

From 1945 to 1965, research was expanded by the Hawaiian Department of Agriculture (HDA) and natural enemies were introduced to control 18 weed species. After the late 1960s, little work was done until 1986 when research on gorse (Markin and Yoshioka 1990) and other weeds was resumed by HDA and the U.S. Forest Service. Of the 21 weed species for which biological control was attempted in Hawaii, 12 have been completely or substantially controlled, a success rate of 57%. Seventy-one species of insects and one fungus were released; of these, 43 insects and the fungus became established (Funasaki et al. 1988; Goeden 1978; Julien 1992). Weeds completely or substantially controlled were blackberry (*Rubus argutus*), crofton weed (*Eupatorium adenophorum*), hamakua pa-makani (*Ageratina riparia*), three-cornered jacks (*Emex australis*), lesser jacks (*E. spinosa*), Koster's curse (*Clidemia hirta*), lantana, two species of prickly-pear (*Opuntia ficus-indica*, *O. cordobensis*), puncturevines (*Tribulus terrestris*, *T. cistoides*), and St.-John's-wort.

In 1975, scientists at the University of Hawaii released a plant pathogen (*Cercosporella* sp. from Jamaica) on Oahu to control the introduced pasture weed hamakua pa-makani. Control was spectacular in zones of high rainfall and optimal temperatures for disease development, with greater than 95% control in the first year. In areas with less than optimal climate, control has been less than 80%. More than 50,000 ha of pastureland have been rehabilitated to full production potential (Trujillo 1985). This was the first, and most successful, control of a weed in the United States by use of an introduced plant pathogen.

Recently, research has resumed on one weed and begun on a second one of Hawaiian forests. Koster's curse, an aggressive, dense shrub 1 to 2 m tall, replaces understory species and seriously threatens ecosystem integrity. In 1952 it occupied fewer than 100 ha in Oahu, but by 1975 it occupied 20,000 ha and, by 1988, 40,000 ha on all the major Hawaiian islands except Lanai (Markin and Burkart 1995). After Koster's curse was controlled in Fiji in the 1930s by a thrips (*Liothrips urichi*), the insect was introduced to Hawaii in 1953; it prevented spread of plants into pastures but did not control those growing in forests. A pyralid moth (*Ategumian atulinalis*) was established in Hawaii in 1966, a buprestid beetle (*Lius poseidon*) in 1988, and a leaf-spot fungus (*Colletotrichum gloeosporioides clidemiae*) in 1988. All damage but, so far, do not completely control the plant. A noctuid moth is cleared for release; 10 additional natural enemies are known on Trinidad that could be candidates for introduction (Markin and Burkart 1995).

Banana poka (*Passiflora mollissima*), a climbing vine native to the Andes of northern South America, was introduced on Kauai before 1920 and today occupies over 40,000 ha of Hawaiian forests, where it spreads through the canopy, killing the trees. For potential control, two moths and a fly introduced from Colombia have been released. Only the pyralid moth is established but is not yet causing much damage to the weed (Markin and Pemberton 1995).

Weeds of Western Rangelands

Research in the northwestern United States and western Canada has concentrated on weeds introduced from Eurasia. Biological control has been attempted for 23 species of these weeds. Seven of these (29%) so far have been completely or substantially controlled in large areas. Additional insects have recently been approved for control of other weeds, and other promising insects are under study (Goeden 1978; Julien 1992; Kelleher and Hulme 1984; Nechols et al. 1995; Rees et al. 1996). Several foreign plant pathogens that appear specific to several of the weed species have been found (Bruckart and Dowler 1986; Défago et al. 1985; Templeton 1982).

By the 1940s, approximately 2 million ha of western United States and Canadian rangelands were densely infested by St.-John's-wort, and the cattle industry was severely threatened. The species was also introduced into Australia in the 1880s; research on biological control began there in 1926. Two species of European leaf beetles (*Chrysolina quadrigemina*, *C. hyperici*) were obtained from Australia and released in California in 1945–1946. Two other insects, imported into California directly from France, were released in 1950: a buprestid beetle (*Agrilus hyperici*), which bores in the roots, and a gall midge (*Zeuxidiplosis giardi*), which forms leafbud galls. Within 10 years, St.-John's-wort was reduced to an occasional roadside plant in California (Holloway 1964), although control is still incomplete in Idaho, Montana, Oregon, and some other northern areas. These insects were released in Canada in the 1950s and in Hawaii in the 1960s (Goeden 1978; Kelleher and Hulme 1984). The geometrid moth (*Aplocera plagiata*) was established in 1976 in Canada (Julien 1992). It was released in the United States in 1989, and is now established in Idaho, Montana, Oregon, and Washington (Rees et al. 1996).

Andres (1985) reported that three of the insect species introduced to control St.-John's-wort—one of the leaf beetles, the buprestid beetle, and the gall midge—were reproducing on the IN *Hypericum concinnum*. Although feeding damage was sometimes severe and plant size was reduced, no reduction in stands was observed and the plant continues to be quite common.

Gorse, a thorny leguminous shrub from western Europe, is naturalized in Australia, British Columbia, Chile, Hawaii, New Zealand, and the northwestern United States where it was planted to form living fences. It competes with forage plants, interferes with forest regeneration, and promotes fires. A seed weevil (*Apion ulicis*) was released in California in 1953 and soon spread to nearly all of the infested area (Markin et al. 1995). This insect destroys 90% to 95% of gorse seeds, but this has produced no noticeable reduction in population density. A European spider mite (*Tetranychus lintearius*) was recently introduced to California from New Zealand to control gorse. Two oecophorid moths also have been introduced: *Agonopterix nervosa* accidentally into Canada and the western United States and *A. ulicitella* into Hawaii. In 1991, a thrips (*Sericothrips staphlinus*) was released in Hawaii (Rees et al. 1996).

Tansy ragwort is a biennial or short-lived perennial herb native to Eurasia and naturalized in North America, South America, Australia, and New Zealand. It occurs in Newfoundland south to New England and the U.S. Pacific Northwest and adjacent Canada. In the west, it is a serious competitor and is poisonous to livestock. Tansy ragwort has been substantially controlled (to 99%) in California, Oregon, Washington (Hawkes 1981; Turner and McEvoy 1995), and British Columbia (Harris et al., 1984). Control has been achieved by a foliage-feeding moth (*Tyria jacobaeae*) from Europe released in 1959 (it was previously tested in Australia in the 1930s but did not establish), a flea beetle (*Longitarsus jacobaeae*) from Rome released in 1969 (Frick and Johnson 1973), and a leaf-mining fly (*Botanophila seneciella*) from France released in 1966 (Rees et al. 1996). Two additional European insects recently released in Australia (Field 1990) may be valuable in the United States as well. A rust pathogen is under study in Europe.

Puncturevine is a prostrate, annual herb with hard, spiny fruit. Native from Africa and the Mediterranean area, through central Asia to China, it is a pest throughout most of the United States but especially from central Texas to southern California and in Hawaii. It has been substantially to completely controlled in areas of warm climate by two species of weevils (*Microlarinus*) (a seed feeder and a stem borer) from Italy released in 1961 (Andres 1978; Andres and Goeden 1995; Maddox 1981). In many areas formerly heavy infested, the plant is almost unknown today, though in colder climates it is still a pest (Rees et al. 1996).

Scotch broom (*Cytisus scoparius*) is a European leguminous shrub introduced as an ornamental and for soil stabilization on road banks. It now displaces IN species and is a weed in replanted forests from British Colombia to California. Two species of moths were accidentally introduced and started giving some control: *Agonopterix nervosa*, found in the 1920s, and *Leucoptera spartifoliella*, found in the 1960s. A seed-feeding weevil (*Apion fuscirostre*) from Italy was tested and released in California in 1964 and is now established in Oregon and Washington. None of these insects gives effective control, although the weevil greatly reduces seed production (Andres and Coombs 1995; Rees et al. 1996).

Poison-hemlock (*Conium maculatum*), a winter annual from the Mediterranean area, is found throughout the United States where it is poisonous to both humans and livestock. A moth (*Agonopterix alstroemeriana*) was accidentally introduced into New York in 1973 and is now established from California to Washington and Montana. It severely defoliates poison-hemlock but is not generally abundant enough to provide control (Rees et al. 1996).

Fourteen species of thistles and knapweeds in the genera *Acroptilon*, *Carduus*, *Centaurea*, *Cirsium*, *Onopordum*, and *Silybum* are NI invasive plants from Eurasia that are major weeds of rangelands and natural areas of the western United States and Canada. Since the 1960s, 18 species of insects have been introduced from Europe (13 have become established) in either or both the United States and Canada to control a complex of 11 of the 14 species (Batra et al. 1981; DeLoach 1991; Frick 1978; Kelleher and Hulme 1984; Rees et al. 1996).

For control of thistles in the genus *Carduus*, a seed-head weevil (*Rhinocyllus conicus*) was first introduced into Canada in 1968 and into the United States in 1969. It attacks, to some degree at least, nine species of thistles but has given best control of musk thistle (*Carduus nutans*), resulting in 90% to 99% reduction in stands in Montana (Rees 1978) and Virginia (Andres and Rees 1995; Kok and Surles 1975) and 80% to 90% reduction in Missouri (Puttler B, pers. com.), and control is beginning in Texas (Boldt and Jackman 1993). It reduces seed production of milk thistle by 50% to 80% in Oregon (Goeden 1995b) and of Italian thistle (*Carduus pycnocephalus*) by up to 55%; it attacks 30% to 90% of the heads of plumeless thistle (*Carduus acanthoides*), but it has not yet reduced stands of these species (Goeden 1995a). A rosette weevil (*Trichosirocalus horridus*) was released in the United States in 1974 and in Canada in 1975. It has provided effective control of musk thistle in Kansas, Missouri, and Virginia. It attacks plumeless thistle but has not yet reduced stands. A root-feeding syrphid fly (*Cheilosia corydon*) was released in 1990; a flea beetle (*Psylliodes chalcomera*) is being tested against musk thistle (Andres and Rees 1995; Rees et al. 1996). A European rust (*Puccinia carduorum*) released in Virginia in 1987, causes early senescence and reduces seed production (Bruckart and Peterson 1990).

Several thistles in the genus *Cirsium*, invaders from Europe, cause much damage in rangelands. For bull thistle (*C. vulgare*), the tephritid fly (*Urophora stylala*) was introduced into British Columbia in 1973 where it infested more than 90% of the heads at two sites by 1978 (Harris and Wilkinson 1984; Rosenthal and Piper 1995a). For Canada thistle (*C. arvense*), three insects were introduced. A leaf-skeletonizing flea beetle (*Altica carduorum*) was released in 1963 in Canada and in 14 American states from 1966–1972 but did not establish anywhere. A weevil (*Ceutorhynchus litura*) whose larvae are crown borers was released in Canada in 1965 and in 10 American states in 1971; it established in Idaho, Montana, Oregon, and Wyoming but provided no measurable control. A stem-galling tephritid fly (*Urophora cardui*) became established in Canada in 1974, in the northwestern United States in 1978, and in Maryland and Virginia. These insects reduce seed production; plants with damaged crowns die during the winter but little reduction in plant density has occurred (Piper and Andres 1995a; Rees et al. 1996).

Knapweeds and starthistles (*Acroptilon*, *Centaurea*) are major rangeland weeds of the northwestern United States and southwestern Canada. Most of the early research on diffuse and spotted knapweeds (*Centaurea diffusa*, *C. maculosa*) was by the IIBC at Delemont, Switzerland, and the AAF Canada (Harris and Myers 1984). AAF Canada established four species of insects in the 1970s in Canada that were later released also in the United States: two tephritid flies (*Urophora affinis*, *U. quadrifasciata*) and a gelechiid moth (*Metzneria paucipunctella*), which feed in flower and seed heads, and a buprestid beetle (*Sphenoptera jugoslavica*), which bores in the stems. Where the two tephritids are established, they have reduced seed crop output by 75% to 95% for diffuse knapweed and 36% to 41% for spotted knapweed. In the 1980s, the CDA also established a cochylid moth (*Agapeta zoegana*) and a weevil (*Cyphocleonus achates*), whose larvae feed in roots and stems, and a rust fungus (*Puccinia jaceae*) (Harris and Myers 1984); most of the insects now have been imported and released in the United States or have dispersed naturally into the country (Piper and Rosenthal 1995; Rees et al. 1996; Rosenthal and Piper 1995b; Story 1995). Two other root-boring moths were released in the 1980s: *Pterolonche inspersa* in Montana and *Pellochrista medullana* in Canada have not established. The seed-head weevil *Bangasternus fausti*, released in the United States in 1990, is now established in Montana, Oregon, and Utah (Dunn and Campobasso 1990). A nematode (*Subanguina picridis*) from the former USSR was established in Canada in 1976 and released experimentally in the United States in 1984 to control Russian knapweed (*Acroptilon repens*), but full-scale releases have not been made (Rosenthal and Piper 1995b). Four additional insects were released on knapweeds in the United States from 1991 to 1993: two tephritid flies (*Chaetorellia acrolophi*, *Terellia virens*) and two weevils (*Larinus minutus*, *L. obtusus*). As the recently established species increase and disperse, and as additional species are released and become established, control should increase substantially in both the United States and Canada.

Five European insect species were established by ARS in California from 1984 to 1992 for control of yellow starthistle (*Centaurea solstitialis*) (Turner et al. 1995; Rees et al. 1996). These were two tephritid flies (*Urophora siruna-seva*, *Chaetorella australis*) and three weevils (*Bangasternus orientalis*, *Eustenopus villosus*, *Larinus curtus*) all of which feed in different ages of flower heads (Turner et al. 1995). The most damaging of these is *U. siruna-seva* which galls 22% of flower heads; *B. orientalis* is the most widespread and abundant. A European rust fungus (*Puccinia jaceae*) is also a good candidate for biological control of yellow starthistle (Bruckart 1989).

Mediterranean sage (*Salvia aethiopis*), a biennial herb native to the Mediterranean area and central Asia, is a toxic weed of rangelands of the northwest that has spread from that region to Arizona and Texas. Two species of weevils (*Phrydiuchus spilmani*, *P. tau*), were introduced in 1971–1973; *P. tau* became established at many sites in California, Colorado, Idaho, and Oregon. At one site in Oregon it reduced *Salvia* cover from 27.7% in 1976 to 1.2% in 1980 (Andres et al. 1995; Rees et al. 1996).

Several weeds—Russian thistle (*Salsola*) and halogeton (*Halogeton glomeratus*) from China and the former USSR—were inaccessible for biological control explorations until recently. Russian thistle, or tumbleweed, a bushy, annual herb from central Asia, has become abundant over wide areas in semidisturbed lands of the western United States. Two moths were released in California: a stem borer (*Coleophora parthenica*) from Egypt, Pakistan, and Turkey in 1973; and a casebearer (*C. klimeschiella*) from Pakistan in 1977. Both dispersed widely and became abundant but did not damage the plant (Goeden and Pemberton 1995).

Skeletonweed (*Chondrilla juncea*), native from the Mediterranean area to central Asia, has been introduced into rangelands and wheat fields in Australia, the northwestern United States, and Argentina. It became a serious pest in Australian wheat fields, but the worst of the three skeletonweed biotypes was almost completely controlled there by the 1970s by a rust (*Puccinia chondrillina*) with help from a mite (*Eriophyes chondrillae*) and a gall fly (*Cystiphora schmidti*), all introduced from Europe (Cullen 1978). The mite and the gall fly, released in Washington State in the 1970s, resulted in excellent suppression of the weed, but the rust reduced only plant vigor and not density (Piper 1985). In California, the rust provided most of the control. The weed remains a serious problem in some areas (Piper and Andres 1995b; Rees et al. 1996; Supkoff et al. 1988).

Leafy spurge is a deep-rooted, perennial herb native from Spain and across Asia to Japan. A serious competitor of forage plants in rangelands and poisonous to livestock, it has invaded most of southern Canada and the northern third of the United States, where it is especially troublesome in Montana, Wyoming, and the Dakotas. Chemical controls are not very effective and are costly. Biological controls also have proven difficult, but a long-term effort in Canada and the United States is now producing good results (Pemberton 1985). Sixteen species of Eurasian insects have been tested and released, four other species are being tested in quarantine, and several others are still under study overseas.

The first species released to control leafy spurge was a large sphingid moth (*Hyles euphorbiae*) released in Canada in 1966; it was ineffective because of heavy attack by ants and a virus. Two species of clearwing moths whose larvae feed in tap roots of leafy spurge were released in the 1970s. One (*Chamaesphecia empiformis*) attacks only cypress spurge (*Euphorbia cyparissias*); the other (*C. tenthrediniformis*) attacks leafy spurge.

Flea beetles in the genus *Aphthona* are perhaps the most promising group of insects found for control of leafy spurge. They tend to feed in gregarious circles, the adults on the foliage and the larvae on root hairs and 1-year-old roots in the top 15 cm of soil. Each species appears to have a unique habitat preference. When released in the proper location, all seem effective in controlling leafy spurge. A release of 500 adults can be expected to reduce cover of leafy spurge from 90% to 2% over 1 ha by the end of the fourth year after release (Rees et al. 1996). *Aphthona nigriscutis* has been the most effective to date, controlling leafy spurge in a 4-km-diameter area of open, well-drained prairies.

Toadflaxes are perennial herbs. Yellow toadflax (*Linaria vulgaris*) is native from Italy to Russia; Dalmatian toadflax (*L. genistifolia* subsp. *dalmatica*), from former Yugoslavia to northern Iran. Dalmatian toadflax is a weed primarily of rangelands and pastures in Colorado, Michigan, Wisconsin, and the northwestern United States but is spreading to other states. Yellow toadflax is weedy throughout the United States and southern Canada. Attack by three accidentally introduced ovary- and seed-capsule-feeding insects (two weevils and a nitidulid beetle) has coincided with the decline of yellow toadflax in western Canada (Nowierski 1995). The two weevils attack yellow toadflax but only the narrow-leaved form of Dalmatian toadflax. The beetle also attacks the broadleaf form of Dalmatian toadflax. The only insect deliberately tested and released was the defoliating moth (*Calophasia lunula*) released in Canada in 1965 and in the United States in 1968, but it attacks primarily the yellow species.

Field bindweed (*Convolvulus arvensis*), native from the Mediterranean area to China, has become naturalized in nearly the entire United States and much of southern Canada. It is a pest primarily of crops but also in rangelands, especially from the Great Plains to western Washington. Many of its natural enemies have been discovered in the western Mediterranean area (Rosenthal and Buckingham 1982). A moth from Italy (*Tyta luctuosa*) (Rosenthal et al. 1988) and a mite from Greece (*Aceria malherbe*) (Rosenthal and Platts 1990) were released in 1987 and 1989 in several states; only the mite established. A very damaging leaf beetle (*Galeruca rufa*) from the Mediterranean area could not be released because it fed on sweet potato and IN *Calystegia* plants in laboratory tests.

Weeds of Southwestern Rangelands

The most damaging weeds of rangelands in the southwestern United States and northern Mexico are IN shrubs or poisonous herbs (DeLoach 1981, 1995; DeLoach et al. 1986; Platt 1959). The importance of these species has increased greatly during the last 150 years as a result of overgrazing, spread of seed by livestock, reduction in range fires, and other factors related to the livestock industry (Buffington and Herbel 1965). The increasing concentration of CO_2 in the atmosphere in the past 150 years also provides a strong competitive advantage for some IN shrubs that are displacing IN grasses (Johnson et al. 1993). Several of the most damaging weeds have related species native to southern South America where they are attacked by IN insects and pathogens. Several of these natural enemies are sufficiently host-specific and could be introduced into North America. Targets included mesquite (*Prosopis*), snakeweeds and broomweeds, willow baccharis (*Baccharis neglecta*) (Boldt and Robbins 1987), and creosotebush (*Larrea tridentata*). Two stem borers were tested for control of snakeweed: a weevil (*Heilipodus ventralis*) and a clearwing moth (*Carmenta haematica*). The weevil was released in New Mexico and Texas but establishment is not confirmed (DeLoach 1995).

Salt-cedars, native to Asia and introduced into the United States in the early 1800s as ornamentals, pose many ecological problems in riparian areas (DeLoach 1990a). An international team of researchers is testing 13 species of promising natural enemies to control these plants. Two species have been recommended for field release in the United States by TAG: a mealybug (*Trabutina mannipara*) from Israel and a leaf beetle (*Diorhabda elongata*) from China. Two other species are being tested in quarantine at Temple, Texas: a psyllid (*Colposcenia aliena*) and a gelechiid leaf tier (*Ornativalva grisea*) from China. A gall midge (*Psectrosema*) has been approved for quarantine testing, and overseas testing has been completed for a foliage-feeding weevil (*Coniatus tamarisci*), both from France, and a pterophorid moth (*Agdistis tamaricis*) and a foliage-feeding weevil (*Cryptocephalus sinaita* subsp. *moricei*) from Israel (DeLoach et al. 1996). Many additional species of apparently host-specific insects attack salt-cedar in the Old World; these could be tested if needed (Gerling and Kugler 1973; Habib and Hassan 1982; Kovalev 1995; Zocchi 1971).

Russian-olive introduced from Israel as an ornamental and windbreak, causes damage very similar to that of salt-cedar, but it does not increase soil salinity, it is little used by white-winged doves or for honey, and it is planted more as an ornamental. It is rapidly invading and is displacing IN species in riparian areas of the west (Knopf and Olson 1984). Opportunistic explorations in China in connection with the salt-cedar project have revealed an exceptionally promising psyllid (*Trioza magnisetosa*) damaging to Russian-olive over wide areas and apparently host specific.

African-rue, though unpalatable, is poisonous to livestock and wildlife. Introduced from North Africa, probably in the 1930s, it now occupies several thousand acres around the commercial airports at Pecos, Texas, and Demming, New Mexico, and along highways leading from there. It could be a candidate for biological control, but exploratory research or testing have not begun because it still occupies a small area. Opportunistic explorations in Turkmenistan revealed a damaging stem borer, a scolytid beetle (*Thamnurgus* aff. *pegani*). This beetle killed half or more of the stems over wide areas and is not known to attack other plants.

Aquatic Weeds in the Southeastern United States

Research on biological control of aquatic weeds focused first on species from South America and later on species from Eurasia and Australia. These weeds occur primarily from coastal North Carolina to eastern Texas. Efficient biological control agents for several of these aquatic weeds have been introduced into the United States and are described in Madsen (Chapter 12, this volume).

Weeds of Other Natural Areas in the United States

Most of the range and pasture weeds discussed above could also be considered weeds of natural areas. They and other weeds introduced into natural areas may displace IN species, cause loss of wildlife habitat, cause plant and animal species to become rare or endangered, degrade recreational areas, or cause human health problems even though they are not agricultural pests. Some of these plants could be controlled by introducing NI natural enemies if conflicts of interest could be resolved or by augmenting the effectiveness of IN control agents already present.

Kudzu (*Pueraria montana*), introduced from eastern Asia for erosion control and forage, has spread widely, damaging forest trees and power lines and smothering IN species (Miller JH, oral presentation). A few promising control agents have been discovered in Japan (Tayutivutikul and Kusigemati 1992a, 1992b; Tayutivutikul and Yano 1989, 1990); a cerambycid beetle recently found in China heavily damages the crowns of the plants (DeLoach, unpublished data).

Melaleuca (*Melaleuca quinquenervia*) from Australia, introduced into southern Florida as an ornamental, has naturalized in recent years to become a serious threat to the Everglades. A recent cooperative project between ARS, CSIRO of Australia, and the National Park Service has already identified several promising insects for biological control (Balciunas et al. 1995).

Purple loosestrife is an herbaceous, perennial weed introduced from Europe growing to 2.7 m tall. A serious invader of wetlands in the northeastern United States and adjacent Canada, it has spread as far as Texas, California, and British Columbia. It has a disastrous impact on vegetation, seriously reduces waterfowl and furbearer productivity, and further threatens several declining animal species (Thompson et al. 1987).

Biological control is the method of choice for control of purple loosestrife (Hight and Drea 1991; Malecki et al. 1993) since Batra and colleagues (1986) found several promising control agents in Europe. Several of these were tested by the IIBC at Delemont, Switzerland, and later in quarantine in the United States (Kok et al. 1992). Three species were released in the field in 1992: two leaf beetles (*Galerucella calmariensis*, *G. pusilla*) and a weevil (*Hylobius transversovittatus*) whose larvae feed in roots and adults feed on foliage. These are well established in five northwestern states and in five North Atlantic states. Also, two species of weevils that feed on flower buds and developing seed capsules were tested: *Nanophyes marmoratus*, released in 1994, is established in New York and Oregon;

N. brevis is approved for release (Rees et al. 1996).

Brazilian peppertree (*Schinus terebinthifolius*), introduced from Brazil into Hawaii and southern Florida as an ornamental, has become a serious weed of natural areas. Two insects established in Hawaii in 1932 and 1954 gave no control (Funasaki et al. 1988). However, Bennett et al. (1990), working at the University of Florida, found at least four additional insects in Brazil that are potential biological control agents. The introduced marijuana (*Cannabis sativa*) also could become a candidate for biological control using insects that attack it in India (Batra 1976) and eastern Siberia (Marikovoski, pers. com.) or a fusarium wilt pathogen from Italy (McCain and Noviello 1985). Habeck (1990) determined that poison ivy possibly could be controlled in Bermuda using natural enemies from the United States. In North America, natural enemies from Japan (DeLoach, unpublished data) might control the plant if conflicts of interest regarding its use as food by birds could be resolved.

Future Directions of Biological Control

Prevention of Entry

The rate of introduction of potentially dangerous pests is increasing with increased trade and travel at a rate far greater than new controls can be developed and at a time when funding for such development is dwindling. Most scientists know the risks of exchanging species between continents, but many tourists and traders do not. Until recently, unwitting or unscrupulous persons still traded weeds and fishes for aquaria. Tourists still try to smuggle fruits, seeds, or attractive plants, which may become weeds or may be laden with pathogens and pests. To improve crops, plant breeders bring in and propagate living materials; some of these may become weeds or harbor pathogens. Shipment of cheap Russian logs from Siberia to the United States was the probable mode of entry of the Asian gypsy moth (*Lymantria dispar*), a serious forest pest.

However, the greatest risk of all may be the introduction of NI species, in enormous numbers, for commercial sale as ornamental or novelty plants (see Reichard, Chapter 15, this volume). In 1993, about 456 million exotic plants were imported through the 16 United States plant introduction facilities, with nearly 80% of these coming through the port of Miami. Port inspectors, overwhelmed, can inspect far less than their goal of 2% of the plants. These imported plants represent a huge pool of potential invaders, directly through their own escape, and indirectly through the insects, pathogens, and other pests they might harbor (Center et al. 1995).

Clearly, to prevent entry of new pests, the front line of defense must be through trade regulation, inspection, education, and public awareness. Center et al. (1995) proposed that all importations of living organisms should be intensively regulated. However, such regulation is unlikely to be implemented in the near term. The importation of NI species is economically lucrative, so legislative attempts to regulate this practice are met by stiff opposition. Restricting importations to seeds might provide some relief but would be strongly contested. Routine fumigation of all plant shipments might eliminate invertebrate stowaways, but the costs and logistics of such a program would be overwhelming. Public education campaigns encouraging use of IN plants and advocating patronage of nurseries providing only IN species have yet to be initiated on any substantive scale. In any event, changing the ways that the general populace uses plants is a slow process (Center et al. 1995). These realities dictate that we deal with the pests as they exist rather than with the cause.

Basic Research Needs

The basic protocol and methodology of biological control of weeds are well established. However, basic research could still make improvements in the most critical stages, although the tedious, time-consuming studies of insect, pathogen, and weed biologies probably will always be necessary because of individual differences in each organism.

Better methods of economic and environmental analyses to resolve various conflicts of interest must be developed and applied at the start of new projects and to select the best projects for research. Similar methods also must be applied to measure the benefit produced as projects are completed.

Much thought has been given in recent years to discover the best locations to find natural enemies and to select the best enemies for testing among the many found. However, the systems developed for selecting these (Goeden 1983; Harris 1973) still need improvements. The methods of host-range testing, though greatly improved since the 1950s (Harris and Zwölfer 1968; Zwölfer and Harris 1971), are still not completely satisfactory and sometimes lead to the rejection of good, safe control agents. The need for development of new methodologies still exists.

The establishment of released organisms in the field is still an area with many unknowns and a high rate of failure. This is particularly distressing because the major portion of the costs has already been invested in the organisms by this stage of the program. Also, far too many sites are lost after releases because of a lack of understanding or economic factors; the landowners or managers may destroy the release sites.

Future Research Opportunities

Several regions have great potential for future successful projects to control weeds in different ecosystems and/or to find natural enemies in areas previously inaccessible. Rich sources of natural enemies are now available in the former USSR and China, where explorations previously could not be made, and in South America. Because of drastically low economic conditions, research can be conducted in Asia at minimum cost, a situation that can be expected to last for only a few years, given the economic growth rate in China and the declining availability of funding in the United States in the last several years.

Pemberton (1990) compiled a list of 25 North American weed species that were introduced from China and/or Japan or that occur naturally there; C.J. DeLoach and P.C. Quimby (pers. com.) added approximately 30 additional species. In northern China, promising insects already have been found: at least 10 insects on salt-cedar, several insects and pathogens on leafy spurge, a crown-boring weevil on Russian thistle, a psyllid on Russian-olive, and several insects on hydrilla, velvetleaf (*Abutilon theophrasti*), and water-milfoil.

Many weeds of the eastern and southeastern United States are native in humid eastern and southern China. Promising control agents are a cerambycid crown borer on kudzu, a gelechiid leaf tier on Chinese privet (*Ligustrum sinense*), a bruchid seed beetle on mimosa (*Albizia julibrissin*), and leaf beetles on smartweeds (*Polygonum*). Preliminary surveys have been made for natural enemies of Chinese tallowtree (*Sapium sebiferum*) and chinaberry (*Melia azedarach*). Other weeds native to China include multiflora rose (*Rosa multiflora*), McCartney rose (*Rosa bracteata*), climbing ferns (*Lygodium japonicum, L. microphyllum*), water-spinach (*Ipomoea aquatica*), tree-of-heaven (Ailanthus altissima) (Anderson 1961), and princesstree (*Paulownia tomentosa*) (Langdon and Johnson 1994). Some of these plants are valuable ornamentals or shade trees in the United States, but seed-feeding insects could reduce their invasiveness without harming their beneficial values.

Several weeds are native to or have close relatives in Japan. Natural enemies are known on poison ivy (*Toxicodendron radicans*) (DeLoach, unpublished data), kudzu, multiflora rose, Japanese honeysuckle (*Lonicera japonica*), McCartney rose, and dock (*Rumex*). Other weeds from southern or southeastern Asia such as itchgrass (*Rottboellia cochinchinensis*) (Ellison and Evans 1990), nutsedge, and pigweeds (*Amaranthus*) possibly could be controlled with natural enemies from Pakistan or Thailand where biological control infrastructures are well developed (Mohyuddin 1981; Napompeth 1992).

Batra (1981) listed 17 species of weeds from Europe being considered by ARS as biological control targets in the northeastern United States. In the southeastern United States, at least 17 species of terrestrial weeds have potential for biological control using natural enemies from the same or related plant species of southern South America (Charudattan and DeLoach 1988; DeLoach 1990b). Preliminary investigations have been made, but more thorough searches are needed to find additional natural enemies.

Conflicts of interest among harmful effects, beneficial functions, and ecological values of some of these weeds probably would preclude biological control. Interaction between biological control workers and ecologists would be desirable to prioritize these weeds for research. These conflicts probably could be resolved so that introduction could proceed. In cases where the weed is also used as a shade tree the standing tree is of value; the introduction of seed-destroying organisms would reduce or halt the dispersal of the weeds without harming ornamental value.

Public Perceptions

Incredibly, in spite of the 50-year history of biological control of weeds in continental North America (93 years in Hawaii), the very successful control of many of them, and the great benefit derived by both agriculture and natural ecosystems, biological control is little known, misunderstood, and even feared by some. The present generation seems to have forgotten (or maybe it never knew) that St.-John's-wort once threatened to destroy the livestock industry and some IN species in California. The older people in Texas remember the plagues of puncturevines but do not know why they disappeared. The general population does not realize that many serious weeds are now permanently controlled in many areas to the point of causing little or no damage, at no cost except for research, and with no environmental pollution. Persons concerned that invasions of NI plants are severely damaging ecosystems do not realize that introduced insects or plant pathogens could control many invaders, or even that control programs are already underway for some of them. Botanists and plant ecologists sometimes do not give credit to the action of insects and plant pathogens in controlling plant populations but contend that populations are controlled only by abiotic factors or that invasive weeds are simply "more aggressive" than the IN plants they replace. This aggressiveness may simply be the result of lack of natural enemies. Every case of successful biological control demonstrates that the simple introduction of one or a few of the natural enemies (insects or pathogens) that feed on the plant in its native range can greatly reduce population size.

The press sometimes mentions "biological control gone awry" (Davis and Buckingham 1988), with an implication of great danger and a cavalier attitude of mad scientists introducing ghastly foreign insects and animals that can later escape and destroy ecosystems. The recent real invasion by "killer bees" and fire ants feeds this hysteria. The bad examples of the introduced mongoose (*Herpestes auropunctatus*) in Hawaii, of the nutria (*Myocastor coypus*) along the Gulf Coast, and of kudzu in the southeastern United States are well known. None of these, however, was introduced by biocontrol workers.

Nothing could be further from the truth than the mad scientist perception. The careful consideration given to the selection of a weed as a candidate for biological control, the excruciatingly careful testing done overseas and in quarantine before release, the safeguards and required review in place at each step of the process, and the consideration given to avoid damage to nontarget plants and to ecosystem health all make biological control one of the most important means of *protecting* ecosystems from the ravages of invading exotics and from broad-spectrum herbicides or mechanical controls that otherwise would be used. Biological control researchers are as much concerned with restoring long-term ecosystem health as with controlling a given weed.

The fact is that invasive, NI weeds presently cause great damage to natural ecosystems, as well as to agriculture and to urban and recreational areas, and harm rare or endangered plants and animals. To do nothing out of concern for causing changes in present "pristine" ecosystems is only to perpetuate the damage. Pristine wildland areas are already essentially nonexistent today. Mechanical or chemical controls are short lived and further damage nontarget plants.

In recent years, increased funding and increasing research are going toward control of weeds that are damaging primarily to ecosystem integrity. However, the biological control scientists must often overcome strong skepticism on the part of administrators as well as environmental advocacy groups who are not familiar with the philosophy, methodology, or past usage of biological control of weeds and who are concerned that introduction of yet another NI organism may cause even greater harm to natural ecosystems.

In the cases where biological control has been successful, plant diversity has increased, ecosystem health has improved, and nontarget IN plants have sustained no more than minor, inconsequential damage (and that in only a very few cases), and no rare or endangered species have been made more rare or more endangered.

Biological control of weeds does have an element of risk. In early years, introductions of control agents were largely unregulated, without oversight, and gave little consideration to possible effects on noneconomic plants. Testing theory was not developed; host-range testing was minimal. The putative worst case mistake, that of temporary damage caused to the sesame plot in Uganda by insects introduced to control lantana, proved to be of little consequence. Nontarget feeding in California on an IN St.-John's-wort by insects released to control the NI species and on IN thistles by insects released to control the NI musk thistle (discussed above) were shown to cause little or no reduction in density of these IN plants.

The bottom line of this discussion is that biological control of weeds is safe, has been widely used and proven over many years, and has controlled so far one-third of the targeted weeds (at least in major parts of their distribution), and control of other species continues to increase. It does entail some risks, but with the safeguards now in place, these are minimal. All other control measures, except for hand weeding, also entail risks. Doing nothing about a serious invading NI weed is not a risk—its destructiveness is a known fact. The biological control achieved is so good in many cases that repeated applications or other control methods apparently will never again be necessary. Such control is inconceivable by any other method.

Conclusions

1. Biological control is the use of undomesticated organisms to reduce density, reproductive capacity, or vigor of weeds to acceptable levels.

2. The introduction of NI agents to control NI weeds provides permanent control, can control the weed in all areas, and is relatively inexpensive. Conflicts of interest between harmful effects and beneficial uses of the weed should be resolved before control begins. Augmentation, a second approach, uses agents already present, is safe, and usually involves no conflicts of interest, but is relatively expensive because repeated treatment is necessary.

3. Biological control has been used worldwide since 1865 in 729 projects against 118 weed species in 51 countries. Since 1902, NI control agents have been released against 21 weeds in Hawaii; since 1945, against 32 weeds in the conterminous United States and 15 in Canada. One third of these weeds has been controlled (control of others appears imminent), greatly benefitting agriculture and natural ecosystems.

4. The philosophy and methodology of biological control are well understood, having been developed and field tested over the past 90 years. A protocol of target weed selection (including the resolution of any conflicts of interest)—overseas exploration, overseas testing, quarantine testing in country of release, release, and monitoring—is well established.

5. Safety of a control program is insured by required oversight and review of test results, by approval and licensing of quarantine facili-

ties and operating procedures, and by issuance of government permits before an NI control agent can be released in the field.

6. Biological control of weeds is a safe procedure. Not a single case is known where nontarget feeding was not predicted by prior testing. In the few cases of feeding on closely related nontarget beneficial or IN plants, damage was minor.

7. Biological control causes no chemical pollution, is highly host specific, and does not damage nontarget plants. It should be the method of choice for controlling NI weeds in natural vegetation where the objective is to control one plant species and leave the others unharmed.

8. The objective of biological control is not weed eradication, but rather reduction in abundance. In practice, it has never eradicated a target or nontarget plant species. Typically, the weed remains occasional or common in the plant community.

9. Biological control is an appropriate option for any NI weed that has minimal benefit, that does not have valuable close relatives, that has natural enemies, and that involves less research cost than damage cost.

10. Many weeds in America could be future targets for biological control; potential enemies are currently under study. Research infrastructure and facilities are in place.

14
Prioritizing Invasive Plants and Planning for Management

Ronald D. Hiebert

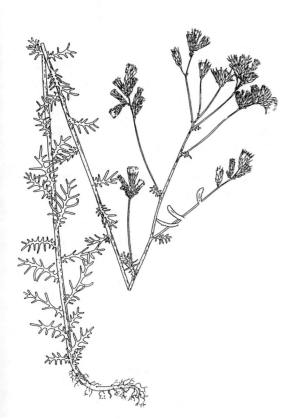

Common crupina
(*Crupina vulgaris*)

The management of invasive nonindigenous (NI) plant species in natural areas or native landscapes is an expensive and complex task for resource managers. Unlike weed control in agricultural situations where the objective is to control all species except the crop, resource managers often want to control one or several undesirable species while causing little or no impact to components of the native community or to natural ecological processes. Managers often are faced with invasion and establishment of hundreds of NI plant species with little knowledge of the effects or potential effects a species may have on the land they manage. It is often unknown if control is feasible or what the cost of a control program may be. Managers need tools to evaluate objectively the kinds and levels of impacts that NI species are causing, their potential for becoming a worse threat, and the feasibility of their control to guide prioritization.

In this chapter, the ecological, economic, and managerial rationales for prioritization of NI plants for control will be considered. The use of decision-making tools by managers will be advocated. A ranking system based on the present level of impacts and the potential of a species to be a threat weighed against the cost and feasibility of control is presented. Directions for application of this decision-making tool are given along with results from applications at various scales. As prioritization of invasive plants is only a first step toward management, guidelines and examples of tools are presented for defining goals and objectives, developing and implementing management plans, and assessing and evaluating results.

Why Prioritize Invasive Plants?

Since the beginning of European expansion, the human-assisted movement of organisms has greatly accelerated. Nearly all nature reserves, grasslands, forests, and parks have been invaded by NI plants and most likely will be invaded by additional species in the future.

Few control programs to date have been successful due to lack of funds or adequate technology (Usher 1988). Resource managers thus must pick their battles wisely and concentrate effort on controlling those species having the largest impact on rare and endemic flora and on landscape-level ecological processes. Studies have shown that only a small proportion of NI species causes significant impacts on nature reserves. For example, Great Smoky Mountains National Park, in North Carolina and Tennessee, has about 1500 species of vascular plants, 400 of which are considered nonindigenous. Only 10 of these appear to be spreading or to pose a threat to park resources (Bratton 1982). Of the 1400 species of vascular plants occurring at Indiana Dunes National Lakeshore, Indiana, 300 are considered nonindigenous but only 14 are thought to be causing or to have the potential to cause significant impacts (Klick et al. 1989). Thus some level of analysis of impacts caused by invasive plants is needed so one can concentrate further efforts on a much reduced list of species. The ecological and managerial/economic reasons for prioritization are further detailed below.

Ecological

Based upon the presently accepted "individualistic" paradigm of community composition, which views the group of species occurring at a particular place at a particular time as the result of chance and the species-specific tolerances and requirements (Soulé 1990), the impact and potential for spread of each species must be considered separately. Based on the above assumptions, the ability of a species to invade an area and spread, and the level and types of impacts, would be expected to vary over time and space. Thus, analysis of impacts must be site-specific and must be periodically repeated.

One cannot predict if NI species will be highly invasive from life history characteristics alone, but some characteristics predispose a species for successful colonization (see Reichard, Chapter 15, this volume; Westman 1990b). Characteristics often found in successful invaders include frequent and high seed production, generalized pollination systems, special adaptations for long distance dispersal, and tolerance of a wide range of conditions (Baker 1965, 1986; Bazzaz 1986). The disturbance regimen (e.g., type, frequency, and intensity) will also modify invasion potential.

Finally, the ecological effects of removal of a species on the composition and function of a system remain poorly understood. Thus one must consider the possible negative effects of control measures.

Managerial/Economic

The cost of control for one invasive species could require a resource manager's entire budget. Thus, as stated above, a manager is forced not only to prioritize among programs but also to prioritize which species to control. To be successful, the manager must be able to acquire additional long-term funding and the cooperation of other land managers in the region. This will require solid information on the level and types of impacts caused by a species on lands

14. Prioritizing Invasive Plants and Planning for Management

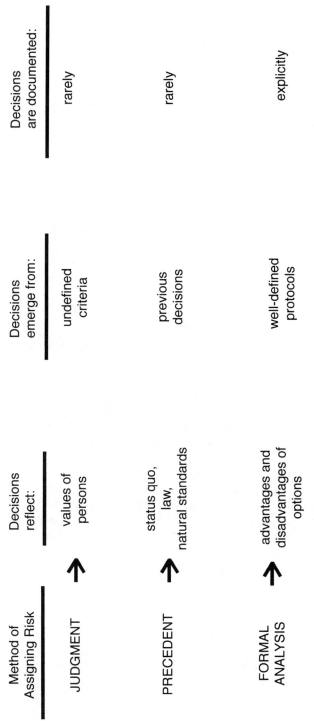

FIGURE 14.1. Different methods of assigning risk or making decisions. (From Kareiva et al. [1991])

within the area of concern. This will often include lands with highly diverse management objectives.

How to Prioritize

Scientists and managers both have an aversion to making difficult environmental decisions. Scientists have been trained to be cautious about predicting what is going to happen based on incomplete information. Managers, often conservative, prefer that mistakes be made through inaction or by the following of standard operating procedures rather than by trying new, innovative plans (Maguire 1991). Delay in action due to inability to make a decision or taking an action that is not information-based can have disastrous consequences for the resource and for the costs of a plant-control program. Any investment in risk analysis or decision analysis can alleviate the problems of scientists risking their professional standards and can give managers strong justification to support their management decisions.

Management decisions are usually made in one of three ways: judgment, precedent, or formal analysis (Fig. 14.1). Granted, many field biologists have an educated idea regarding which invasive plants are causing major impacts and which pose threats. Managers have a strong intuition as to how well different management scenarios fit their land-management objectives, what the cost of control efforts would be to an overall management program, and the probable reactions of neighbors to various control actions. The judgment of experienced and well-educated professionals is irreplaceable in decision making. However, decisions made by judgment alone are rarely based on defined criteria, are subject to personal biases, provide no documentation of the reasoning process leading to the decision, and include no assurance that the full array of factors was considered.

Decisions based on precedent may be easier to defend, but they often are not responsive to variation in NI species performance that may occur at diffferent times and in different areas. Nor does this approach encourage development of new and innovative technologies and procedures to deal with complex problems.

The consistent application of an analytical tool (i.e., formal analysis) of some type (e.g., to decide which plant species to control or how to control them) can alleviate most problems associated with making decisions based upon judgment or precedent alone. An analytical approach assures that consistent attention is given to the range of factors related to the problem at hand and to the consequences of alternative decisions. An analytical framework documents the procedures and the reasons for the decisions made, thus reducing the aversion to making decisions. An analytical approach provides defendable justification for decisions made and strong support for program backing.

Steps in the Decision-Making Process

In making even the simplest decision, the mind follows a distinct series of steps. These logical, basic steps in the decision-making process are preparation, incubation, inspiration, and verification (Evans 1986). Interpolated into a generic pragmatic problem-solving situation, the stages are identifying the problem, analyzing the problem, defining goals and objectives, identifying and evaluating alternative solutions, developing and implementing a management plan, assessing and evaluating the management plan, and educating participants (Coughlan and Armour 1992).

The National Park Service (NPS) follows these basic steps in its guidelines for integrated pest management (IPM). It defines IPM as "a process for determining if pest management is needed; when management should be initiated; where and at what frequency treatments should be applied; and what physical, biological, or chemical strategies should be employed and how effective these treatments are in achieving management objectives" (USDI, NPS 1991). As applied to the management of NI plants, NPS policy calls for control in areas zoned as natural. However, due to the high cost of control of invasive plant species and the possible negative side-effects of treatment, NPS policy calls for priority to be given to those plants that cause significant impact to park resources and are easiest to control. Thus if policy is to be followed, park managers are directed to submit each step of the decision-making process to formal analysis.

Steps in the IPM process are as follows:

1. Identifying the problem. Are NI species present in natural zones within the park? At a minimum this requires a survey of the park and determination if a species is indigenous to the area or is nonindigenous. However, the mere presence of an NI plant species does not indicate a problem. It must be determined if a species is causing an impact or has a high potential to interfere with the purpose for which a piece of land is managed.

2. Analyzing the problem. To guide wise management decisions, one would ideally want to know the abundance and distribution, the life history in detail, and the type and magnitude of effects that each NI species has on elements and processes of the system of concern. In most cases, some of this information is lacking. Therefore, it would take years of research to gather the information. Additionally, by the time the research is completed it may be too late to save the resource of concern. Thus it is necessary to conduct the best analysis possible in a timely manner based upon existing information. If data are available, the following three questions require answers. Which species presently appear to be causing an impact? Which species have a high probability for causing more serious impacts if not treated soon? Which have a high potential to cause an impact in the near future? In addition, one needs sufficient information to be able to state the problem clearly.

3. Defining goals and objectives. Management goals regarding invasive species are often not explicitly stated. It is important to clearly define realistic goals and objectives based on the resources and the land-management objectives. Goals and objectives should not only be realistic but they should reflect the values of the organization or organizations (Coughlan and Armour 1992). Goals may range from providing high-quality forage for livestock, to maintaining wildlife habitat, to preserving an endangered species, to preserving a historic landscape. Goals may be stated in fairly general terms, but objectives must be explicitly stated and should be quantifiable. Objectives may include the establishment of thresholds of change in current conditions before a certain action may be taken.

4. Identifying and selecting alternatives. In most cases, identification of a problem and identification of possible solutions are directly linked. This, the most creative portion of decision making, usually benefits from open group participation. This is especially true in the development of control programs for invasive plants in natural or other uncultivated landscapes as little precedent has been set. Besides finding alternatives having a high probability for control of the target species, alternatives must consider issues such as acceptance and support of all parties, effects on nontarget resources, and the costs in manpower and dollars.

5. Developing and implementing a management plan. The selected treatment or management strategy should be an overall approach to prevention or control of undesired invasive plant species based upon a solid ecological understanding. Any invasive-plant management program should take the holistic view, recognizing the interactions among the pest species, natural enemies, competitors, and various disturbances. One should also consider the possible negative effects of various treatments on other desired species, the environment, and human health.

6. Assessing and evaluating the plan. As the science of control of undesired plant species in natural areas is in its infancy, it is extremely important to document community responses to control methods. This allows objective evaluation of success or failure and also guides managers in the modification of control programs.

7. Educating participants. Education at various levels is an important aspect of any successful control program. First, educating land-management workers about the basis of a control program adds their assistance in recognizing problems, unknown populations, etc. Second, education can develop public support for control programs. Finally, sharing of information among fellow scientists and land managers should increase the knowledge base upon which to build new and improved control programs.

Decision-Making Tools

Numerous decision-making tools (Table 14.1) have been developed to solve problems having the potential of causing risk to human health and life. Other processes have been developed to aid in making decisions about problems affecting the environment. The best known of the latter is the environmental impact assessment process required for certain actions by the National Environmental Protection Act (NEPA). Other tools aid in a benefit/cost analysis of alternative management actions or strategies. Some processes are quantitative; some are qualitative. Some are costly and take a long time to complete; others are inexpensive and relatively fast. Some work better for large-scale situations; some, for small-scale situations. Some are better for bringing about consensus of various interest groups or experts with different opinions. Some lend themselves better to a particular stage of the decision-making process than do others. Major decision-making tools and processes are summarized in Table 14.1.

Environmental Assessment

The National Environmental Protection Act of 1969 was intended to help managers make wise decisions about use of federal resources. If a plant-control program occurs on federal lands or is federally funded, one is required to provide environmental documentation to support the decision or management action. Thus a manager should view the process as

TABLE 14.1. Examples of analytical techniques and their possible use in the problem-solving steps.

Analytical technique	Problem-solving steps	References
Nominal group	Problem analysis; definition of goals and objectives; identification of alternatives; evaluation of alternatives; development of management plan	Delbecq et al. (1975); VanGundy (1984); Brightman (1988)
Backstep (cause and effect)	Problem identification; problem analysis	Erickson (1981); Armour and Williamson (1988)
Five W's	Problem analysis; development of management plan	VanGundy (1984)
Problem boundary	Definition of goals and objectives	Brightman (1988)
Delphi	Identification of alternatives	Delbecq et al. (1975); Crance (1987)
Brainstorming	Identification of alternatives	VanGundy (1984); Napier and Gershenfeld (1985)
Goal attainment	Definition of goals and objectives; evaluation of alternatives; monitoring and evaluation	Brightman (1988)
Ordinal ranking	Evaluation of alternatives	Erickson (1981)
Decision trees	Evaluation of alternatives	Behn and Vaupel (1982); Maguire (1986); Maguire et al. (1987)
Precedent diagrams (PERT)	Development of management plan	Erickson (1981)
Flowcharts	Development of management plan	Erickson (1981)

From Coughlan and Armour (1992).

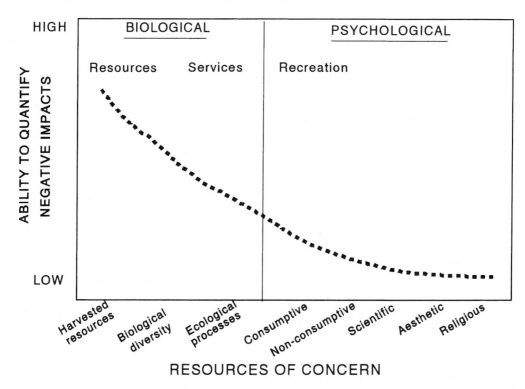

FIGURE 14.2. Extents to which effects of nonindigenous species are amenable to economic quantification. (Fom OTA [1993])

a decision-making tool rather than just as tedious paperwork. Through NEPA, reasonable management alternatives and impacts to the environment are analyzed. The process also allows public involvement. The decision makers will weigh information against concerns required by NEPA (e.g., impacts to cultural or historic resources, flood plain communities, threatened and endangered species, and wetlands) and other factors identified as significant by the manager and the general public. Also considered are which alternative actions will cause the least environmental impacts and how to best mitigate unavoidable impacts.

For plant-control programs, an environmental assessment (EA) or an environmental impact statement (EIS) will be required. The purpose of an EA or EIS is to inform management and the public about the environmental consequences of a project action, to identify potential impacts, and to formulate mitigation measures. In a plant-control program, an EA would be invoked at the stage when research demonstrates that the invasive plant is modifying resource availability or use and when a control action has been proposed. The process could be effectively used to make best predictions as to impacts of no action, a delay in action, and various action alternatives; to identify needed mitigation measures (e.g., planting desired species to replace the species removed); and to inform the public of the intended action and the rationale for its proposal.

Both EAs and EISs are written by an individual or team with the responsibility of gathering all available and pertinent information (through literature searches and communication with experts). The potential impacts and feasible mitigation measures are presented for all alternatives identified including no action. The agency proposing the project normally selects a preferred alternative based upon this analysis.

Benefit/Cost Analysis

Projecting economic costs of a proposed plant-control program requires solid scientific knowledge of present impacts and reliable predictions of future impacts that a species may cause; a good idea of costs and feasibility of control are also necessary. Due to the heavy reliance on market value in economic analysis, it is difficult to place a market price on resources such as aesthetics, biodiversity, or recreation. For example, it is much easier to calculate an economic cost of a plant pest on a cultivated crop such as corn or the loss of forage value to rangeland than it is place a cost on the modification of a natural community in a park or the loss of biodiversity in a preserve (Fig. 14.2).

Benefit/cost analysis (BCA) is a method of weighing decisions by comparing the ratio of benefits to the economic costs of a proposed action. Its best use is to compare the cost of no action to action alternatives and to prioritize among plant control projects when there are insufficient funds to do them all, or when other methods have been insufficient to drive a decision (OTA 1993). Of course, BCA is of little use if one does not know the impacts, cannot predict future impacts, or cannot quantify the cost in dollars of the plant invasion or of control measures.

Expert Opinion Techniques

Expert opinion decision-making processes are based upon the assumption that the opinions of a group of experts are better than the opinion of a single expert and that in light of incomplete information, a group of experts can better predict the probable outcome of a certain action than can analysis of current literature and theory alone (Crance 1987). Two basic techniques sharing common protocols are used: nominal group techniques (NGTs) and the Delphi method. The methods differ most significantly in that NGTs involve an interactive group structure whereas the Delphi method seeks the individual opinions of experts with no face-to-face interaction. The Delphi method has the advantage over NGTs because it allows a group of individuals to reach consensus without ever meeting (Green et al. 1990). It eliminates committee activities and helps to reduce certain psychological factors such as dominance of a discussion by one or more participants, the bandwagon effect, and unwillingness to abandon a previous opinion. Also, because opinions will never be publicly linked to a single individual, one can get a more candid response from panel experts with the Delphi method.

Both techniques require participation of three groups: the decision makers wanting the information; the group or individual designing the meeting or questionnaire, receiving and analyzing responses, and preparing a report; and the group of experts being asked to respond to specific questions. The process usually involves a minimum of four steps starting with identification of experts, orientation of experts to the problem and objectives, several iterations of responses and refinement of responses, and preparation of a report. Both techniques produce qualitative (subjective) outputs. They are normally much faster and less expensive to perform than EIA or quantitative risk-assessment techniques, discussed below.

To date, neither NGTs nor the Delphi method have been employed to make decisions concerning the managment of invasive plants or to choose among alternative management actions. These techniques have been used in corporate planning and in the field of renewable resources (Crance 1987), to identify social science aspects of natural resources management (Parker et al. 1993), and to gain expert opinion on effects of reintroduction of wolves on prey species and large-game hunting in the Greater Yellowstone Ecosystem (Lime et al. 1993). There is no obvious reason why these methods could not be effectively applied to decisions concerning assessment and management of plant invasions.

Other Analytical Tools

Numerous other decision-making tools have been developed and used to make various types of management decisions. The strengths and weaknesses of major categories of decision-making tools are summarized in Table 14.2. Few if any of these tools have been applied to rank invasive plants or to aid in the development of programs to control invasive plants. This does not mean that some or all could not be applied to various stages of the decision-making process. For example, the Delphi method has been applied to the development of habitat suitability indices for threatened and endangered aquatic organisms (Crance 1987). Similar techniques could be used to set limits of acceptable change or conditions that would trigger a management action for invasive plants. Self (1986) developed a ranking system for control of NI species at Point Reyes National Seashore, California. Other less formalized systems have been developed and employed to rank NI plants in Everglades National Park, Florida, and in Great Smoky Mountains National Park, North Carolina and Tennessee (Bratton 1982). These initial attempts inspired development, testing, and application of a generalized system for ranking NI plants. This system is presented in detail below (Hiebert and Stubbendieck 1993).

A Generalized Nonindigenous Plant Ranking System

This system was developed to support NPS policy of giving priority to management and control of NI invasive plants that cause or have the potential to cause a major impact (and are easy to control) and low priority to those that cause little impact (and are difficult to control). Based upon ecological principles and management concerns, the system allows for the ranking of species based on their present level of resource impact and on their innate ability to function as pests. This ranking can then be weighed against the feasibility or ease of control. An additional factor included in the system is the urgency of action or the

TABLE 14.2. Characteristics of various analytical techniques.

Analytical technique	Inter-disciplinary approach	Quanti-tative	Documents logic	Handles many ideas	Illusion infalli-bility	High leadership skills required	Time consuming	Complexity	Task oriented	Reduces conflict potential	Meets need for social interaction
Nominal group	H	H	H	H	M	M	M	L	H	H	L
Backstep	H	L	H	H	M	L	M	M	H	M	M
Five W's	H	L	H	H	L	L	L	L	H	M	H
Problem boundary	M	L	M	M	L	M	L	L	M	M	M
Goal attainment	H	H	H	M	H	M	M	M	H	M	M
Brainstorming	H	L	L	H	M	H	L	L	H	M	H
Delphi	H	M	H	L	M	H	H	M	H	H	L
Ordinal ranking	M	M	H	M	M	L	L	L	H	M	M
Decision trees	L	H	H	L	H	H	M	H	H	M	L
Flowcharts	M	L	H	H	M	M	M	M	H	H	L
Precedent diagrams	M	M	H	M	M	M	M	H	H	H	L

Letters refer to performance (L = low; M = medium; H = high) regarding the characteristics. From Coughlan and Armour (1992).

financial and resource cost of delay in action. At a minimum, if applied by a knowledgeable ecologist(s), the system should allow for placing species into categories: innocuous, disruptive, or having a high potential of being disruptive. This enables land managers to concentrate effort on a few species and to weigh the cost of various management alternatives, including no action. The information accumulated in applying the system can guide the development of control programs and can provide solid documentation to support management decisions.

This system has been reviewed and applied to parks of various sizes and in different biomes: Indiana Dunes National Lakeshore, Indiana (Klick et al. 1989); six small shortgrass, midgrass, and tallgrass prairies (Stubbendieck et al. 1992, Stumpf et al. 1995); Olympic National Park, Washington (Olsen et al. 1991); and the state of Minnesota (MDNR 1991). The system, or a modification thereof, is being considered by the National Biological Service to rank invasive plants on U.S. Department of the Interior lands nationwide. The system can also be modified for use on lands with other management objectives such as forage for domestic livestock or habitat for wildlife.

System Description

This ranking system uses numerical ratings, is written in outline format, and is divided into two main sections: (I) Significance of Impact and (II) Feasibility of Control or Management (Appendix 14.1). Although the actual numerical rating is not important, Stubbendieck et al. (1992) considered a species with a combined score of 50 out of 100 for significance of impact to be seriously disruptive and needing appropriate attention. The rating for significance of impact can then be weighed against the feasibility of control (the higher the score, the higher the feasibility of control) by management. To further assist management in prioritizing species for control, the cost of delay in action is considered in the urgency ranking. A detailed description follows.

Significance of Impact

A. Current Level of Impact

This section concentrates on ranking an NI species based on the present degree and extent of impact caused by the species. Element 1 addresses species response to the disturbance regimen. If the species is found only in sites recently or frequently disturbed, the species is not considered a serious threat. If the species is found in mature, undisturbed natural communities, it is considered a serious threat. Element 2 addresses how many populations (stands) are found in the park and the size of the populations. Element 3 rates a species based on its effects on ecological processes and structure of native communities. Element 4 addresses which park resources are threatened. Finally, element 5 addresses the visual impact as perceived by an ecologist.

B. Innate Ability of Species to Become a Pest

This section ranks a species based on the life history traits that preadapt it to become a problem and on its known impacts in other areas. Important life history characteristics include potential rate of increase, adaptations for long-distance dispersal, and the breadth of habitats in which the species can colonize and thrive. Element 1 is essentially a screening device. If the species cannot reproduce in the area, it most likely will not pose much of a threat. Species that probably will not reproduce in an area are horticultural species transferred from areas with different environmental conditions. Element 2 addresses how a species reproduces. The assumption is that vegetative reproduction allows an adapted ecotype to be maintained, resulting in local spread. Sexual reproduction allows for maintenance of genetic variation, production of propagules for long-distance dispersal, and the possibility of forming highly adapted gene combinations. If the species can reproduce both vegetatively and sexually, pest scoring will obviously be relatively high.

Elements 3, 4, and 5 address factors determining the intrinsic rate of population increase of a species (i.e., number of seeds produced and regularity of this seed production). Element 6 deals with dispersal ability. This factor can usually be rated based on presence or absence of special adaptations for seed or fruit dispersal, for example, wings for wind dispersal, air spaces for water dispersal, or bristles for animal dispersal. Element 7 asks if the species needs bare (disturbed) soil to germinate or if it can germinate in a relatively closed (undisturbed) community. Element 8 looks at what the species can do once it has colonized an area. Is the species able to overgrow and outcompete indigenous (IN) species? Finally, scientists should not ignore effects of the species recorded in other natural areas.

Feasibility of Control or Management

Less is known about the feasibility of managing NI plants in natural areas relative to the impacts they have on natural systems. Most research efforts concerning plant invasions have been in agricultural systems where the goal is to control all but one species while not adversely affecting the single-crop species. In natural areas, the goal is to control one or a few species while not affecting diverse assemblages of IN species. However, many factors, discussed below, will affect the funds and effort required for successful control.

A. Abundance Within Park

The straightforward logic applied here is that the larger the number of populations and the greater their size, the larger the effort required to control the species.

B. Ease of Control

This section deals not only with life history characteristics affecting the level of effort needed to control the species but also with the probability of success if unlimited funds and personnel were available. Element 1 addresses the seed bank, which directly influences the required duration of a control program. Information on seed longevity in soil is not available for many species, making this element hard to score. However, a best estimate should be made based on available information. Element 2 addresses vegetative reproduction of the species, which influences the number and kinds of treatments required to control the species and whether the underground parts of the plant must be removed; it also dictates the protocol for disposal of plant material. Element 3 addresses not only the level of effort required but also the kind(s) of control measures necessary. It follows the preferred steps of the NPS Integrated Pest Management Program in that mechanical treatment is preferred over chemical treatment. Element 4 deals with the presence or absence of propagules adjacent to the park and the probability of propagules being dispersed into the park. Consideration should be given to the ability of managers to control the species outside park boundaries through cooperative control programs.

C. Side Effects of Chemical/Mechanical Control Measures

As stated earlier, researchers must consider the effects that eradication or control measures will have on the system being restored or preserved. Will the treatment open up areas for the same species to recolonize or for invasion by other NI species with equal or greater impact? In some cases, no action may be preferable.

D. Effectiveness of Community Management

Controlling NI species through sound ecological management is by far the preferred control method. In some cases, controlling trampling by visitors, restoring historical fire regimes, or reinstating shoreline processes or natural hydrological regimens will shift the competitive edge to the desired IN species.

E. Biological Control

Biological control is ecologically feasible for many NI species. However, due to the high costs of developing well-tested biological control agents (see DeLoach, Chapter 13, this volume) it is economically feasible only for NI species causing major impacts over broad geographical areas and normally only if the species cause economic and ecological impact. Similarly, biological control is not feasible if the species to be controlled has some economic value. Abundance of closely related IN species in the area where the NI species is to be controlled also lowers the feasibility because of possible negative side effects. The responsibility for conducting long-term studies involved with selecting and screening possible control agents lies with the U.S. Department of Agriculture.

Urgency

After various species are ranked according to their levels of impact and their feasibilities of control or management, the species with the highest scores should be addressed first. The cost of delaying an action either because of finances or because of impact to the natural resources of the park is a valid criterion to use when making this final decision.

System Application

The description below is for a single land-management unit such as a park, preserve, or forest unit. Examples and suggestions follow on how to modify the method for application at other scales.

Before one can apply the system to rank species within a given unit, one must be equipped with certain baseline data. One must know and understand the management objectives. In the case of an NPS natural area, a general goal, as stated in the NPS Organic Act of 1916, is to preserve the indigenous/historic vegetation and associated processes. For an area where the main objective is maintenance of wildlife habitat, one must know the target wildlife species and the preferred forage/cover as well as the plant species that have direct or indirect negative impacts. Second, one should know which plant species grow in the area and which are a desired part of the preferred vegetation. It is also important for the person applying the system to have a good general knowledge of present and historical ecological processes occurring in the managed area. A vegetation map is helpful.

Using the above data, one can create a list of species known to occur or having the potential to occur in the area. The area should be surveyed for distribution and abundance of these species. Observations should be made of any other features that will aid in ranking the species.

The next step is to conduct a literature search for data on any undesirable invasive species found in the survey. This search may reveal information on life history traits of the various species (e.g., number of seeds per plant, mode of reproduction, longevity of seeds in seed bank) and on ecology of the species in their IN and NI habitats. The literature may also give information that will aid in ranking the species as to ease of control.

Armed with a good understanding of the land-management objectives, the ecology of the area, the distribution and abundance of the target species, and information on species ecology, one is ready to rank the species. Although the system is designed to facilitate ranking, unequivocal choices will not always be possible. Users should make their best guesses based on their own knowledge and observations and on available literature.

For surveys and rankings completed in NPS areas, results are submitted in a report including a list of NI species present and a species abstract for those considered presently to pose a significant threat to park resources. Klick et al. (1989) included management recommendations and maps showing the local distributions of NI species.

The ranking system was designed for use in small (100ha or less) to large (1000s of ha) natural areas. National Park Service areas surveyed to date range from an 80ha historic site (Homestead National Monument, Nebraska) to a 364,000ha park (Olympic National Park, Washington). The system has been modified to rank NI plants and animals for the state of Minnesota. The state was divided into natural resource districts. The approach was to have district ecologists rank NI plants and animals in their district based on their knowledge and field surveys. Results for the districts were then synthesized into a single ranking for the state through consensus-building processes. This approach could be followed and taken to a higher level by ranking species at the state or ecoregion level and synthesizing these rankings into an overall ranking for a biome (e.g., Great Plains) or the nation.

Present thinking is that the best scale for analyzing and managing invasive plants is regional. Analysis at this level adds complexity in that multiple land-management objectives ranging from raising cultivated crops for maximum financial return to preserving natural areas are usually involved. Thus, application of the ranking system should be preceded and followed by consensus-building procedures to insure that all key participants are informed. After ranking is completed managers should seek consensus on priorities and cooperation in management.

Application of Results to Management

Ranking invasive plants for management and control based on their current or potential level of impact and their feasibility of control completes only the problem-identification and problem-analysis stages of the decision-making process. Developing and implementing a management plan must follow. However, much of the information needed to develop such a plan is provided by the ranking system. Stubbendieck et al. (1992) and Stumpf et al. (1995) plotted ranking scores relative to significance of impact and feasibility of control (Fig. 14.3). This made it easier to visualize high-priority or low-priority species. It also better focused the invasive-

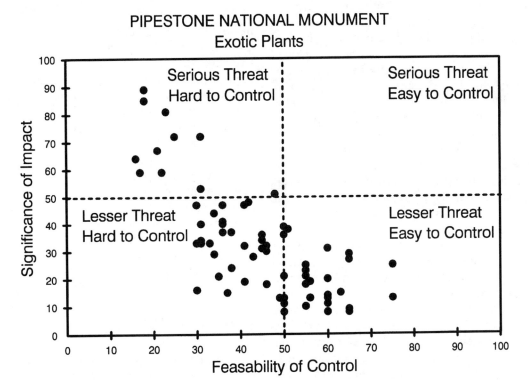

FIGURE 14.3. Significance of impact versus feasibility of control or management for nonindigenous species at Pipestone National Monument, Minnesota. (From Stubbendieck et al. [1992])

plant problem; the majority of invasive plants were categorized as innocuous. Key gaps in information will become apparent by using the graphing approach (e.g., longevity of seeds in seed bank) and will guide the direction of future research.

If ranking has been done at a regional scale and a multiagency and multidisciplinary advisory group has been established, the following steps toward implementation of a management plan are suggested.

Refine Goals and Objectives

As new information is now available, specific objectives can be refined. Because overall goals may differ among agencies, a search for objectives fulfilling multiple goals should be sought. This task can be accomplished by an advisory group or task force.

Identify and Evaluate Alternatives

A good way to start the process of identifying alternatives would be brainstorming exercises considering all possible solutions from the advisory group or task force. Creativity and original thought should be encouraged; rules should be drafted to minimize criticism of others' ideas. Brainstorming can also be employed to develop criteria upon which to weigh the relative validity of alternatives generated. Other NGTs or the Delphi technique can also be effectively employed to generate alternatives and to reduce a list to feasible ones. Other screening techniques, some as simple as listing advantages and disadvantages of each alternative, can be used to reduce the list. However, many evaluation techniques developed in economics and psychology could readily be employed to evaluate alternatives for management of invasive plants. Graham et al. (1991) developed a risk assessment technique for regional-scale application emphasizing consideration of spatial characteristics. Kangas (1994) described the use of the analytic hierarchy process (AHP) as a way to get public involvement in forest planning and to attain compromise between conflicting land-management objectives. Other techniques for evaluating alternatives are listed in Table 14.1, compared in Table 14.2, and assessed in Coughlan and Armour (1992). Of course, the NEPA process can be employed at this stage of planning.

Develop a Management Plan

The management plan brings together all information into an integrated approach for action. Because it is important to generate support from all involved parties and to identify all probable sources of funding, education is an important component throughout the process but is especially critical at this stage. The approach recommended by VanGundy (1984) is to involve all those affected throughout the planning process, to stress benefits to be gained from proposed actions, and to gain support from key opinion leaders. The seven recommended steps in developing a management plan are (1) to develop and evaluate goals and objectives, (2) to assess resources, (3) to define activities and tasks, (4) to assign responsibilities, (5) to estimate time for each activity, (6) to schedule activities, and (7) to assess implementation strengths and weaknesses. The management of invasive plants at a regional scale will most likely be an expensive and long-term proposition, and so it is important that the plan be done well and in sufficient detail so all participants know the objectives, the schedule of activities, the responsibilities for funds, personnel, and equipment, the indicators of success, and the associated monitoring requirements to measure success; all participants should agree to and approve the final plan. The plan should also identify key research that can fill gaps concerning the life histories of target species, assess the effects of invasion on regional natural resources, and test various novel control techniques.

Conclusions

1. Most natural areas and native landscapes throughout the globe have been invaded by NI plant species. Managing these invasions and preventing further ones

present a complex and expensive problem for land managers. Few successful plant-management programs exist because of lack of technology, funding, and adequate planning.

2. Based upon the presently accepted "individualistic" paradigm of community composition, which views the group of species occurring at a particular place at a particular time as the result of chance and species-specific tolerances, the effects and potential for spread of each species must be considered.

3. As a rule, only a few NI plants in a given area cause significant impacts to natural resources. Thus some type of analytical tool is needed to identify which species are innocuous, which have a high potential to cause impacts in the future, and which are currently causing significant impacts.

4. Management decisions are usually made in one of three ways: judgment, precedent, or formal analysis. Decisions based on judgment alone are rarely based on defined criteria; they provide no documentation of the decision-making process used, are subject to personal biases, and give no assurance that a full array of factors was considered. Decisions based on precedent may be easier to defend but are not responsive to variation in ecological interactions over time and space and may hinder application of innovative solutions. Decisions based on formal analysis alleviate these problems by providing a logical process for consideration of a full range of factors.

5. Logical steps in the decision-and-planning process are identifying the problem, analyzing the problem, identifying and evaluating alternatives, developing a management plan, and implementing, monitoring, and evaluating the plan.

6. Education should be a key component throughout the decision-and-planning process. Educating land-management staff creates allies to sell the program, to identify problems in the field, to locate target species populations, and to assist in program implementation. Education of key participants in the region of concern fosters partnerships and financial and logistical support.

7. Numerous decision-making tools have been developed in other fields to aid in making sound decisions, to foster consensus among persons with various interests and values, and to provide rationale for decisions made. Few such tools have been employed to make decisions concerning management of invasive plants. However, most could be adapted for this use.

8. Examples of decision-making tools that most land managers are familiar with and that can be applied to management of invasive plants are the NEPA process, benefit-cost analysis, and various expert opinion techniques.

9. A system has been developed and tested by Hiebert and Stubbendieck (1993) to prioritize NI plants for management in natural areas. The system ranks species based on their present level of impact and their innate potential to become pests. The system separately rates the feasibility of control of each species to allow managers to weigh the threat or impact against the cost and feasibility of control.

10. Application of the system allows managers to classify species as being innocuous, as possessing a high potential to be pests, or as currently causing significant impacts. It also identifies gaps in existing key information. This allows the manager to concentrate further efforts on a small fraction of the total number of NI plant species.

11. The ranking system is used as a tool in the problem-identification and problem-analysis stages of the decision-making process. It also identifies key research needs and provides basic information to identify and evaluate alternatives.

12. A final management plan integrates many different types of information. Policy makers now think that analysis and management of plant invasions are best addressed at the regional scale. Therefore, all key parties in the defined region should be active participants in planning, implementing, and evaluating of an invasive-plant-management program.

APPENDIX 14.1. A system for ranking nonindigenous plants.

I. SIGNIFICANCE OF IMPACT
   A. Current Level of Impact
      1. Distribution relative to disturbance regimen
         a. Found only within sites disturbed within the last 3 years or sites regularly disturbed   0
         b. Found in sites disturbed within the last 10 years   1
         c. Found in mid-successional sites disturbed 11 to 50 years before present (BP)   2
         d. Found in late-successional sites disturbed 51 to 100 years BP   5
         e. Found in high quality natural areas with no known major disturbance for 100 years   10
      2. Abundance
         a. Number of populations (stands)
            1) few; scattered (<6)   1
            2) intermediate number; patchy (6–10)   3
            3) several; widespread and dense (>10)   5
         b. areal extent of population
            1) <5 hectares   1
            2) 5–10 hectares   2
            3) 11–50 hectares   3
            4) >50 hectares   5
      3. Effect of natural processes/character
         a. Plant species having little or no effect   0
         b. Delays establishment of native species in disturbed sites up to 10 years   3
         c. Long-term (more than 10 years) modification or retardation of succession   7
         d. Invades and modifies existing native communities   10
         e. Invades and replaces native communities   15
      4. Significance of threat
         a. Threat to secondary resources negligible   0
         b. Threat to areas' secondary (successional) resources   2
         c. Endangerment to areas' secondary (successional) resources   4
         d. Threat to areas' primary resources   8
         e. Endangerment to areas' primary resources   10
      5. Level of visual impact to an ecologist
         a. Little or no visual impact on landscape   0
         b. Minor visual impact on natural landscape   2
         c. Significant visual impact on natural landscape   4
         d. Major visual impact on natural landscape   5
                                          Total possible = 50
   B. Innate Ability of Species to be a Pest
      1. Ability of complete reproductive cycle in area of concern
         a. Not observed to complete reproductive cycle   0
         b. Observed to complete reproductive cycle   5
      2. Mode of reproduction
         a. Reproduces almost entirely by vegetative means   1
         b. Reproduces only by seeds   3

APPENDIX 14.1. *Continued*

c. Reproduces vegetatively and by seed	5
3. Vegetative reproduction	
a. No vegetative reproduction	0
b. Vegetative reproduction rate maintains population	1
c. Vegetative reproduction rate results in moderate increase in population size	3
d. Vegetative reproduction rate results in rapid increase in population size	5
4. Frequency of sexual reproduction for mature plant	
a. Almost never reproduces sexually in area	0
b. Once every 5 or more years	1
c. Every other year	3
d. One or more times a year	5
5. Number of seeds per plant	
a. Few (0–10)	1
b. Moderate (11–1000)	3
c. Many-seeded (>1000)	5
6. Dispersal ability	
a. Little potential for long-distance dispersal	0
b. Great potential for long-distance dispersal	5
7. Germination requirements	
a. Requires open soil and disturbance to germinate	0
b. Can germinate in vegetated areas but in a narrow range or in special conditions	3
c. Can germinate in existing vegetation in a wide range of conditions	5
8. Competitive ability	
a. Poor competitor for limiting factors	0
b. Moderately competitive for limiting factors	3
c. Highly competitive for limiting factors	5
9. Known level of impact in natural areas	
a. Not known to cause impacts in any other natural area	0
b. Known to cause impacts in natural areas, but in other habitats and different climate zones	1
c. Known to cause low impact in natural areas in similar habitats and climate zones	3
d. Known to cause moderate impact in natural areas in similar habitats and climate zones	5
e. Known to cause high impact in natural areas in similar habitats and climate zones	10
Total possible = 50	
II. FEASIBILITY OF CONTROL	
A. Abundance Within Park	
1. Number of populations (stands)	
a. Several; widespread and dense	1
b. Intermediate number; patchy	3
c. Few; scattered	5
2. Areal extent of populations	
a. >50 hectares	1

APPENDIX 14.1. *Continued*

b. 11–50 hectares	2
c. 5–10 hectares	3
d. <5 hectares	5

B. Ease of Control
  1. Seed banks
    a. Seeds remain viable in the soil for at least 3 years   0
    b. Seeds remain viable in the soil for 2 to 3 years   5
    c. Seeds viable in the soil for 1 year or less   15
  2. Vegetative regeneration
    a. Any plant part is a viable propagule   0
    b. Sprouts from roots and/or stumps   5
    c. No resprouting following removal of aboveground growth   10
  3. Level of effort required
    a. Repeated chemical and/or mechanical control measures required   1
    b. One or two chemical and/or mechanical treatments required   5
    c. Can be controlled with one chemical treatment   10
    d. Effective control can be achieved with mechanical treatment   15
  4. Abundance and proximity of propagules near park
    a. Many sources of propagules near park   0
    b. Few sources of propagules near park, but these are readily dispersed   5
    c. Few sources of propagules near park, but these are not readily dispersed   10
    d. No sources of propagules are in close proximity   15

C. Side-Effects of Chemical/Mechanical Control Measures
  1. Control measures will cause major impacts to community   0
  2. Control measures will cause moderate impacts to community   5
  3. Control measures will have little or no impact on community   15

D. Effectiveness of Community Management
  1. Routine management of community and/or restoration or preservation practices (e.g., prescribed burning, flooding, controlled disturbance) effectively controls target species   10
  2. Cultural techniques (burning, flooding) can be used to control target species   5
  3. Neither 1 nor 2 is effective   0

E. Biological Control
  1. Biological control not feasible (not practical, possible, or probable)   0
  2. Potential may exist for biological control   5
  3. Biological control feasible   10

Total possible = 100

III. URGENCY
  1. Delay in action will result in large increase in effort required for successful control.   High
  2. Delay in action will result in moderate increase in effort required for successful control.   Medium
  3. Delay in action will result in little increase in effort required for successful control.   Low

Section IV
Regulation and Advocacy

15
Prevention of Invasive Plant Introductions on National and Local Levels

Sarah E. Reichard

> If you can look into the seeds of time, and say which grain will grow and which will not, speak then to me . . .
> —*Macbeth*, Act I, iii

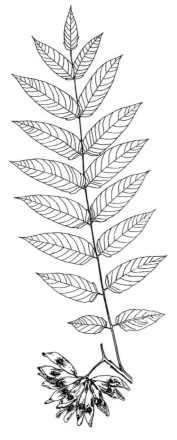

Tree-of-heaven
(Ailanthus altissima)

Banquo's metaphorical plea to the witches of Shakespeare's play is echoed literally by managers of natural areas today. Invasive nonindigenous (NI) plants have an increasing presence in natural areas worldwide; it appears that there are virtually no nature reserves in the world that are immune from this problem (Usher 1988). As resource managers are faced with dwindling resources allocated to an increasing number of management problems, the ability to predict which introduced plant will become invasive ("which grain will grow and which will not") would be of great value. We are beginning to develop such prescient ability.

Just as preventive medicine is gaining recognition as an important part of human health care, we must realize that sensible preventive steps are an important component of environmental health care. For instance, we can use a number of ecological and life history traits of species, such as their taxonomy, habitat breadth, reproductive biology, and success in past introductions, to make educated estimates of which plant species will invade. Such information can be used to develop a more restrictive policy on introduction of species into the United States, thereby slowing the rate of new introductions, and can also suggest a "triage" approach, allowing managers to prioritize which species to control. Identifying which species to target at an early stage of invasion may be critical to effective control (Hobbs and Humphries 1995).

This chapter explores accidental and, especially, intentional plant introductions. I discuss current laws regulating intentional introductions in the United States, the ways these laws might be changed, and some of information we will need to enact reasonable changes. Finally, I explore some immediate ways to reduce intentional and accidental invasions into natural areas on the level of the reserve rather than of the continent.

Modes of Species Entry

Pollen and other plant remains have been found in archeological sites dating to some of the earliest civilizations (Bottema 1995; Miller 1995). Although many plants may have been collected near the sites by early peoples, it is apparent that some were also collected from distant locales and were traded or carried for various uses such as food, technology, and medicine (Fritz 1994). With the advent of global travel, the human-mediated transport of plants, both accidental and intentional, outside their natural ranges is occurring at a rate orders of magnitude larger than natural rates of movement. Most of these introductions fail to become self-sustaining populations, but many do establish in their new habitats.

Many invasive NI species, both herbaceous and woody, have been introduced accidentally by humans, their domesticated animals, or their machinery (Baker 1986; Muenscher 1955; OTA 1993). Agricultural practices were responsible for some of the earliest introductions, especially via weed-contaminated crop seed (Baker 1986; Mack 1991). Domestic animals may carry seeds in their fur (Muhlenbach 1979), they may ingest seeds that remain viable in manure (Baker 1986; Kruger et al. 1986), or seeds may be brought into natural areas with straw used for pack animals (Kruger et al. 1986). Many species were probably introduced in ship's ballast (Baker 1986; Kruger et al. 1986); soil, sand, and rocks incidentally containing seeds were loaded onto ships in Europe and then exchanged for cargo in North America. Studies of these "ballast floras" in the late 1800s indicated that at least 386 species, mostly European in origin, were associated with ballast grounds in New York and New Jersey alone (Muhlenbach 1979). Changes in laws regarding discharge of ballast materials have eliminated this avenue of introduction. An accidentally introduced species may have come from a region with a very different climate than the one to which it has been dispersed; its resource requirements may not be met and its chances of survival thus may be low.

Many invasive NI species have resulted from intentional introductions for landscape purposes (Baker 1986; Foy et al. 1983; Luken and Thieret 1995; Reichard n.d.), medicine (Foy et al. 1983), forage (Foy et al. 1983), and timber plantations (Kruger et al. 1986). Some plant species, such as multiflora rose (*Rosa multiflora*), were actually introduced with the intention that they would spread to control soil erosion. Regardless of whether a species is introduced in the hope that it will spread beyond the point of planting or that it will persist only at the point of planting, the resource requirements are considered and the probability of persistence is high. The most frequent reason for original introductions in North America, especially of woody plants, is for landscaping (Fig. 15.1). Botanical gardens, nurseries, and individuals commonly import new species to meet demands of the public for new landscaping materials. The horticulture industry is an important one, with gross sales of $5.3 billion in 1992 (USDCES 1994). Throughout its history this industry has constantly tried to cope with waxing and waning interest in specific plants. Horticulture professionals, however, are only a part of the total introduction picture associated with the industry. Plant enthusiasts often bring in plants, even those declared noxious by the federal government; 82% of the seeds of noxious species intercepted at points of entry into the

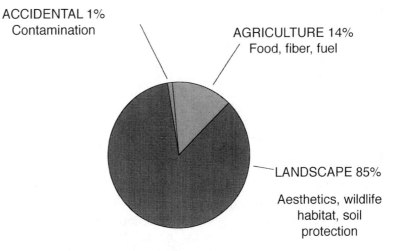

FIGURE 15.1. Reasons for original introduction of woody plants.

United States were in personal baggage (OTA 1993).

Current Laws

Preventing introductions of invasive plant species is the most powerful tool we have to manage invasions. Regulations regarding allowable levels of weed contaminants in crop seed and changes in ballast-discharge protocols have reduced many accidental introductions. Most woody invasive plants, however, are intentionally introduced for landscape and agricultural purposes (Fig. 15.1), and thus, with better knowledge about invasive species and more inclusive laws, invasions by these species might have been prevented.

In the United States, the 1974 Federal Noxious Weed Act provides the main authority for restriction of "weeds," currently prohibiting import of 94 listed species. The act has been less than successful in preventing introductions of invasive plant species for several reasons, primarily because it has been applied mostly to agricultural pest plants. It has also been applied to too few species; at least 750 species meeting the act's definition of "noxious" still remain unlisted (OTA 1993). The purpose of listing is to restrict movement of a known problem species, with action taken only after a species is shown to present a significant risk. The act does not require evaluation or regulation of "new" intentionally introduced species at the time of introduction. Proposed changes in U.S. Department of Agriculture Animal and Plant Health Inspection Services (USDA-APHIS) policy would mandate screening for early detection and assessment of incipient infestations, not only for weeds of commodity-based systems but those of natural ecosystems as well (Westbrooks n.d.). The policy has yet to be implemented.

There are four possible policy strategies that could be used at the United States border to screen out potential problem species (Ruesink et al. 1995): (1) all species can be admitted unless proven to be invasive and thus specifically prohibited (a "dirty list" approach); (2) all species can be prohibited unless known to be noninvasive and listed as such (a "clean list" approach); (3) each species can be tested for invasiveness before a decision is made regarding its entry; or (4) an informed estimate of invasive potential can be made based on available information about other invasive species. Currently, the United States, in accordance with the Federal Noxious Weed Act, allows all plants into the country as long as they are not on a brief list of "dirty" species that harbor agricultural pests or are known weeds. This is consistent with strategy 1 above. Because there are very few restrictions limiting what can be introduced, botanical gardens, nurseries, agricultural organizations, and individuals can import plants and exchange seeds at will.

The dirty list approach is commonly used to prevent interstate transport of harmful species, mostly fishes and wildlife, but there is a growing realization that more systematic and formal methods need to be developed (OTA 1993). Changing to a "clean" (i.e., allowed) list approach by which all species would be banned from importation unless they could be proven to be pest-free and to lack weed potential is one such method; species could be placed on the clean list and no further reviews would be needed for subsequent introduction of those species. The clean list approach is analogous to the U.S. Food and Drug Administration's regulatory system for approving drugs for human use. A previous attempt, in 1973, by the U.S. Fish and Wildlife Service (USFWS) to use this approach for NI fishes and wildlife was withdrawn after the USFWS received more than 5500 negative comments (OTA 1993).

The clean list approach shares, with the third strategy, the need to hold the material in quarantine for an extended period of time, a politically and legally unfeasible and expensive solution for goods in trade. Holding the species in quarantine to evaluate invasiveness may reveal little anyway, unless it is held for decades. Many invasive species have a considerable "lag period" between introduction and commencement of invasion (Hobbs and Humphries 1995; Scott and Panetta 1993).

This leaves the fourth strategy, an approach based on evaluation of previous introductions, to guide the screening process. Approval for a species would be withheld until it was determined to have a low or high potential for invasiveness and could be placed on a clean or dirty list, with some questionable species perhaps held for further determination of invasive qualities. This would provide a more conservative approach and therefore more protection for the environment than under current laws and would be fair to those wishing to introduce species by allowing at least some species into the country quickly. Costs of such a program of evaluation could be borne by those seeking to introduce a species, in exchange for distribution rights (a sort of patent) for a specified time period. The costs and time necessary for evaluation might also serve to encourage selectivity when choosing species to import.

Developing Predictive Methods

Model Types

The ability to recognize potential invaders of natural areas depends on identification of traits that are positively correlated with invasion success and that can be measured without the actual invasion occurring. Such predictive traits may be attributes of species (e.g., polyploidy) or may be more broadly defined ecological attributes assessing habitats, ranges, and past histories of species and of related species. Recently, there have been several attempts to develop models useful in predicting which plants are likely to invade natural areas. Six such models are briefly discussed below; the characters used are discussed below under "Possible Predictors of Invasiveness."

The oldest approach for predicting invasive success is the compilation of lists of characters of "weedy" species (Baker 1965, 1974; Roy 1990). The lists generally have come from observations of colonizing species (e.g., agricultural weeds establishing on bare soil). One problem with trying to use lists of traits is that any given invasive species will lack some of the traits, confounding predictions based on them. Prediction should ideally consider combinations of species' traits as well as ecological and biogeographical information because such information is generally easier to acquire.

A more refined version of character lists provides decision paths based on characteristics of previous invaders and/or ecological factors. One such path model, produced to evaluate plant introductions into Australia (Panetta 1993), used a combination of life history characters, bioclimatic information, taxonomic relations, and life forms to screen species (Table 15.1). It is easy to use and the

TABLE 15.1. Summary of traits used in six predictive models.[1]

	Pinus species in South Africa (Richardson et al. 1990)	Annual species in Great Britain (Perrins et al. 1992b)	Introductions to Australia (Panetta 1993)	South African species in Australia (Scott and Panetta 1993)	Pinus species (Rejmánek and Richardson n.d.)	Woody species in North America (Reichard n.d.)
Variables						
No. of species used	60	49	N/A	242	24	349
Correctly predicted	N/A	71%–78%	N/A	N/A	68%–100%	76%–86%
Ecological attributes						
Geographic origin		−				+
Bioclimatic match			+			
Climatic/latitude range				+		*
Problem in native range				+		
Native congeners				−		
Congener a problem in native range				+		
Congener a problem in introduced range			+			
Invades elsewhere			+	+		+
Fire tolerance	+				−	
Degree of serotiny	+					
Mycorrhizal associations		−				
Species attributes						
Plant height		−			−	
Plant longevity						
Leaf longevity						+
Leaf area		−				
Vegetative reproduction			+			+
Reproductive system						+
Agamospermy						*
Self-compatibility		−				*
Polyploidy						−
Juvenile period/growth rate	+	+			+	+
Flowering period		+				−
Pollination mechanism		−				
Fruiting period		−				+
Seed crop variability/mast year	+				+	
Number of seeds produced		−				*
Seed bank type/longevity	+	+				
Seed dispersal mechanism			+		*	−
Seed mass/size	+	+			+	−
Seed-wing index	+				−	
Seed germination requirements		−				+
Percent germination					−	−

[1] N/A indicates the variable is not applicable to the specific model; + indicates a trait with predictive value; − indicates a trait with no predictive value; * indicates discussion of the trait but no formal analysis.

information needed is relatively easy to find. With this path model some species would be rejected as a poor risk, some would be accepted, and some would need further evaluation.

Another approach for developing a predictive model is to define "invasion windows" of specific ecosystems. In susceptible ecosystems these windows would generally occur after disturbances such as fires or floods. Characteristics allowing a species to exploit such a window are then defined, and future invaders with these characteristics are identified. This approach was used in the South African fynbos (Richardson et al. 1990), an ecosystem characterized by high-intensity fires at 10 to 40 year intervals. Five functional groups were identified using the traits of *Pinus* species (Table 15.1); the groupings reflect various levels of fire resistance and stress tolerance. This approach appears to work well for ecosystems that have easily defined windows of opportunity for invasion.

Equally instructive for examining the traits of successful invaders is characterization of species that do not invade or that are very minor invaders. While some characters generally make sense for successful invaders (e.g., production of abundant fruit) it is difficult to determine which characters contribute to invasive ability and which are reflective of the source pool of plants (e.g., landscape plants may be bred to produce much colorful fruit, but that trait may not actually affect invasive success). Thus, in this instance one needs to compare invaders with a peer group of noninvasive species to see if they actually do produce more fruit. Such comparisons generally require a statistical component.

Focusing on 242 South African species already introduced into Australia and showing various levels of invasiveness, Scott and Panetta (1993) divided the species according to weediness level, examined several characteristics (Table 15.1), and performed multiple logistic regressions to identify predictive variables. By examining the status of agricultural weeds in South Africa, they found that they could rather reliably predict which of these would become weeds in Australian agriculture; they did not find suitable predictors for plants that invade systems where commodity production is not the management goal.

Annual plants in Great Britain were evaluated in an effort to derive weed potential from plant traits (Perrins et al. 1992b). Forty-nine species were divided into "weedy" and "non-weedy" groups, scored for the characteristics (Table 15.1), and then submitted for statistical analyses including principal coordinates analysis, hierarchical clustering, stepwise discriminant analysis, histograms, and comparison of seven cases of weedy and non-weedy congeners. Only the discriminant analysis and congener comparisons identified predictive traits. Discriminant analysis finds a linear combination of characters that maximizes variation between the groups and minimizes variation within the groups. The character combinations can be used to predict invasive success. Three different species sets identified different predictive characters in discriminant analysis, but four characters—maximum seed width, seed bank type, relative growth rate, and month of flowering onset—were found in all three sets, with correct predictions in cross-validation ranging from 71% to 78%. In the congener comparison the weedy species flowered longer in six of the seven cases, with no other differences.

Rejmánek and Richardson (1996) used discriminant analysis of 10 life history traits (Table 15.1) of 24 pine species—12 known to be invasive and 12 known not to spread where cultivated—to find predictive traits. They found that three traits—mean seed mass, minimum juvenile period, and mean interval between seed crops—were adequate to maximize differences between the groups. In repeated cross-validations they correctly predicted 68% to 100% of the species as invasive or not invasive. They tested the results on angiosperms with good predictive results.

Finally, I analyzed attributes of 349 woody plant species (235 invaders and 114 non-invaders) introduced into North America over 70 years ago (Reichard n.d.). Two statistical techniques were used to predict invasive success in the groups: discriminant analysis and classification and regression trees (CART), which, like discriminant analysis, uses data structure to classify species as invasive or noninvasive. Unlike discriminant analysis, however, CART evaluates the characters individually rather than linearly to create a hierarchical binary tree (Brieman et al. 1984). Thus, the single best distinguishing character is selected, then the next, and so on. In both analyses different subsets of randomly selected species were used to create the models, and then the models were validated on the remaining species. Both methods had high predictive value (76% to 86%) using essentially the same characters (Table 15.1). From the results of these statistical models I then created a simple-to-use decision tree based on ecological, geographical, and species information. This approach also had high predictive value, admitting no invaders outright, denying admission to 89%, and recommending that the remaining 11% be examined further for invasive potential before admission.

Possible Predictors of Invasiveness

Table 15.1 provides a list of the characters each model used to predict invasive success as well as characters evaluated that were universally not able to identify invasive NI species. Some of the latter are believed to be traits

Geographic Origin and Bioclimatic Matches

Correspondence of climatic conditions such as precipitation, timing of precipitation, and minimum and maximum temperature between native and introduced ranges has been suggested as a way to assess which species may be invasive in a given range. Such climate matching has long been used to select species suitable for agriculture, horticulture, and forestry. Homeoclimatic matching may be of greatest value for predicting invasiveness in smaller, more uniform land masses such as islands, but in a large area such as North America, which encompasses virtually every possible climate type, it may be of less utility for prescreening. Climatic matching analyses may be best used to provide an estimate of potential range for a species already invading or predicted by other methods to invade, although factors such as soils, land use, and interactions with existing species must be simultaneously considered (Panetta and Mitchell 1991).

Climatic and Latitudinal Range

A wide latitudinal range, indicating that a species has a broad adaptability to a number of ecological conditions, may be of greater significance in predicting invasive success than homeoclimatic matching. Such a range may also mean that a species is more likely to be accidentally introduced. South African species invasive in Australia have been shown to have wider ranges of latitude in their home area than noninvaders (Scott and Panetta 1993). Woody invasive plants in southeastern Australia are also derived from a greater variety of climates than are noninvasive plants (Mulvaney 1991). In Australia, a comparison of three invasive NI species of *Echium* found that the weediest of the species had much more widespread native ranges than the less invasive species (Forcella et al. 1986). Woody invaders in North America have on average 6° more latitude in their native ranges than do noninvaders, a statistically significant amount (Reichard n.d.). An analysis of brome grasses (*Bromus*) indicated a significant correlation between latitudinal range and the number of continents invaded (Roy et al. 1991). Rejmánek (1995) concluded that species in Asteraceae and Poaceae native to Europe and northern Africa and invasive in North America had latitudinal ranges about 10° greater than those not invasive there. Although such ranges appear to be useful predictors of invasive potential, a serious limitation is that reliable information about them is difficult to find.

Taxonomic Relationships

Taxonomic relationships may also be useful in estimating invasive potential of a species. Panetta (1993) used such relationships as a part of his scoring system for assessing proposed plant introductions into Australia, finding that 87% of the noxious plant species of Australia have weedy congeners. Daehler and Strong (1993) also claimed that "it is still reasonable to predict that an area successfully invaded by species A will be susceptible to invasion by closely related species B."

Certainly, some groups do tend to invade repeatedly. Species of Poaceae account for 10 of the 18 worst weeds in the world (Holm et al. 1979), and Asteraceae, Fabaceae, Rosaceae, and Solanaceae are responsible for a large proportion of the world's weed flora (Holm et al. 1979). Many introduced weeds in California are members of Asteraceae, Brassicaceae, and Poaceae (Stebbins 1965). These families are among the most species-rich in the world; their ubiquity as sources of weeds may in fact be less a matter of success than of proportion. Still, the Ericaceae are also a species-rich family but are noticeably absent from the invasive flora of the world. And in North America, although 53 families are represented in the invasive woody plant flora, seven families are responsible for 66% of the 76 most damaging species (Reichard n.d.). Two of these, Rosaceae and Fabaceae (*sensu lato*), are species-rich, as are, to a lesser extent,

Myrtaceae, but the other four families—Caprifoliaceae, Oleaceae, Salicaceae, and Tamaricaceae—are not. Similarly, Scott and Panetta (1993) decided that the large number of introduced species of Asteraceae in Australia was the result of the size of the family, and that, in fact, other families considered highly invasive in Australia made only an average contribution to the introduced flora.

In general, what all the above families have in common is high reproductive success, including high seed production, effective seed dispersal, quick seed germination, and ability to reproduce vegetatively. Thus, knowing that a species is in one of these "weedy" families may be a shorthand way of recognizing invasive characteristics. A species in one of these families or in a genus that has one or more highly invasive members may not necessarily be a priority for exclusion or control but should certainly be flagged for future monitoring.

Historical Performance

The past performance of a species often provides a useful indication of its future performance. In my survey of 235 invasive and 114 noninvasive woody NI species in North America, I found that 54% of the successful invaders were also invasive in other places; only 15% of the species not invasive in North America were invasive in the native range. The predictive power of such data has been noted for Australian plants (Panetta 1993; Scott and Panetta 1993), for Hawaiian passerine birds (Moulton and Pimm 1986), and for many vertebrates (Ehrlich 1989) and insects (Crawley 1987). The tendency to invade repeatedly is likely due to various species characteristics that facilitate invasion; like taxonomic relationship, it may provide a shortcut to predicting invasiveness.

Information about other parts of the world where a species may invade can be derived by consulting floras of areas with climates similar to the one in which the invasion is occurring, by checking computerized abstract services (e.g., BIOSIS, AGRICOLA, or TREECD) for abstracts that mention the species as escaping, or by checking compendia of "weedy" species (e.g., Holm et al. 1979). The advent of worldwide communications such as the Internet, however, may allow the fastest and most reliable way to check the status of a species around the world. One bulletin board, the World Weeds Database, maintained at Oxford University, has already proved a worthy forum for querying about a species' invasion and spread in other parts of the world. Current and future technology with the World Wide Web will make an accessible and complete database of invasive plant species a reality in the near future.

Fire Tolerance/Degree of Serotiny

Many ecosystems require regular fires to maintain communities. In such ecosystems, the ability to tolerate fires by means of thick bark and accumulations of seed banks, especially in the canopy (e.g., serotinous cones in pines), are positive attributes for an invasive NI species (Richardson et al. 1990).

Plant Height and Leaf Characteristics

Plant height does not appear to provide predictive power for invasive success in either annual species (Perrins et al. 1992b) or *Pinus* species (Rejmánek and Richardson 1996); neither does plant longevity (Rejmánek and Richardson 1996). Leaf longevity (e.g., evergreen versus deciduous) has some predictive power for North American invasive woody plants, mostly for specific bioclimatic regions of the continent (Reichard n.d.). Significantly more invasive NI species in North America are semievergreen (having photosynthetic activity in stem tissue or abscission of some of the leaves), but the numbers are still low, with 14% of the invasive species and 4% of the noninvasive species having this trait. Leaf area does not have predictive value for annual species (Perrins et al. 1992b).

Vegetative Reproduction

Methods of vegetative reproduction, such as root sprouts or soil layering, also can quickly increase a population and may allow faster recovery following a disturbance. A population that increases through vegetative reproduction is simultaneously producing new plantlets capable of sexual reproduction. In North America, 44% of woody invaders reproduce vegetatively, but only 25% of noninvasive woody species do so (Reichard n.d.). Panetta (1993) also used vegetative reproduction as a criterion for evaluating NI species. In southeastern Australia, significantly more invasive woody species reproduce vegetatively than do noninvasive ones (Mulvaney 1991).

Reproductive System

The type of reproductive system affects the ability of a single colonizing individual to result in an invasion or may determine success in a variable environment. Perfect flowers (male and female structures in the same flower), if the plant is self-compatible (see below), have the greatest reproductive flexibility; plants with such flowers would appear to be better adapted to invade when in isolated populations. Dioecious species (male and female flowers on separate plants) may have some advantages because the male and female plants have different physiological and seasonal requirements (Freeman et al. 1976) and thus may have greater survival in a patchy environment. In woody angiosperms I found no differences in reproductive system for invasive and noninvasive NI species (Reichard n.d.), but self-compatibility and agamospermy (producing seed without fertilization) were significantly more common in the invasive species. The latter two traits were not included in the models, however, because the information was incomplete for too many species.

Polyploidy

Polyploidy, the condition of having more than two sets of chromosomes in the nucleus of a somatic cell, may influence invasive ability by increasing self-compatibility, seed number, and seedling vigor and other traits (Levin 1980). Many of the world's worst weeds are polyploids (Barrett 1982), but a number of these species are in Asteraceae and Poaceae, families noted for their high levels of polyploidy (Heywood 1993). Is polyploidy facilitating invasion, or is some other factor in these families doing so, with polyploidy merely an artifact? In North American woody plants I found no difference in the proportion of polyploids among invaders and noninvaders (Reichard n.d.). Annual "weedy" species also did not have higher chromosome numbers (indicating polyploidy) than non-weedy annuals (Perrins et al. 1992b). In southeastern Australia, however, there apparently was a correlation between increasing chromosome number and the probability of invasiveness (Mulvaney 1991).

Juvenile Period/Growth rate

A short period between seed germination and the onset of reproduction is associated with many invasive species; Baker (1974) suggested that his "ideal" weed would have rapid growth through the vegetative stage to reproduction. A short juvenile period means a fast population growth rate; it also may allow a colonist to begin reproducing before it can be detected. Annuals, often with a juvenile period of only days, make up 22% of the herbaceous species invading natural areas in North America; biennials, with a 1-year juvenile period, are another 12% (derived from an unpublished list of the National Coalition of Exotic Pest Plant Councils). Growth rate, which can be linked to juvenile period, was a predictor of "weedy" status in annuals in Great Britain (Perrins et al. 1992b). Even among woody perennials, however, short juvenile periods are associated with invasive ability. In North America, for instance, woody invaders begin to reproduce on average about 3 years before noninvasive species (Reichard n.d.). Invasive pines have shorter juvenile periods than noninvasive ones (Rejmánek and Richardson 1996; Richardson et al. 1990). A short juvenile period can be used to evaluate a species' reproductive ability before introduction and also to prioritize control within natural areas. A species with a short juvenile period will be able to increase populations quickly and thus must be quickly controlled.

Flowering and Fruiting Periods

Long flowering periods may mean a greater accessibility to pollinators and a greater chance of seed set. Perrins et al. (1992b) found that the month of onset of flowering consistently distinguished weedy from nonweedy annual species and that the length of the period of flowering was longer in weedy species compared to their non-weedy congeners. In woody plants the flowering period is nearly 2 months longer in invasive NI species than in noninvasive (Reichard n.d.). Nevertheless, this character was not selected by either the discriminant analysis function or the classification and regression trees as useful in distinguishing invaders from noninvaders (Reichard n.d.). A long fruiting period, however, allowing greater opportunity for seed dispersal, was employed by the models. Baker (1974) stated that a characteristic of the ideal weed is continuous seed production for as long as growing conditions permit. The length of flowering and fruiting periods is strongly correlated (Reichard n.d.), so it is difficult to tell which may be more critical for invasive ability. It is likely that a longer fruiting season would be typical of an invasive species.

Seed Crop Production and Variability

High fecundity is a characteristic of colonizers, including many NI species, and was listed by Baker (1974) as a character ideal for weeds. Large seed crops ensure that at least some of the progeny will establish each year. This trait, however, is exceedingly difficult and time consuming to measure; although knowing the size of seed crops may elucidate the biology of invaders, this trait is limited in value for prediction policy. Some species vary also in their annual seed production; Richardson et al. (1990) and Rejmánek and Richardson (1996) found that species with consistently large seed crops are more likely to become invasive than those that infrequently have large crops.

Seeds and Their Dispersal Mechanisms

An effective method of seed dispersal is critical to successful invasion. Rapidly dispersed species ultimately have larger final distributions (Forcella 1985a). Although a number of invasive NI species are vertebrate-dispersed (Rejmánek 1995), especially in tropical and subtropical areas, many highly successful invasive NI species are wind dispersed, for example, Australian pine (*Casuarina*) and tree-of-heaven (*Ailanthus altissima*). Among North American woody NI species, including the 76 considered to be natural area pests, there was no clear tendency for invasive plants to be dispersed by biotic agents versus abiotic agents (Reichard n.d.). A similar study of woody NI species in southeastern Australia also found no differences in dispersal mechanisms between invaders and noninvaders (Mulvaney 1991).

Seed size or mass may affect both dispersal and seedling vigor. Small seeds are lighter, which may mean increased distance in wind-dispersed species. Species with smaller seeds may also be able to produce more seeds per plant (discussed above). Large seeds have more food-storage tissue, which may provide greater resources for seedlings. I found no differences in seed sizes of North American invasive and noninvasive woody NI species, but Rejmánek and Richardson (1996) and Richardson et al. (1990) found seed mass to be a useful predictor of invasive ability in pines.

Seed Germination and Longevity

Many plant species germinate in response to a precise set of conditions that overcome dormancy. Baker (1974) suggested that an ideal weed would have germination requirements that could be fulfilled in many environments. (If seeds lack well-developed dormancy mechanisms, they may germinate in a greater variety of places and situations.) In North America, 51% of invasive NI species do not require any seed pretreatment such as cold-chilling or scarification, but only 30% of noninvasive species lack dormancy mechanisms, a significant difference (Reichard n.d.). The most "weedy" species of *Echium* in Australia germinated far more quickly than two less weedy species (Forcella et al. 1986). No differences between invasive and noninvasive species have been found in the percent of seeds germinating in either *Pinus* (Rejmánek and Richardson 1996) or woody plants in general (Reichard n.d.).

Seed banking effectively increases crop size by allowing seeds to accumulate in the soil or canopy until a propitious time for germination occurs. Serotinous cones, which represent a canopy seed bank, are more common in invasive than in noninvasive pines of the South African fynbos (Richardson et al. 1990), and the type of seed bank (according to Thompson and Grime 1979) was selected as a predictor of weediness in annuals (Perrins et al. 1992b). Invasive species may also have greater seed longevity than noninvasive ones, a trait allowing them to remain in the soil seed bank for long periods of time, germinating as conditions became hospitable (Mulvaney 1991).

Using Model Information

Ultimately, assessing the risk of invasiveness of a species should be done in tandem with assessing the risk of specific effects on ecosystems. None of the models discussed specifically assesses environmental costs of introduction, although several (Perrins et al. 1992b; Reichard n.d.; Richardson et al. 1990; Scott and Panetta 1993) do attempt prediction based on analyses of species impacting primarily natural systems. Such information should be used to develop more restrictive introduction policies now, knowing that we may be able to improve policies through time.

Models to predict potential invasion of plant species into natural communities will likely be more difficult to refine than those developed for agricultural communities, simply because natural areas express more spatial and temporal variation than commodity-production systems. Still, the models discussed here have been successful, and continuing research into the nature of invasions and invasive species will only refine them.

Finally, policy changes occur in response to public awareness of a problem. The significance of plant invasions as an environmental problem must be repeatedly presented to the public. An informed public can also help in early reporting of new invasions. Avenues of education include talks to local garden clubs and environmental interest groups, brochures, articles in newsletters and local newspapers, and signs in natural areas reminding visitors of potential routes of introduction. Building public support for policy changes is a time-consuming but essential process.

Monitoring for New Invasions

Even if changes are made in policy related to plant introductions, invasions will still continue as species slip through systems of inspection and exclusion. It is imperative that some form of monitoring system be in place so that invasions can be detected in the earliest phases. Once an invasive species becomes well entrenched, its control is expensive and virtually impossible (Hobbs and Humphries 1995). Early detection and action are necessary so that we can move from managing well-established invasions to preventing further invasions.

Monitoring requires alert resource managers, an efficient strategy for checking areas of susceptibility, and, most of all, an effective protocol to report invasions and share information. First, the "ground troops"—those out in natural areas doing surveys and maintenance work—must be trained to recognize indigenous (IN) species and to identify anomalous species. Teaching them only the common weeds of an area will not help them recognize a new invader. Second, resource managers should identify "hot spots" of invasion such as roadsides, hiking trails, and riparian corridors in their preserves and should establish a schedule of monitoring these areas. The frequency of monitoring depends somewhat on location of the reserve and on the type of ecosystems involved. Areas invaded by slower-to-mature plants such as woody or herbaceous perennials might be checked once a year or less, while those invaded primarily by annuals might be monitored twice a year, particularly in spring, when last year's introductions may begin germinating, and then later in summer, when recent introductions begin to germinate and many plants are in flower and easy to spot.

Finally, if a new species is found to be invading we must consider the most difficult aspect of monitoring: reporting the information so that others nearby may be alert for the species' presence. Currently no agency takes responsibility for listing new invaders found in a region or for notifying other land managers of their presence. New USDA-APHIS policy mandates that the federal government should coordinate communication and cooperation among relevant public agencies about detection and control of weed species, but this policy is not yet implemented (Westbrooks n.d.). State noxious weed boards might also be appropriate agencies. However, in these days of reduced budgets it is doubtful that these agencies or any others will step forward in the absence of adequate funding. Instead, we may perhaps have to rely on the Internet with its informal quick and easy communication, and on the growing number of nongovernment groups such as the Exotic Pest Plant Councils.

Detection of new invasive NI species or new populations of existing ones is only the first step in controlling incipient invasions. Resource managers must also develop emergency procedures to eradicate plants quickly in satellite populations. This can be as basic as assigning a person to both detection and eradication duties. Still, management activities may need to be prioritized (see Hiebert, Chapter 14, this volume).

Preventing Invasions on the Local Scale

There are several steps that resource managers can take to reduce the probability that invasive NI species will be introduced to their preserves. First, they can make sure that any landscape plantings in the preserve utilize IN species. Many reserves can trace initial introductions to intentional plantings within reserves (Macdonald 1990). If there is a visitor's center, plantings of IN species can be used both to soften lines of the building and to educate visitors about the species they may see. If there are additional buildings—offices, maintenance sheds, or residences—IN species should also be used for landscaping. One caveat in planting IN species in a natural area is, however, that locally collected material should be used when available. Not only is there the possibility of contaminating local genotypes of a species with nonlocal ones (Gehring and Linhart 1992), but if plants from outside are brought in, the probability of introducing seeds of NI species along with transported soil increases. If there are no propagation facilities available, existing plants may be transplanted; greatest survival will likely be with young plants transplanted while dormant. Rare species should not be moved for landscaping lest they die in the transplant.

Second, local invasions into natural areas may be avoided by sensible hygiene. Tourist vehicles may be a major vector of weed seeds (Clifford 1959; Schmidt 1989; Wace 1977); their access to sensitive areas should be limited. Visitors may also spread seeds unintentionally by bringing them in on boats, hiking boots and other hiking equipment, and pack animals. At every entrance to a lake or hiking trail there should be signs warning visitors of the role they may play in damaging the very beauty they seek to enjoy. Receptacles should be provided for disposal of seeds found when equipment is checked. Those who work in natural areas should remember that they and their vehicles may also be vectors of seeds. Mowers, trucks, bulldozers, and other pieces of equipment should be cleaned frequently, especially if they have been used in areas known to be infested with invasive species. Boots should be checked every day.

Finally, resource managers can reduce introduction of NI species by creating a buffer zone around the preserve comprised mostly of

IN or NI species with a low probability of invading. Many species invading natural areas are dispersed from surrounding landscape plantings. The width of the buffer would depend, ideally, on the predominant method of dispersal of the invasive plants of concern (with bird-dispersed seeds requiring perhaps greater width than wind-dispersed) and, practically, on population density and land use of the area. If it is impractical to create a buffer zone of IN species around the entire preserve, emphasis could be placed on dispersal corridors such as rivers, streams, and entry roads. Building a buffer zone may involve the cooperation of surrounding landowners; they may be induced to participate if educated about environmental problems caused by invasive species and perhaps also provided with some IN species.

Conclusions

1. Current laws regarding species introductions are inadequate to stem the tide of invasion. These laws recognize problem species only after they reach the point where containment is expensive and likely to be ineffective. Sweeping changes are necessary if we are to curb the entry of ecosystem-damaging species. Such changes will require the ability to recognize invasive species before they actually invade.

2. Models generated in different parts of the world and using a variety of species have found that good predictors of invasiveness include wide latitudinal range, prior history of invasions, vegetative reproduction, short juvenile period, long fruiting period, and extended seed longevity. These models have been successful in predicting invasions at rates exceeding the 50% expected by random assignment to invasive or noninvasive categories.

3. Access to information about performance of a species in areas to which it was introduced is extremely useful in predicting its behavior in other areas where it may be introduced. To that end, global Internet bulletin boards and databases should be developed and maintained.

4. Species should be prohibited from introduction until it is shown that they have a low probability of invasiveness. The importer should pay the costs of evaluation in exchange for distribution rights for a period of time. Once a species has been evaluated it would then go on a "clean" or "dirty" list; future introductions of the species would not need reevaluation.

5. For more inclusive legislation to be accepted regarding introductions, the public must be informed about the long-lasting environmental effects of invasive species. Those persons knowledgeable about invasions should take every opportunity to educate. In natural areas, posted signs should tell visitors about the problem.

6. Monitoring of invasion hot spots such as roads into natural areas, hiking trails, and riparian corridors should allow for early detection of new invaders, which is essential if control is to be effective. Information on new invaders should be shared among resource managers in a region. Managers should know which species or types of species may be likely to invade based on past experiences; they should monitor for them and be prepared to control them.

7. Resource managers should endeavor to protect reserves from the introduction of seeds of invasive plants. This may involve limiting vehicle access, cleaning service vehicles, requesting visitors to clean footwear and equipment, and encouraging nearby landowners to eliminate invasive species from their property.

Acknowledgments. I am grateful to all of the participants in the Aspen Global Change Institute's 1994 session "Biological Invasions as Global Change" for stimulating discussions. I especially thank Randy Westbrooks of USDA-APHIS and John Randall of The Nature Conservancy for their encouragement and many useful comments over the years.

16
Exotic Pest Plant Councils: Cooperating to Assess and Control Invasive Nonindigenous Plant Species

Faith Thompson Campbell

> Although we need quantitative measurements of the effects of various stages of L[ythrum] *salicaria* invasion on the structure, function, and productivity of North American wetland habitats, the replacement of a native wetland plant community by a monospecific stand of an exotic weed does not need a refined assessment to demonstrate that a local ecological disaster has occurred.
>
> —Thompson et al. 1987

Faya-tree
(*Myrica faya*)

The impact of invasive nonindigenous (NI) plants in the United States has been sufficiently well established to justify action even in the absence of specific information about particular species. We must be practical. Public and private land-managing agencies and scientific research institutions will never have sufficient funds to study every introduced plant species or every ecosystem in which such a plant has become established. We should also be humble; we may never fully understand the invasion process, particularly for each of the hundreds of potentially invasive species in each of our many ecosystems. One truth is clear: as time passes, many species will spread to new areas or increase in density if controlling actions are delayed.

The Exotic Pest Plant Councils (EPPCs) have been formed to improve management of invasive NI plant species. These councils promote research on both the process of invasion and control methods. However, they also act based on the best information available at the time. EPPC members have learned from experience the principle enunciated by many scientists (Ashton and Mitchell 1989; Hobbs R, oral presentation; Kummerow 1992; Macdonald et al. 1989; McEvoy PB, oral presentation; Windle P, oral presentation) that delay allows the problems caused by invasions to become almost irreversible.

Inaction, or action that is too little or too late, is thus both ecologically and economically expensive. (Ashton and Mitchell 1989)

Such is the nature of weeds that a small, tractable problem can easily become a huge and impossibly expensive problem. (Kummerow 1992)

It is a paradoxical and revolutionary concept for many that simply leaving a natural area alone, or "letting Nature take its course", would actually result in the loss of these species and ecosystems, along with wasting the efforts expended in the first place to designate them for protection. (Dudley and Collins 1995)

Evidence of damaging plant invasions can be found everywhere in the United States with the possible exception of Alaska. People's awareness is heightened in some areas compared to others. However, this awareness may be better explained by factors other than true ecological impact. A conspicuous change, for example, change from a saw grass (*Cladium jamaicense*) marsh to a thicket of trees, may be more likely to get people's attention than a more subtle change in species composition of the herbaceous flora of a forest floor. An invasion that proves costly to economic interests is much more likely to be recorded and to stimulate control efforts than is one that does not. But conspicuousness and economic impact are separate from ecological effects.

For example, cheat grass (*Bromus tectorum*) is so widespread that its control is considered impractical. Furthermore, the grass has some value as forage for livestock, so it is not so disruptive of the grazing industry as are other rangeland herbaceous weeds. These factors, not an absence of ecosystem effects, explain why cheat grass is not listed as a noxious weed by federal or state agencies and why reporting of its occurrence is spotty. The EPPCs, not allowing these factors to determine their decisions, include cheat grass in their lists and educational efforts because of its foreign origin, ecosystem impacts, and area covered.

Measuring the Impact of Invasions

Many parameters are often used to evaluate particular invasions or to compare invasions in different places. These measurement tools include the numbers of NI plants established in the flora, the proportion of NI species in the total flora, the rate of expansion, the impact of individual species in transforming ecosystem functions, the vulnerability of an ecosystem to stress, the size of area invaded, and the economic impact. One should also consider whether the NI species equals a high proportion of the biomass, for example, 90% of the shrub layer at a particular site. Used in isolation, each parameter has too narrow a focus. As I will demonstrate, too great a reliance on any one or even on a few can undermine the validity of the evaluation.

Numbers

Attention to the numbers of NI species and to the proportion of the overall regional flora they represent is sporadic and may not reflect actual damages. For example, contrast the varying significance attached to these numbers in discussions of invasions in Hawaii, southern Florida, the grasslands of the North American Intermountain West, and the eastern deciduous forest biome.

Hawaii, southern Florida, and, to a lesser extent, the grasslands of the Intermountain West, have been widely cited as examples of areas severely affected by invasive NI plant species (Carlquist 1980; Loope and Mueller-Dombois 1989; Mack 1989; Schmitz and Brown 1994). The eastern deciduous forest is rarely so described. Regarding Hawaii and southern Florida, one often sees references to the high numbers and proportion of NI species as a component of the total flora. For example, about 800 plant species are considered invasive in Hawaii (Loope and Mueller-

Dombois 1989). Although the number of problem weed species in the Intermountain West exceeds 60 (USDI, BLM 1994; Vail 1994) and may exceed 100 (Rice P, pers. com.), this is almost never considered to be a relevant factor in determining the impact of invasions there. Instead, the focus is usually on ecosystem changes and economic losses due to impacts of one or a few species on the livestock industry. The eastern deciduous forest has also been invaded by large numbers of species—more than 200 by my count—but these invasions have nevertheless received little attention.

I suggest that the reasons for lack of attention to numbers of NI species in the Intermountain West and eastern deciduous forests are not necessarily related to true ecological impacts. In these cases, extraneous factors result in less human interest in the numbers. The takeover may not be so conspicuous a change, for example, a change from one set of herbaceous species to another (Schwegman 1988). There may be no immediate measurable economic effect, or the economic costs of dozens of species may be "swamped" by the overwhelming costs associated with one species. Or the conversion may have begun sufficiently long ago, and may have become so complete, that no one can know the composition of presettlement vegetation, for example, the grasslands of central California.

The numbers of NI plants established in the flora or the proportion of NI plants relative to the total number of plant species in the region represent one aspect of an invasion problem, but these parameters must not be used exclusively. Nor should a low number of NI species or low proportions of NI species in the flora be considered evidence that there is no problem.

The entire United States is subject to invasion. At least 2000 plant species of foreign origin have been identified growing outside cultivation in the country (OTA 1993); the number may be 3500 (Morse L, pers. com.). Unfortunately, there exists no summary for the United States that identifies problem introductions and the geographic extent of these problems. Various organizations are compiling data, usually limited to lands under their jurisdictions. Among them are The Nature Conservancy, the U.S. National Park Service, the U.S. Fish and Wildlife Service, and the U.S. Department of Defense. The U.S. Department of Agriculture maintains a database on agricultural pests, but these data have proved to be very incomplete based on a sample of widespread "weeds" of natural areas.

On behalf of the EPPCs, I have also been compiling a list through a mail survey of EPPCs, native plant societies, natural heritage programs, and academic botanists. The survey includes Hawaii but no other Pacific islands or those in the Caribbean. Three hundred fifty species have been annotated by botanists, resource managers, or knowledgeable amateurs as serious invaders in one or more states. Thus, "only" 10% to 15% of introduced species has been so far identified as serious. I consider this estimate to be low because of the relatively few scientists studying the problem. As these scientists are focusing on some species to obtain information, many more species are expanding their ranges virtually undetected.

The highest proportion of introduced plants relative to the overall flora is found in tropical and subtropical areas. In Hawaii, the figure is greater than 45% (OTA 1993). Florida follows closely, with about 40% of the flora outside cultivation consisting of introduced plant species (Schmitz and Brown 1994). However, relatively high proportions are also found in states near ports where the plants could enter: 29% in New England and 28% in Illinois (OTA 1993). California has such a large IN flora that the proportion represented by NI species is not high, but this does not diminish their damaging impacts. More than 2.8 million ha in the state are heavily infested with yellow starthistle (*Centaurea solstitialis*) (Beck 1993), a spiny plant that discourages recreational use of the land on which it grows. Furthermore, chemical constituents in the plant cause horses to lose jaw muscle control, with death as the ultimate result (Westbrooks 1993).

Rate of Spread

Although a plant species with a rapidly expanding range certainly demands attention, rates of invasion are extremely difficult to use in evaluating a species' impacts. This is true, first, because many if not all invasive plants seemingly have a lag period between establishment and explosive expansion (Brookreson n.d.; Hobbs R, oral presentation; Vitousek PM, oral presentation). Whether or not this lag is a real phenomenon or only a perceptual one stemming from inadequate attention by researchers may never be determined. The important result, however, is that people are lulled into complacency. In the early stages, while the species is relatively easy to control, attention is focused elsewhere. If the species does become an aggressive invader, it usually escapes recognition until it has occupied an impressive area and has established numerous outlying populations that are foci of further dispersal. By that time, successful control requires a major undertaking with serious economic and ecological costs.

Unfortunately, scientists have not yet developed reliable systems for predicting which species that have not yet erupted will prove to be dangerous invaders (see however, Reichard, Chapter 15, this volume). One factor does seem to be highly indicative: species that have been invasive elsewhere (Reichard S, oral presentation; Whiteaker L, pers. com.). While scientists continue developing and testing predictive models, resource managers should, at a minimum, become familiar with those species already showing invasive behavior in the managers' own or similar ecosystems. In addition, national and state policymakers should initiate voluntary or regulatory actions to control intentional transfer and planting of such species.

The repercussions, if such preventive actions are not taken, are clearly illustrated by purple loosestrife (*Lythrum salicaria*) and tropical soda-apple (*Solanum viarum*). Purple loosestrife entered the United States probably by the 1830s. Despite the recognition some decades ago that the species is invasive in wetlands, it continued to be deliberately dispersed by the beekeeping and horticultural industries. Now purple loosestrife is at least in all Canadian provinces and 41 conterminous states. Once a population is present, it is capable of rapid spread without further assistance. It spread to 15,000 ha of wetlands in 68 counties in Minnesota over the course of about 80 years (Skinner et al. 1994). This dispersal is assisted by prolific production of seeds—a single mature plant can produce annually more than 2.5 million seeds—which remain viable for many years (Malecki et al. 1993). The seeds are also easily transported in water, soil, or on animals' fur or feet. Establishment is greatly enhanced by disturbance, and seedling densities can approach 20,000 plants per square meter (Malecki et al. 1993). However, few if any wetlands in North America south of the Arctic can be considered safe from disturbance (Thompson et al. 1987).

In the United States, tropical soda-apple, a weed of pastures and woodlands, was first detected in Florida. Cattle farmers battled their own infestations individually for several years before state and federal agricultural agencies became aware of the problem. The weed is spread when cattle and wildlife ingest the seeds. Interstate sale of cattle and composted manure has facilitated spread of tropical soda-apple in about a decade to more than 400,000 ha in four states and to Puerto Rico (Fig. 16.1). In July 1995, the U.S. Department of Agriculture listed the species as a federal noxious weed (*Federal Register*, Vol. 60, No. 133 [12 July 1995], pp. 35829–35831); however, it did not impose a quarantine on cattle from infested counties so, even now, continued human-assisted dispersal appears likely.

Ecosystem Impacts

Altered Disturbance

As Heywood (1989) noted, often one species can have an overwhelming effect on ecosystem properties. Among the ecosystem impacts

FIGURE 16.1. Current range of tropical soda-apple (*Solanum viarum*) as indicated by the National Agricultural Pest Information System.

most often mentioned is a change in the fire regimen. Introduced grasses can promote more frequent or hotter fires contributing to destruction of indigenous (IN) plants and wildlife habitat. Examples include broomsedge (*Andropogon*) in Hawaii and cheat grass and salt-cedar (*Tamarix*) in the Intermountain West and desert Southwest (Billings 1994; Lovich 1994; Macdonald et al. 1989; Vail 1994).

A second type of ecosystem disruption attributed to various NI plants is a change in soil chemistry. Fayatree (*Myrica faya*) in Hawaii fixes nitrogen on cinder and ash pockets, making them more easily invaded by other plant species and potentially changing pathways of succession (Macdonald et al. 1989; Whiteaker L, pers. com.). In California, iceplant (*Mesembryanthemum crystallinum*) deposits salt on soil, excluding IN plants; iceplant also shades out competitors (Macdonald et al. 1989). Salt-cedar trees also exude salt from their leaves, suppressing germination of IN species (Lovich 1994).

Changes in the structure of an ecosystem can be quite conspicuous and drastic. Melaleuca (*Melaleuca quinquenervia*) transforms the everglades ecosystem of southern Florida from a primarily saw grass marsh to a dense monoculture of tightly packed trees (Fig. 16.2). By trapping silt in their roots, salt-cedar trees and Bermuda grass (*Cynodon dactylon*) (Dudley T, oral presentation) can prevent periodic scouring of the river or creek bottom on which IN plants depend. Salt-cedar trees also change the stream channel. The North American intermountain perennial grass/shrub ecosystems have been converted into annual grasslands by overgrazing, fire, and invasion of cheat grass and other species.

Altered Wildlife Habitat

Nonindigenous plants sometimes deprive IN animals of food, shelter, or nest sites. Wildlife species diversity in the everglades is decreased 60% to 80% when the saw grass marsh is replaced by melaleuca (Ewel 1986). The endangered snail kite and wood stork are among the birds that do not adapt to the new ecosystem (Maffei 1994). On Florida's coasts, the roots of Australian-pine (*Casuarina*) trap turtle hatchlings. When purple loosestrife displaces indigenous wetland vegetation, especially cattails (*Typha* spp.), the conservation value of the wetland is greatly diminished. Waterfowl do not feed on the plant (Skinner et al. 1994). The seeds of purple loosestrife also are inedible for most birds (Brookreson n.d.). The vegetation becomes too dense for nesting waterfowl to penetrate (SDSWPC 1993). Muskrats need cattails to build their nests. Rawinski and Malecki (1984; cited in Malecki et al. 1993) found that, in a national wildlife refuge in New York, muskrats and long-billed marsh wrens used cattail stands almost exclusively; red-winged blackbirds preferred purple loosestrife (SDSWPC 1993; Skinner et al. 1994; Thompson et al. 1987). Fishes such as northern pike lose spawning habitat (Hight and Drea 1991; SDSWPC 1993; Skinner et al. 1994). Other species harmed by purple loosestrife include frogs, salamanders, snakes, toads, and turtles, which potentially lose shallow-water breeding and/or feeding sites (SDSWPC 1993).

Cheat grass makes poor fodder for such IN wildlife species as deer and antelope

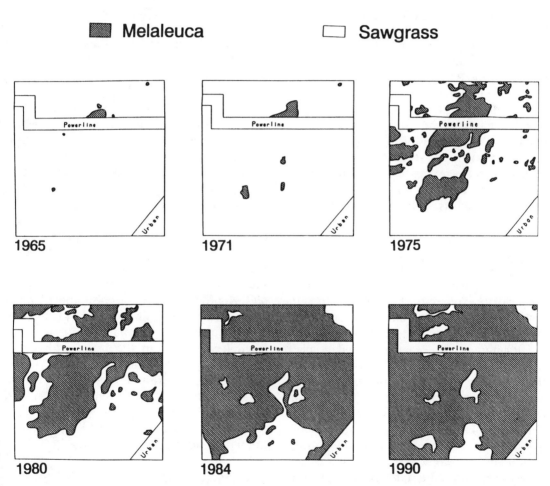

FIGURE 16.2. Areal expansion of melaleuca (*Melaleuca quinquenervia*) over a 25-year period in a 1 mile² section of land in Dade County, Florida. Area between parallel lines indicates a powerline corridor. Triangular area in the lower right indicates an urban area. (Data are from Laroche and Ferriter [1992])

(Whisenant S, oral presentation). Rodents and lagomorphs—the prey base for raptors in the Snake River Birds of Prey National Conservation Area in southwestern Idaho—prefer the IN shrub-perennial grass habitats. Habitats made up of NI annual grasses support no lizard populations.

Knapweeds (*Centaurea*) reduce plant diversity and possibly animal diversity as well (Rice et al. 1992). Elk in Montana need grass in winter and early spring, not NI herbaceous species (Kummerow 1992). In the desert Southwest, salt-cedar trees lower water tables, depriving desert wildlife of already scarce water (Macdonald et al. 1989).

Various European brooms, particularly Scotch broom (*Cytisus scoparius*) and French broom (*Genista monspessulana* = *Cytisus monspessulanus*), are widespread west of the northern Sierra and Cascade ranges from California into British Columbia. In California, Scotch and French brooms have invaded more than 400,000 ha. Deer in the Sierra Nevada will not eat brooms (Bossard and Rejmánek 1994). Erosion from land infested by spotted knapweed (*Centaurea maculosa*) can be double that covered by IN bunchgrasses, reducing water infiltration and increasing siltation that affects fish populations (Beck 1993).

Competitive Interactions

By competitive interaction, invading NI plants can lead to extinction of IN plants. In remnants of the highly endangered prairies of Oregon's Willamette Valley, brooms threaten the habitat of the Federal Candidate plant Kincaid's lupine (*Lupinus sulphureus kincaidii*), the food plant for two subspecies of the candidate Oregon silver spot butterfly (Isaacson D, pers. com.). In Idaho, competition with cheat grass has reduced IN annuals such as gymnosteris (*Gymnosteris nudicaulis*) from "abundant" to rarely seen; this species and others are now on the state sensitive plant list (Rosenstreter 1994). Competition with filaree (*Erodium cicutarium*), the invasion of which is encouraged by cheat grass, is reducing populations of the Federal Candidate 1 species Aase's onion (*Allium aasae*) (Rosenstreter 1994).

In Golden Gate National Recreation Area, California, two IN species of manzanita (*Arctostaphylos*) and an IN ceanothus (*Ceanothus*) cannot reproduce because they are being overgrown by German ivy (*Senecio mikanioides*). Dense German ivy infestations are smothering vegetation along nearly the full length of the California coast. Large monotypic stands of purple loosestrife jeopardize several rare species, including bog turtle, dwarf spike-rush (*Eleocharis parvula*), and Long's bulrush (*Scirpus longii*) (Malecki et al. 1993).

At the tenth annual symposium of the Florida EPPC in 1995, Kathy Craddock Burks of the Florida Department of Environmental Protection described numerous examples of rare IN plants harmed by invading aliens. Some impacts are direct; examples are crowding or shading out of rare dune vegetation by Brazilian-pepper (*Schinus terebinthifolius*) and Australian-pines (*Casuarina*). A rare Florida species, inkberry (*Scaevola plumieri*), is similarly threatened by an introduced congener from Hawaii, beach naupaka (*S. sericea*). Japanese climbing fern (*Lygodium japonicum*) is threatening the habitat of 32 listed plant species in Appalachicola National Forest. On Eglin Air Force Base, 60 listed plant species are threatened by invasions of Cogon grass (*Imperata brasiliensis*) and Japanese honeysuckle (*Lonicera japonica*) as well as by feral pigs. Nonindigenous plants also impede management intended to benefit rare plants. Prescribed burning to enhance five federally listed fern species in the pine rocklands of Dade County, and a rare bromeliad (*Tillandsia*) on bald-cypress (*Taxodium*) in Loxahatchee Strand, has been hindered by favorable response to fire of Burma reed (*Neyraudia reynaudiana*) and Old World climbing fern (*Lygodium microphyllum*).

Area Infested

The total area infested by a particular species is probably more a reflection of how widespread the suitable habitats are than of the species' invasiveness or ecological impact. This is especially true after considerable time has elapsed since first introduction. When attempting to measure a species' "invasiveness," it is probably more relevant to ask what proportion of the suitable habitat has been occupied by the NI plant species.

Information about the extent of various species' current range is poor or difficult to obtain. As noted above, the U.S. Department of Agriculture's pest management database has not proved to be reliable. Other institutions' list-compiling exercises will be fragmented by land ownership considerations and will apparently not include complete data on areas affected. Nevertheless, there are certainly many species that are widely spread across the United States.

The champion is surely cheat grass. This species alone is present in an area of more than 40 million ha (Mack, cited in Rosenstreter 1994)—an area almost as big as the state of California. Its wide establishment was greatly assisted by overgrazing of livestock, which devastated the vulnerable IN bunchgrasses (Mack 1989). By now, however, cheat grass is the dominant species on up to 25 million ha (Rich TD, oral presentation). Assisted by its association with fire, cheat grass almost certainly will continue increasing its relative population density within its present range and will also likely spread to new areas.

Other rangeland weeds are also reported to be quite widespread. The reports reflect the species' ability to invade natural grasslands and artificial pastures *and* the interest of researchers in invaders with immediate economic impacts. In 70 years, spotted knapweed has spread to 2.9 million ha in nine states and two Canadian provinces (Beck 1993). In Montana, the state of introduction, infestations cover a total of 1.9 million ha (Beck 1993), the size of Delaware. "[T]he foothill grasslands in areas such as western Montana's Missoula and Bitterroot Valleys are ... in more peril than any other vegetation type in the West" (Bedunah 1992). Yellow starthistle, introduced into California in 1869, has spread to more than 3.8 million ha in 10 states and two provinces (Beck 1993), an area larger than Maryland, Delaware, and the District of Columbia combined.

Less information is available on the trees, shrubs, and vines that invade wooded areas. Tree-of-heaven (*Ailanthus altissima*), the most widespread woody invasive, is found across the country wherever moisture allows (Reichard S, oral presentation). Bush honeysuckles (*Lonicera*) are widespread shrub invaders in the eastern deciduous forest. One of them, Amur honeysuckle (*L. maackii*) is found in ecosystems stretching from central Maine to northern Georgia to North Dakota and eastern Texas (Reichard S, oral presentation). This region equals one-third of the area of the conterminous United States.

Other species, while more limited by broad ecological requirements, can still dominate large geographic areas. Purple loosestrife invades sunny freshwater wetlands. It reaches its highest densities in the northern half of the United States and in southern Canada, where it easily invades wet meadows and shallow and deep freshwater marshes, but is now present across the entire continent (USDI, FWS 1995) with a range of 192,000 ha (Meyerdirk D, pers. com.).

Some species are not so widespread because they can grow only in restricted habitats. This does not diminish the seriousness of their impacts. Melaleuca is present on up to 198,000 ha (Center et al. 1994), nearly 25% of Florida's remaining everglades ecosystem. Australian-pines occupy 33% of surveyed undeveloped coast on the Gulf side of the Florida peninsula and 46% of the surveyed undeveloped Atlantic coast (Johnson 1994).

A rough total for areas invaded nationwide was compiled by considering just the few species for which area data are available and disregarding all rangeland weeds other than cheat grass (because of the likelihood of overlapping infestations). This total exceeded 43 million ha, a figure representing 5% of the area of the conterminous United States. The Department of the Interior reported that NI plants (excluding cheat grass) infest 6.5 million ha of federal lands and are spreading at a rate of at least 1.2 million ha per year

(Whitson T, oral presentation). At this rate, weeds cover a new area the size of Delaware every 6 months.

These quantitative data do not include any of the vines, shrubs, and herbaceous species invading the forests constituting the natural vegetation in the eastern third of the United States.

Economic Impacts

Losses or control costs associated with NI plant species are not necessarily linked to either the overall numbers of NI species in a region or the proportion of that region's IN flora that these species represent. Nor are attempts often made to place economic valuations on ecosystem impacts. Such data are usually available primarily for crop or livestock forage losses. Occasionally, other associated costs are calculated.

For example, the Great Plains was considered by Mack (1989) to be relatively less vulnerable to invasion; only 13% of its flora is of foreign origin (OTA 1993). However, one of those species has amassed considerable economic repercussions. Leafy spurge (*Euphorbia esula*) has infested more than 0.6 million ha in the northern Great Plains and another 0.4 million ha across the country. Direct and indirect losses to livestock production as a result of leafy spurge, nationwide, were estimated at $110 million in 1990 (Beck 1993).

The Congressional Office of Technology Assessment estimates that NI plants—including Bermuda grass, Canada thistle (*Cirsium arvense*), Johnson grass (*Sorghum halepense*), shattercane (*Sorghum bicolor*), and summer-cypress *Kochia scoparia*, (Bridges 1992)—cost American farmers between $3.6 and $5.4 billion annually in the forms of reduced quantity and quality of production and higher costs for herbicides and other control measures (OTA 1993). A study in Minnesota concluded that purple loosestrife threatens the state's $14 million wild rice crop (Skinner et al. 1994).

In recent years, the United States Animal and Plant Health Inspection Service (APHIS) has spent more than $9 million annually on efforts to control or eradicate NI plants in the United States. The Forest Service and Bureau of Land Management together have spent an additional $2.5 million annually on control of NI plants, primarily on public rangelands west of the Dakotas (Kaiser pers. com.; Waters pers com.). Projects to control weed infestations in national parks would cost a total of $80 million if they were fully funded. Due to financial constraints, the National Park Service has allocated only $6 million to this task over 4 years; fewer than 10% of control projects have received funds (Johnston G, pers. com.). The U.S. Fish and Wildlife Service spends about $3 million annually controlling "pests," including animals, on National Wildlife refuges (Furniss S, pers. com.). The Corps of Engineers devotes $10 million to research and control of aquatic weeds (Cofrancesco A, pers. com.).

Individual states and counties also spend millions of dollars to control invasive NI plants. In Florida, state and county agencies and water management districts spent at least $30 million per year to combat problem-causing NI aquatic plants (Schmitz DC, pers. com.). The state of Washington and its counties spend more than $6 million each year (Penders L, pers. com.).

One associated cost sometimes reported is increased fire suppression effort as a result of the greater flammability of certain NI plants. The Bureau of Land Management in Idaho spends an average of $4 million annually on fire suppression. Although the precise amount attributable to cheat grass is not stated, "Cheatgrass has increased the size and intensity of wildfires. Rangeland burned in Idaho since 1980 is 1.8 million acres" (Vail 1994). In Florida, as new houses are built near wetlands containing melaleuca stands, crown fires fueled by the tree have become an economic and safety issue. In addition, more than $1 billion in losses were associated with a 1985 melaleuca fire under electrical transmission lines, causing more than 2.3 million people to lose service (Laroche 1994). An explosive fire in 1936 that burned the town of Bandon, Oregon, to the ground has been blamed in part on gorse (*Ulex europaeus*) infestations surrounding the town. Dense stands of purple loosestrife also increase the cost of roadside ditch maintenance (Skinner et al. 1994).

Ecosystem costs are usually calculated in terms of money spent to acquire the area or in terms of lost recreational opportunities. For example, purple loosestrife potentially could destroy Minnesota wetlands acquired or otherwise protected at a cost of more than $25 million; 140,000 waterfowl hunters and 2.5 million anglers would have fewer opportunities to pursue their sports (Skinner et al. 1994).

More Subtle Impacts

The impacts outlined above are relatively dramatic and often impinge on human activity. It is much more difficult and it is rarely attempted to establish an economic cost for NI species invading natural areas not utilized for commodity production. Furthermore, more subtle or long-term changes are often not measured. For example, few investigators study the long-term impact on plant communities themselves, as opposed to impacts on associated—usually vertebrate—animals. Are these impacts unimportant ecologically? Or are they just overlooked because humans have a bias toward short-term economic considerations and charismatic megafauna?

Nonindigenous plant species can prevent the growth of seedlings of IN species. On Theodore Roosevelt Island, a small island in the Potomac River under management by the National Park Service, Japanese honeysuckle has prevented regeneration of American elm (*Ulmus americana*), wild black cherry (*Prunus serotina*), and yellow-poplar (*Liriodendron tulipfera*). English ivy (*Hedera helix*) has smothered IN wildflowers. Both honeysuckle and ivy suppress established trees by shading (Thomas 1980). Banana poka (*Passiflora mollissima*) has the same effect in Hawaiian forests. Until studies are done, would it not be prudent to operate on the assumption that other invasive NI vines similarly suppress IN plants?

Any mechanism that threatens the ecological integrity of a protected area, including the relative abundance of plant species, should be considered a serious conservation problem. Often, protecting the biotic communities was one of the reasons for designation of the protected area in the first place. Where this goal was not explicitly stated, it remains an important part of protection. A survey of all units in the national park system conducted in 1986–1987 found NI plants to be the most common threat to our national parks' resources. This response, while alarming, is not surprising given the already noted nationwide extent of the problem of plant invasions. The most intensive and longest-lasting control programs have been carried out in parks in Hawaii and Florida. However, the third most costly control program now under way is in Great Smoky Mountains National Park, Tennessee and North Carolina. Biologists in Great Smoky Mountains National Park have targeted 33 NI plants, including oriental bittersweet (*Celastrus orbiculatus*) and a privet (*Ligustrum*). The programs in the Everglades, Great Smokies, Haleakala, and Hawaii Volcanoes national parks each cost more than $250,000 per year. Smaller amounts are being spent by several parks in the Intermountain West or on the Great Plains, such as Canyonlands, Capitol Reef, Death Valley, Roosevelt, and Yellowstone national parks, to battle the rangeland weeds and salt-cedar found throughout the region. Other park units conducting weed-control programs are Golden Gate and Gateway national recreation areas, Indiana Dunes National Lakeshore, and Redwood National Park (Cacek T, pers. com.)

Overall Significance of Invasions

Vitousek (oral presentation) concluded that biological invasion is one of seven "global changes" caused directly by humans that are changing Earth *now*; Billings (1994) concurred. People are now the principal vectors of species' dispersal; Noble (1989) went so far as to say that natural long-distance dispersal capabilities are no longer relevant. Furthermore, Vitousek (oral presentation) considered biological invasion to be second to direct land-use changes as an anthropogenic cause of loss of biological diversity. In another measure of its significance, Vitousek (oral presentation) regarded biological invasion as irreversible in many cases—almost as irreversible as complete extinction of a species.

It is often hard to separate the impacts of a disturbance—grazing, stream channelization, dams, or development—from those of an associated plant invader (Ramakrishnan and Vitousek 1989). Vitousek (oral presentation) maintained that the two factors are so closely linked that they should be considered together as a syndrome. Certainly, attempting to restore a community by combatting only one of the synergistic factors is not likely to lead to success (see Luken, Chapter 11, this volume). Halting or modifying the disturbance activities is politically difficult. However, because of the ubiquitous availability of potential invaders (Hobbs R, oral presentation; Noble 1989; Vitousek PM, oral presentation), even success at curbing anthropogenic disturbances will probably not curb invasions now. One needs to address the invading plants themselves. This was the approach recommended in the EPPC's comments on the draft management plan for the Snake River National Conservation Area (see "Role of Exotic Pest Plant Councils," below).

One specific example is the failure of attempts to restore IN perennial grasses through changed grazing patterns and seeding of crested wheat grass (*Agropyron cristatum*). Cheat grass persisted despite these efforts (Young 1994). Salt-cedar is another example. This tree is an extremely aggressive invader and is very difficult to eliminate once it is established. In the opinion of Lovich (1994), habitats dominated by salt-cedar are likely to remain so, even if other causes of habitat degradation are halted. Purple loosestrife is probably a third species that, once established, will persist despite efforts to protect the wetland from other anthropogenic perturbations.

The Need for Coordination

From the point of view of conservation, priority should be given to protected areas and important wildlife habitats. However, public and private protected areas are often surrounded by infested areas. Public and private land-managing agencies will never have adequate funds to carry out a perpetual holding action intended to stop weeds at the property line. Furthermore, they must prepare for invasions by new species. Important political alliances and economies of scale can be realized by cooperating with those motivated by concerns other than conservation. These allies would include people struggling to control invasive NI plants that cause economic losses to ongoing commercial activities such as livestock grazing, farming, and use of water bodies.

To protect parks, wildlife refuges, and public lands generally, resource managers will need the active cooperation of those responsible for managing neighboring lands and waters. In addition, they will need sensible national policies that effectively prevent entry to the country of as many invasive NI species as possible and containment programs directed at the first outbreaks of species that have entered the country despite these precautions.

In the absence of complete information, such a program will need methodologies to evaluate risks. There will never be sufficient funds or enough scientists to monitor every introduced species to see whether it reaches the threshold of outbreak (especially given the perceived lag, which can be long). Resources to do lengthy and complex ecosystem studies to prove more subtle impacts will also remain scarce.

The cohesive program described above does not now exist. In 1993 the Congressional Office of Technology Assessment described efforts in the United States to counter the effects of invasive NI species as "a largely uncoordinated patchwork of laws, regulations, policies, and programs" (OTA 1993).

It was in the absence of a cohesive national program that the EPPCs were formed. Their task can be described as promoting creation of new barriers that will obstruct the importation and transport of invasive NI species. The EPPCs will consider and employ the full armament: laws and regulations; encouraging more responsible behavior by those who import or transport plants, including the establishment of legal liability; technical fixes (washing, inspections, management, etc.); and mechanical, chemical, and biological methods of combatting NI species already established in natural ecosystems. The choice of method depends on the specific factors of a particular species at a particular site.

Role of Exotic Pest Plant Councils

The original EPPC, that in Florida, was formed to coordinate control efforts by many players. Finding the overall effort to be lagging, the EPPCs have moved increasingly into advocacy of policy because few other organizations are addressing threats to natural areas from invading plants. Therefore, the purpose of the EPPCs is to raise awareness wherever invading plants pose a problem; to promote cooperative efforts to find the most effective means of addressing each regional or local manifestation of the problem; and to promote and develop comprehensive national programs incorporating prevention, exclusion, early detection, and eradication or control. Exotic Pest Plant Councils build on each member's strengths and prevent duplication of effort. They integrate management investigations with managers, ecosystem impacts with ecologists, and legislation with policy-makers.

The Florida EPPC (FLEPPC) was formed in 1982. Early on, FLEPPC adopted a priority list of four species: Australian pine, Brazilian pepper, giant sensitive catclaw (then called *Mimosa pilleta*), and melaleuca. This action followed launching of efforts to protect the East Everglades area from invasion by melaleuca and to eradicate NI plants from the "Hole in the Donut" portion of Everglades National Park.

In succeeding years, much of the effort has been devoted to melaleuca. In 1990, FLEPPC established the Melaleuca Task Force, made up of professionals from member agencies. The task force developed a management plan, which was updated in 1994 (Laroche 1994). The plan established the goal of protecting the integrity of Florida's natural ecosystems from biological degradation by melaleuca. The plan's objectives were to eliminate melaleuca from Florida's natural ecosystems; to achieve overall reduction in melaleuca throughout Florida such that maintaining natural areas is economically feasible; to provide public information to encourage support for management; and to coordinate with and support goals of the Everglades Restoration Task Force. Among the new recommendations in 1994 were calls for increased funding for biocontrol evaluation and subsequent release and for a quarantine facility. Another recommendation called for utilizing the resources of EPPC and others to lobby at state and federal levels. Operational recommendations focused on continued melaleuca elimination programs in all publicly owned natural areas in order to establish a melaleuca-free buffer zone on those areas' boundaries. To carry out these recommendations, member agencies were asked to identify a task leader for each function. The South Florida Water Management District (SFWMD) took the lead in investigating control techniques for mature and seedling trees.

Subsequently, a staff person at SFWMD developed proposed protocols and submitted an application to the responsible regulatory agency for experimental applications of herbicides to control melaleuca. Other agencies participating in experiments were the National Park Service, Florida Department of Agriculture and Consumer Services, Florida Department of Environmental Protection, Dade County Parks and Recreation Department, and pesticide registrants. The SFWMD applied the pesticide; each agency assigned a staff person to assist in monitoring.

Once a small pilot project was completed, a larger, operational study was proposed in which agencies would spray up to 120 ha, monitor for pesticide residues in surface and ground water and soil, and monitor the impact on nontarget species. A consulting firm carried out an independent evaluation of pesticide impacts on melaleuca (including reinfestation), and on nontarget species (Forest Resources Management 1992).

The SFWMD also wrote a draft cost analysis of melaleuca management in water conservation areas for FLEPPC in June 1992. The analysis said that the program was not truly integrated because neither biocontrol nor fire and flooding were yet being used. This resulted in a great reliance on herbicides. The analysis recommended concerted removal programs in particular sites for 3 years to eliminate all melaleuca from the area. It discussed the appropriateness or lack thereof of heavy equipment and herbicide application methods. The analysis concluded that biocontrol agents would be the cornerstone of effective long-term management, especially by killing seedlings to prevent reinfestation.

The draft cost analysis also described the weaknesses of various measurement techniques for determining the extent of melaleuca infestations. It recommended no single approach; instead, it suggested mapping various parts of the Everglades region using the method most suitable for specific conditions.

The analysis reported the results of a retrospective study of eight heavily infested 1-square-mile plots using sets of aerial section photos. By digitizing melaleuca coverage in each photo, adjusting to exclude areas previously cleared for farming or development, calculating the percentage of melaleuca infestation in each site for each year, and performing a regression analysis, a mean growth equation of 4% per year was developed (Fig. 16.3).

California EPPC (CALEPPC) carries out research directly through its own working groups. These volunteers receive the coopera-

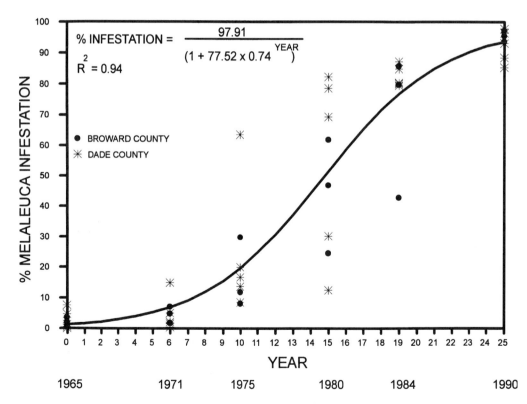

FIGURE 16.3. Rate of melaleuca (*Melaleuca quinquenervia*) expansion in permanent plots in Broward and Dade counties, Florida. (From Laroche and Ferriter [1992])

tion of state agencies, which often provide some staff and equipment. In turn, the state agencies get prompt access to the research results. At present, groups are researching French broom, German ivy, and salt-cedar; cooperating agencies include state parks, the U.S. National Park Service, and probably the U.S. Bureau of Land Management (BLM) (Bossard CC, pers. com.).

One role of the EPPCs is to share our members' expertise with others who manage NI plants, thereby leading to well-thought-out efforts with a higher chance of success. For example, the EPPCs of California, Florida, and the Pacific Northwest provided comments on BLM's draft management plan for the Snake River Birds of Prey National Conservation Area, specifically as it pertained to invasive annual NI plants therein. The EPPCs urged the BLM to modify its proposed actions to reduce the emphasis on fire management and to increase emphasis on restoring IN plant communities and on ecosystem dynamics and processes. The EPPCs also suggested that the bureau address livestock grazing as a probable contributing factor to invasion by cheat grass and other NI plants. The need for systematic and frequent inventories to evaluate the extent of infestations, detect new problems, and determine the effectiveness of management measures was noted. The EPPCs urged a focus on detecting and controlling outlier populations.

In areas of heavy infestation of cheat grass, the EPPCs recommended that the BLM attempt active restoration using burning, herbicide spraying, and mowing to reduce seed crops of NI plants, followed by planting of plugs rather than seeding of IN grasses. Since about 72% of wildfires are human caused (draft plan), the EPPCs recommended that the BLM place much greater emphasis on preventing as opposed to suppressing fires.

As noted above, public and private land-managing agencies will never have adequate funds to protect their lands when surrounding areas continue to be infested. Environmental constraints on pesticide use and other factors may also restrict choices for management. In any case, it is important to limit invasions by new species. Therefore, the EPPCs give high priority to improving prevention and exclusion efforts at U.S. borders with regard to imports and to regulation of deliberate transport, for example, interstate trade or use in roadside or wildlife plantings.

From this perspective, the EPPCs are not satisfied with either the Federal Noxious Weed Act (FNWA) of 1974 or its implementation by APHIS. The EPPCs have joined other organizations in supporting a proposal to strengthen the act. The proposed amendments seek to ensure: (1) that quarantine requirements and inspection systems effectively exclude as many invasive NI species as possible; specifically, APHIS would apply the Act to control wildland as well as agronomic weed species; (2) that APHIS staff have the knowledge and resources to identify and attempt to eradicate rapidly the first outbreaks of weeds entering the country despite quarantine precautions; (3) that APHIS has authority to prohibit intentional interstate shipment of listed plants so as to slow spread of invasive plants already in the country; and (4) that non-USDA experts have a greater voice in recommending NI species to be listed.

The EPPCs are collaborating with managers of national parks and wildlife refuges to explain to the public the environmental values lost to "weeds," the importance of launching and maintaining aggressive control programs, and the need for adequate funding.

The EPPCs are also supporting adequate funding and staffing to allow the Agricultural Research Service and Corps of Engineers to conduct the research and testing necessary to develop biological control programs and other components of site-specific, effective, and environmentally sound control strategies.

Public understanding is a crucial factor in determining the potential for success in reaching our goal to preserve biodiversity from the threat posed by invading NI plants. Therefore, the EPPCs will coordinate the preparation of articles, videos, and other educational materials for dissemination through scientific, conservation, and general-interest outlets.

In efforts to educate the public and policymakers, the EPPCs face several hurdles. Some are weaknesses in perception. Most people have only a very general knowledge of biology. They cannot recognize one plant species from another; they do not know which ones are indigenous (Miller M, oral presentation; Schwegman 1988). So they literally do not see the difference between a pristine and an invaded community.

The concept of invading NI species is complex. Efforts to simplify the educational message by focussing on one species have not been as successful as one might hope in creating a public understanding of the need to address the underlying causes across the board. This problem is common to environmental efforts generally.

Addressing invading NI species also demands balancing of conflicting benefits or risks. For example, many introduced species are useful (Reichard S, oral presentation) and familiar. Some, such as purple loosestrife and Japanese honeysuckle, are attractive. People must be persuaded to stop planting these and other invasive plants in order to protect the surrounding environment.

Introduced species do not elicit the same concern as chemical pollution because they are not perceived to be a threat to human health and well-being (Vitousek PM, oral presentation; Westbrooks R, oral presentation). Two consequences follow: Other environmental problems receive the lion's share of attention and resources; and people are reluctant to use herbicides in combatting a threat they do not appreciate.

As I noted above, the perceived time lag between establishment of a species and its demonstrated invasiveness contributes to uncertainty as to which species will be invasive and to delay of control measures. To top it off, human nature, displayed in many policy arenas, resists perceiving a problem until the situation has reached crisis proportions.

Conclusions

1. Taken individually, the numbers of NI species, the percent of the flora that is nonindigenous, the total area invaded, public attention, and economic impact are not necessarily good measures of ecological effects. Measurements of "ecosystem effects" should be based on broader foundations than the demonstrated impact on a few vertebrates.

2. Nonindigenous plant species have invaded a wide range of ecosystems in at least 49 of the 50 states. Our knowledge of the extent of such invasions and their impact on the IN community is poor due to lack of study and the difficulties of studying invasions.

3. Nonindigenous plant species are often not perceived as invasive until they are widespread; consequently, successful control requires a major undertaking with large economic costs. I recommend, first, implementation of programs aimed at early detection and monitoring of potentially invasive species and, second, rapid eradication efforts at the initial signs of spread. Information should also be exchanged with other knowledgeable people in the United States and abroad; considerable weight should be given to their experience with the same or similar species in reaching management decisions.

4. Protection of natural areas from invasion by NI species requires a comprehensive national program that includes measures to exclude, detect, and eradicate NI species. We lack such a program at present. Constituencies other than conservationists would also benefit from better management of NI species. Therefore, conservationists, scientists, and public and private agencies should work together with other affected interests to adopt and implement effective laws, programs, and voluntary ethical standards. Research is a necessary component of such a comprehensive program, but control efforts cannot be put on hold pending resolution of all scientific questions.

5. Coordination of efforts is also beneficial at regional, state, and local operational levels. Voluntary coordinating bodies such as the Exotic Pest Plant Councils can integrate members' research and control programs and undertake additional cooperative projects.

Acknowledgments. I thank the many people who assisted me in the preparation of this chapter. First thanks go to the leaders of the Exotic Pest Plant Councils: Carla Bossard, Brian Bowen, Dan Thayer, and Lou Whiteaker. Randy Westbrooks, Howard Singletary, George Beck, and John Randall have taught me much over the years. Like many others, I first became involved with alien species in the context of Hawaii; thanks there go to Rob Milne, Dan Taylor, Tim Tunnison, Larry Katahira, Lloyd Loope, Art Madeiros, and Ron Nagata. Finally, thanks to John Schwegman, who has the soul to care about the disappearance of woodland herbs.

17
Team Arundo: Interagency Cooperation to Control Giant Cane (*Arundo donax*)

Paul R. Frandsen

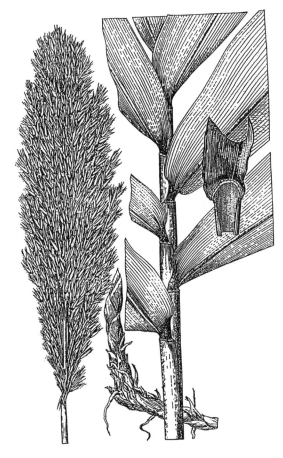

Giant cane
(*Arundo donax*)

To manage large-scale plant invasions effectively, a regional approach involving several agencies and many people is often required. In southern California such a cooperative gathering of agencies has focused efforts on problems created when riparian habitats are invaded by giant cane (*Arundo donax*). This gathering of personnel is called Team Arundo.

Historical Setting

As the Spanish settled the New World, they brought their technology, animals, and plants to help them survive in the foreign and sometimes hostile environments. They needed a lightweight building material, a source of firewood, food for their cattle, and a means of protecting the soil from wind and water erosion. To satisfy these needs, they introduced a grass, arundo or giant cane (*Arundo donax*). Missions along the old trading routes were built using giant cane as the lath to hold the plaster; leaves of the plant were also used as thatch for the construction of roofs. Giant cane escaped and spread slowly from the early ranchos. As more settlers came to enjoy the climate, they planted more cane.

It was not until the last 40 years that major shifts in plant communities in southern California became evident to those who understood the intricate web of life along rivers in this arid region. Water diversions allowed year-round flow rather than in only the winter and spring. Novel habitats created in riparian areas apparently favored the development of dense stands of giant cane (Rieger and Kreager 1989). In addition to riparian habitats, giant cane also grows in irrigation ditches, streams, and near seeps (Hoshovsky 1993a).

Plant Biology

Giant cane can reach a height of 6 m and may grow up to 6.3 cm/day under optimum conditions (Rieger and Kreager 1989). It has hollow, segmented culms 1 to 4 cm in diameter. A creeping rootstock bears fibrous roots, which penetrate up to 4.9 m deep in the sand. Although giant cane grows on many different soil types (Hoshovsky 1993a), it is most frequently found on sandbars very close to the river edge (Rieger and Kreager 1989). Growth can occur year-round. The seeds are not viable; the plant reproduces by vegetative pieces as small as a pencil or by "spears" coming from the root mass.

Southern California rivers wash, scour, and deposit sand and small rock after winter storms. The native riparian forest is typically a jumble of cottonwood (*Populus*), mule fat (*Baccharis*), willow (*Salix*), and other plants. These species have various mechanisms for persistence under such a disturbance regimen. Giant cane interferes with regeneration of the native riparian forest by growing more rapidly than indigenous (IN) species, by shading out these species, and then by burning during the fire season (Rieger and Kreager 1989). Bell (1994) presented a conceptual model of plant succession as influenced by giant cane. Invasion by cane alters fire frequency, which in turn favors dense regeneration of cane. Indigenous woody plants are eliminated. Each year more acres of rivers wash, become filled with giant cane, and then catch fire. Simultaneously, development projects put semi-treated wastewater into the rivers, providing giant cane with a constant source of water that did not exist in earlier historical periods.

There is no known biological agent that kills giant cane in the invaded region; however, in some areas prescribed grazing by goats has produced some control (Hoshovsky 1993a). Stands become so thick that birds cannot fly through the dense growth. Pure stands lack a diversified structure, so the birds that depend on the multistory riparian forest for nest sites cannot find suitable habitat.

In addition to changing the conservation value of riparian areas, giant cane also modifies the disturbance regimen. Dense, monotypic stands create debris dams that have damaged bridges and caused extra maintenance costs for removal of the dams. In irrigation ditches, water flow may be impeded by giant cane.

The plant, thought to be a native of the Mediterranean region, has an invaded range extending from central California into Mexico, Arizona, and Nevada; colonies have also been observed in central Oregon. While it normally thrives in river bottoms, it has been found on hillsides and is used as an ornamental.

The Problem

As giant cane expanded its southern Californian geographical range following fires and floods, it presumably eliminated certain endangered species through either direct competition or habitat destruction. This extirpation of endangered species, in turn, increased the cost of doing business for developers and public agencies. In some instances the developers were required to mitigate environmental impacts set in motion by invasion of giant cane. The resulting expense to developers prompted much controversy and generated pressure from state and federal resource agencies and from local governments. Trips to Washington, D.C., were made, letters were written, and contentious meetings took place all in an effort to facilitate development projects. Mitigation activities were directed at small plots where chances for long-term habitat protection were minimal.

At the time, no one was able to perceive the extent and magnitude of effects associated with giant cane. Those who fought annual fires accepted the cost of fighting the fires to save homes but did not take any action to eliminate the basic cause of the fires. Flood control districts paid to remove giant cane but in many instances had to restore habitats after management activities. Water agencies lost large amounts of water to transpiration and thus had to increase their purchase of imported water. It has been estimated that giant cane uses about three times more water than the native vegetation (Iverson 1994). Finally, giant cane damaged bridges by modifying water-flow patterns and impeded recreational access to local rivers.

No single agency could correct its own particular problem with giant cane while, at the same time, avoiding conflict with other agencies. For example, fire suppression personnel were not allowed to create firebreaks because park managers assumed that this damaged habitat. Transportation agencies were not allowed to remove giant cane when it grew over bridge railings so they cut the tops of it, thus stimulating release of basal stems. In large agencies, such as the U.S. Army Corps of Engineers, different administrative sections provided contradictory instructions for maintenance of flood structures while not negatively affecting IN plants. The decision to use heavy equipment to clear floodways of giant cane required permits from and consultation with the U.S. Army Corps of Engineers, the U.S. Fish and Wildlife Service, the California Department of Fish and Game, the California Regional Water Quality Board, and various county flood-control districts. Environmental groups that might protest removal of the cane had to be given notice. In one case, an agency was told to mitigate removal of giant cane as if it were a preferred species. The first 48-acre site where giant cane was removed required state and federal permits and a posting with a state clearing house. This process took a year.

The Solution? Team Arundo

Telephone calls were made to the local battalion chief of the U.S. Army Corps of Engineers and to the regional ecologist of The Nature Conservancy to see if either person thought giant cane was a serious resource-management problem. The ecologist responded and planned the first meeting with all interested parties. There were 18 representatives from city, county, state, and federal agencies, a conservation group, The Nature Conservancy, and Monsanto Corporation. This assemblage initially had no name, no mission, no agreement, and no unity. I hosted the meetings, set agendas, invited potential members, and determined what each person or organization required to participate in this venture. The name we chose for the project was the Santa Ana River Habitat Recovery Project. Our group assumed the name of Team Arundo.

It was obvious from the outset that members of Team Arundo had little in common. Six months elapsed before members started communication with each other. This new, face-to-face interaction revealed new needs along the river, generated further misunderstandings, and made clear just how little we knew about up-river processes. An attempt to write a memoramdum failed, but we continued to meet. Team Arundo became an ad hoc group dealing with emerging resource-management problems along the Santa Ana River.

The first ever cooperative Team Arundo project involved hand removal of giant cane from a 5-acre site. From this followed a day-long workshop attended by more than 100 people dealing with nonindigenous (NI) plants and their management, creating a guide to the removal of giant cane, initiating a riverwide planning effort, developing maps to show where giant cane occurred, looking for various uses of the plant, conducting a study on water uptake by the plant, sharing of resources among agencies, and educating the public about the effort. As an ancillary benefit, friendships began to develop among members, relationships began to improve, and face-to-face dealings started to lessen the tension.

A 50-acre fire that cleared a large population of giant cane helped to focus further actions. In an effort to eradicate this population while plants were reduced in stature, Monsanto Corporation provided technical support, paid for a helicopter to spray herbicide, developed a public announcement, and provided contacts in the herbicide industry. One county sent a fleet of trucks to help with the spraying. The city of Riverside provided staff, vehicles, and funds. As a result of this project, federal and state agencies began to rethink their river-management strategies.

After 2 years, more planning has been completed and a subcommittee has been formed to address management issues in the upper river. Heads of various agencies have embraced the concept of giant-cane control. More agencies have sent representatives to Team Arundo. The entire arena of river management has changed for the better. In Riverside County alone more than 100 acres of riparian areas are under management, and many other projects are in the planning stages.

Negotiations are nearly complete for creation of the largest mitigation bank in southern California—the Santa Ana River. The U.S. Environmental Protection Agency, the U.S. Army Corps of Engineers, the U.S. Fish and Wildlife Service, the California Department of Fish and Game, and the Regional Park and Open-Space District (lead agency) are working on an agreement for integrated management of the river.

From Team Arundo has emerged a new approach to river management that offers potential for protection of endangered species, water conservation, flood control, fire protection, and increased recreational access. Central to the success of Team Arundo was an understanding of the ecological impacts of giant cane coupled with development of improved techniques to manage the plant.

The Future

The organization known as Team Arundo has achieved many of the goals established at its inception: better understanding of the Santa Ana River and its ecology; public acceptance of herbicides and heavy equipment as tools for river restoration; increased state and federal flexibility regarding management practice; and better cooperation among various competing interests. Much work still needs to be done because giant cane occurs throughout many parts of the river basin.

Control of giant cane was a relatively simple and straightforward task. However, the development of procedures to implement control of giant cane was much more complex. Permits are now being prepared for management of a larger section of the river, thus simplifying the permit-acquisition process. New rules and procedures formed by large administrative groups such as the U.S. Army Corps of Engineers have better addressed local needs.

If one is to be successful in managing a particular NI plant, a broad-based effort involving all stake-holders is necessary. Nonindigenous plants and animals are fundamentally problems requiring the management by people through involvement of the community, the press, and conservationists. I try very hard to make agency representatives part of the team, thus changing positions of power to positions involving a common goal. Pressure to be more responsible to all parties involved at all levels of government can help in implementing a management plan. Homework is required in building this base of support in and out of an organization, for seeking the needs of each partner, and for providing consistent, positive leadership. Funding for Team Arundo now totals nearly $850,000.

Conclusions

1. Giant cane (*Arundo donax*) has invaded many riparian areas in southern California with corresponding negative effects on en-

dangered IN species. It also increases intensity of fires and thwarts flood-control efforts.

2. Because the Santa Ana River encompasses a variety of jurisdictions and because various agencies have different management goals for the river, it was historically difficult to achieve control of giant cane and restoration of the riparian community.

3. A loose association—Team Arundo—including county, state, and federal agencies, conservation groups, The Nature Conservancy, and industry was formed to reclaim the Santa Ana River from giant cane.

4. Team Arundo facilitated cooperation among diverse groups with diverse management goals. It created educational materials. It served to focus funding on the basic cause of many problems along the river. It developed novel management strategies. And it allowed restoration to occur throughout the Santa Ana drainage area.

18
A Multiagency Containment Program for Miconia (*Miconia calvescens*), an Invasive Tree in Hawaiian Rain Forests

Patrick Conant, Arthur C. Medeiros, and Lloyd L. Loope

Invasiveness of a species in a particular part of the world can often be successfully predicted by the species' invasive behavior in another part of the world with similar environmental conditions. When naturalized miconia (*Miconia calvescens*; Melastomataceae) was discovered on Maui, Hawaii, in 1990, biologists responsible for protecting native ecosystems paid attention because of the infamous reputation of this species in ecologically similar Tahiti (French Polynesia), where it has come to extensively dominate forests since its introduction there in 1937. This species was introduced to the Hawaiian Islands as an ornamental in the early 1960s without attracting any great attention and was sold at nurseries for 2 decades. As of 1996, miconia had aroused a battle cry throughout the state of Hawaii, and cooperative efforts for its control were underway on four islands.

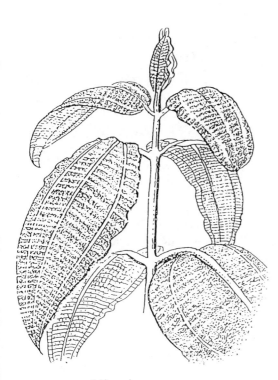

Miconia
(*Miconia calvescens*)

Distribution and Ecology of Miconia

Miconia, a tree up to 10 to 15 m tall, native to the neotropics, is widespread and overwhelmingly dominant on Tahiti, an island with rain forest habitat very similar to that of the Hawaiian Islands and other high islands of the Pacific. First introduced to Tahiti at the

Papeari Botanical Garden in 1937, the species now occurs on over 65% of the 1045 km² island and forms dense, monospecific stands on more than 25% of the island (Meyer 1996). Ecosystems become completely transformed as miconia gains dominance, due to its creation of deep shade that no indigenous (IN) plant species can tolerate (Meyer 1996). In Tahiti, 70 to 100 IN species, including 35 to 45 species endemic to French Polynesia, are directly threatened by invasion of miconia into forests (Meyer and Florence, unpublished data). This invader has now spread to three other islands of French Polynesia. It is also naturalized to a lesser extent in Sri Lanka and Jamaica (Meyer 1996).

The indigenous range of miconia extends from 20°N in Mexico, Guatemala, and Belize to 20°S in Brazil and Argentina (Meyer 1994). The upper elevation limit of the species in its indigenous range is 1830 m in Ecuador (Wurdack 1980). Meyer (1994) determined that the form with very large leaves (to greater than 1 m in length) with purple undersides occurs only in Mexico, Guatemala, Belize, and Costa Rica; specimens examined by Meyer were collected at elevations between 45 m and 1400 m. In Tahiti, it forms dense stands at elevations as high as 1300 m (Meyer 1994). The potential upper elevation range of miconia is an important consideration for the Hawaiian Islands, where large areas of relatively intact IN rain forest occur between 1000 m and 2000 m (Cuddihy and Stone 1990; Jacobi and Scott 1985).

Although apparently a gap invader in its indigenous habitat, and abundant only very locally if at all, miconia is ideally suited for invasion of wet, tropical oceanic islands. Under rain forest conditions, it grows rapidly, produces abundant seed, accumulates a large, persistent soil seed bank, is efficiently dispersed by birds, and tolerates shade (Meyer 1996). Saplings can grow at a rate of 1 m per year; trees in ideal conditions can flower and fruit after only 4 to 5 years of growth at a height of 4 to 5 m. A single mature tree is capable of producing millions of tiny (ca. 0.6 mm diameter) bird-dispersed seeds each year. Two of the genera of nonindigenous (NI) birds believed to disperse miconia in Tahiti (Gaubert 1992) occur also in Hawaii (*Zosterops* and *Pycnonotus*). In Tahiti, miconia forms dense, monotypic stands in areas where mean annual rainfall exceeds 2000 mm. It appears to reduce biological diversity in most habitats receiving 1800 to 2000 mm or more of annual precipitation, including high-elevation rain forests.

Through perusal of botanical garden records and discussions with knowledgeable horticulturists, we have been able to determine that miconia was present in the Hawaiian Islands on Oahu (a single tree in Wahiawa Botanical Garden) by 1961. It had reached the island of Hawaii by 1964, and Maui by the late 1960s or early 1970s. It may have reached Kauai by the early 1980s, based on the recent (1995) discovery there of a fruiting tree more than 10 m tall.

Agencies and Citizen Groups Involved in Miconia Control in Hawaii

In spite of the fact that Hawaii is plagued with plant invasions to a greater extent than most places in the world, a coordinated system designating agency responsibilities is lacking (Smith 1985; Tunison et al. 1992). Agencies involved in pest control are chronically underfunded, and each has its own priorities. The fact that the state of Hawaii is spread over eight major islands tends to discourage unified, statewide campaigns against any particular pest. Each island has its own perceived problems with invasive weeds. And within each island the meagre resources of each individual agency for pest control are typically devoted to a few species considered by that agency as top priorities. A certain amount of cooperation and coordination has existed all along, of course, but the alarm caused by the realization of potential damage from miconia is stimulating united action.

The island of Maui is where efforts in the Hawaiian Islands were first mobilized against the invader. About 20 years after the plant's apparent introduction at Helani Gardens near Hana in northeastern East Maui, National Park Service biologists became aware in 1990 that it was present on Maui, and an alarm was raised (Gagne et al. 1992). Seven populations were found in the general vicinity, but prospects seemed good for eradication because all were easily accessible. More than 20,000 plants were removed from 1991 to 1993 with assistance of volunteers. However, a much larger concentration of miconia, discovered by a Maui state forester in September 1993, occurs in several foci within a 150 ha area on a 500-year-old lava flow up slope (100 to 350 m elevation) and west of Helani Gardens. By the time of its discovery, the population contained an estimated 1000 fruiting trees (Hobdy R and Medeiros AC, pers. obs.). A multiagency effort at eradication has been mobilized by the Melastome Action Committee and the East Maui Watershed Partnership, both of which include representatives from federal, state, county, and private entities.

The Melastome Action Committee (MAC) was formed in mid-1991 through the initiative of a conservation coordinator of a private organization (Maui Land and Pineapple Company) and the local representative of the Maui County Resource Conservation and Development Office of the U.S. Department of Agriculture. Formation of the committee was spurred by the need to address the severe threats to conservation lands posed by miconia and by a glorybush (*Tibouchina herbacea*), both members of the melastome family. The committee name recognizes that many of the 15 species of Melastomataceae thus far introduced to Hawaii have become aggressive weeds. The MAC has met periodically since, with participation of the Hawaii Department of Land and Natural Resources (HDLNR), the Hawaii Department of Agriculture (HDOA), The Nature Conservancy (TNC), the East Maui Irrigation Company (EMI), the University of Hawaii, the National Park Service (NPS), the U.S. Forest Service, and (since November 1993) the National Biological Service (NBS). Activities of the committee include public education, lobbying at the Hawaii legislature and Maui County for funding for weed control programs, and facilitating cooperative chemical, mechanical, and biological control programs.

Whereas the MAC was formed specifically to address the threat from miconia, another state-federal-private alliance, the East Maui Watershed Partnership (EMWP), was formed in late-1991 with the objective of managing the shared watershed "to protect this area from nonindigenous (NI) pest animals, weeds and other threats" for sustained supply of water and conservation of biological diversity. Members include HDLNR, HDOA, TNC, EMI, NPS, Maui County, and Hana and Haleakala ranches. By 1994, the EMWP had begun to recognize miconia as the greatest obstacle to accomplishing their mission. The EMWP and MAC have worked jointly and effectively to develop a strategy, obtain funds, and implement miconia control on Maui. Individuals of various agencies have stepped in to fill essential niches in the control effort. Major commitments are being made by HDLNR (overseeing control within largest populations), TNC (eradication within peripheral populations), and NBS (monitoring effects of control of miconia populations).

Stimulated largely by increased awareness of the threat from miconia generated initially by concern on Maui, important efforts are underway on other islands. A chapter of the MAC was formed in 1995 to address the problem on the island of Hawaii, where the situation is much more difficult to grasp because miconia is probably much more widespread there than on other Hawaiian islands. An HDOA employee, working with the Sierra Club and other volunteers, has been exploring Oahu for miconia and eliminating fruiting trees, saplings, and seedlings. When a private citizen reported miconia on Kauai in 1995, employees from HDLNR, HDOA, TNC, and the National Tropical Botanical Garden canvassed the invaded area for additional plants; information from local residents was helpful.

Meanwhile, the HDOA has mounted an effort to explore biological control methods. Biological control appears promising in the long run, but no group advocates inaction in miconia containment or eradication while awaiting biocontrol success.

Strategies and Tactics

Containment Versus Eradication

In formulation of strategies for combatting miconia in Hawaii, it has been difficult to settle on the term for the goal to be achieved—whether it is *eradication* or *containment*. Because of the life history characteristics of miconia, with 4 to 5 years and 3 m of height growth separating seedlings from fruiting trees, eradication is clearly possible for some small localized populations and may be possible for entire islands. Working against eradication is the longevity of the seed bank (more than 3 years); eradication will clearly require sustained commitment, but so will containment. Some participants working at the statewide level have advocated referring to the statewide effort as a containment program (Holt RA, pers. com.). We advocate aiming for eradication at a locality or island level, with the realization that it may be only an ideal. Unfortunately, as long as there is a seed source of miconia in the state, there will always be a strong possibility of infestation of new areas or reinfestation of areas from which the plant has been eradicated. It can easily be moved intra- or inter-island on dirty boots.

Aerial Spraying

The largest miconia population on Maui, discovered from the air in 1993, was initially inaccessible on the ground because of the extremely rough terrain of a 500-year-old lava flow. As a holding action to limit seed production, a Hughes 500-D helicopter with an attached compressed-air-powered sprayer spot-sprayed herbicide on larger, fruiting trees beginning in early 1994. This specific-delivery, spot-spray system had been developed to control illegal marijuana cultivation in remote mountain areas. The herbicide (Garlon 4®, ester formulation of triclopyr) was applied with surfactant and blue dye (Turfmark®). The dye assists the pilot in judging application rate and identifying treated plants. This program has been implemented by HDLNR, with technical assistance from HDOA. National Biological Services researchers conducted monitoring to assess effects of the spraying. In the initial trials, about 70% of sprayed individuals were killed; others lost leaves and aborted flowers and green fruits, yet recovered and fruited in the next fruiting season (A. Medeiros, unpublished data). Vegetation plots and tagged individuals are being monitored to determine survival of sprayed miconia, effects on nontarget plants, and long-term succession after localized canopy disturbance.

Removal of Miconia by Ground Crews and Volunteers

Hand removal (uprooting) is an effective method of removing plants less than about 3 m tall. Adventitious rooting of uprooted individuals occurs occasionally but is rare. If larger individuals cannot be uprooted and are cut down, the stump must be treated with an herbicide (e.g., Garlon 4®) or it will resprout. Garlon 4 has been shown to be highly effective for this purpose in tests conducted by TNC and NPS.

In large concentrations of the tree, canopy removal often results in a spectacular germination of the miconia seed bank; the seedlings may cover substantial areas in clearings. About 18 months after germination, there can be up to 500 to 1000 seedlings/m^2, with the tallest about 0.7 m tall. These can effectively be dealt with by spraying with Garlon 3A®. A second (and, perhaps, third) treatment will be required after another 1 to 2 years to destroy the remaining seed bank (although seedling numbers in subsequent generations are reduced).

The greatest problem for on-the-ground control and monitoring of the largest miconia population on Maui is one of access. The terrain is extremely rough on the 500-year-old lava flow, making walking very difficult, even on a cleared path. Traversing 100 meters in some areas through a vegetation cover can take more than an hour. An old, partially overgrown bulldozer track proved the best access route initially; new access routes are now being developed to allow crews to reach control sites rapidly and safely. Work is to be done by contract at a cost of $7000 to $10,000 per mile.

Preventing Spread of Miconia Seeds

An important factor in mechanical and chemical control is the seed bank associated with miconia stands, which necessitates monitoring and removal of emergent seedlings for more than 3 years. Another problem is the potential transfer of the seeds to other sites in soil on the boots and equipment of crews engaged in assessment and control efforts. This problem was dramatized in an incident experienced by the NBS research program on Maui. After a miconia seedling was found in a rain forest vegetation plot several miles distant from the nearest known population, it was clear that the seed was dispersed by a researcher using gear worn at both sites, even though gear was cleaned by hand-washing between uses. After this incident, those working with control of the plant were encouraged to wear conspicuously marked footwear and other "dedicated," gear, that is, gear used only for work involving miconia.

The seed dispersal problem greatly complicates the issue of using volunteers. Whenever miconia control is undertaken, a supervisor must be responsible for seeing that safeguards are taken seriously. Whenever bulldozers and other vehicles are used, they must be pressure washed immediately afterwards.

Biological Control

An important prospect for long-term control of miconia in areas where it is invasive is classical biological control. The HDOA began efforts to pursue this option in mid-1993. Exploratory entomologist Robert Burkhart visited Costa Rica, Brazil, Uruguay, Paraguay, Argentina, Trinidad, and Tobago. More than 70 species of potential biological control agents were found, including fungi, weevils, leaf beetles, and numerous species of butterflies and moths. Unfortunately, because of delays in quarantine when these insects were brought into the United States, populations of some species died before arrival in Hawaii. Later, under quarantine conditions in Hawaii, most remaining insect biocontrol candidates eventually perished, due primarily to low numbers of individuals imported and the lack of needed specific conditions at the Honolulu quarantine facility. To circumvent this latter problem, contract work within the IN range of miconia should also be considered.

This first exploratory trip nonetheless documented good collecting sites and distribution of natural enemies. Exploration needs to be done in the northern part of the IN range. Several fungi survived shipment back to Hawaii, including the leaf spot fungus *Cocostroma myconae*. Unfortunately, none of them could be successfully cultured. Burkhart reported *C. myconae* had the most impact of those pathogens he saw in Costa Rica and Brazil. He believes that the prospect of finding effective pathogens for biocontrol of miconia may be more promising than finding effective insect natural enemies (Burkhart R, pers. com.). Further exploration has in fact been done by HDOA in the northern range of miconia in Guatemala, and cooperators in Brazil have sent diseased miconia specimens to the Plant Pathogen Quarantine Facility in Honolulu (Killgore E, pers. com.). Examination and assessment of the pathogens for use as biocontrol agents is ongoing. Prospects for funding of the biocontrol program are now uncertain but are improved by the interest of the government of French Polynesia in cooperatively supporting biocontrol work with HDOA.

Success of biological control of invasive plants has had mixed results. Julien (1992) estimated that between 60% and 75% of the releases of classical biocontrol agents are ineffective. Fortunately, the Hawaiian flora lacks IN melastome species, so that opportunities for bringing in generalist melastome feeders may considerably enhance the chances for success against miconia. But even assuming best case scenarios for biocontrol, success is probably at least a decade in the future.

Education

Public education, a crucial aspect of the containment and eradication program, has been aided by the distinctiveness of the plant, with its huge leaves with purple undersides. Major articles on miconia first appeared in the *Maui News* and the *Honolulu Star-Bulletin* in May and June of 1991. A colorful and factual "wanted" poster on the plant was prepared and produced in large numbers. Since it became available in late 1991, this poster has been distributed widely to educate people and solicit reports of miconia locations.

In April 1993, Jean-Yves Meyer, a French graduate student conducting studies on miconia in Tahiti, visited Hawaii; he was shown *Miconia* populations on Maui and Hawaii and made contacts with numerous agencies and individuals. In June 1994, Arthur Medeiros of NBS was sent to Tahiti and, assisted by Meyer, obtained good photographic documentation of the situation in Tahiti and French Polynesia. This documentation of potential damage by miconia has proved invaluable in convincing doubters of the need for prompt action. Meyer completed his Ph.D. thesis on miconia in late 1994 and now works in Tahiti for the Ministry for the Environment; he presented a keynote talk on miconia at the 1995 Hawaii Conservation Conference in July. Medeiros has made numerous presentations based on his Tahiti visit. In October and November of 1995, major stories on miconia appeared in the *Boston Globe* and in *U.S. News and World Report*.

A public service announcement on the threat from miconia was prepared under the direction of the MAC. It is shown repeatedly by several TV stations; it led to the crucial first report of the plant from the island of Kauai.

Prospects for Success

There are many potential invasive plant control projects in Hawaii. Control of miconia is very high in priority because (1) its potential impact on crucial watershed lands and on biological diversity in Hawaii—even at relatively high elevations—is enormous; (2) the largest populations are at densities where containment and eradication are possible; and (3) the biology and structure of the plant make containment or eradication feasible at this time, at least on some islands and parts of other islands. Another very significant factor is that this is the first large cooperative interagency NI plant control effort in Hawaii; its success or failure will be crucial in determining the feasibility of such efforts.

Over time, balance between ground and aerial control and biocontrol is clearly essential. Currently the emphasis is on physically removing or killing plants with herbicide because of the perception that we are near the point of no return with proliferating miconia populations. No one is advocating putting all resources into biocontrol at this time, but it is very important that biocontrol be pursued aggressively. Although we have advocated aiming for eradication at a locality or island level, we fully realize that it may be only an ideal for more dispersed populations. Unfortunately, as long as there is a seed source of miconia in Hawaii, there will always be a strong possibility of infesting new areas or reinfesting areas from which the plant has been eradicated. Biological control must be pursued, but is at least 10 years away. Mechanical and chemical control must be considered the only proven control methods for the foreseeable future.

Appendix. Selected Plant Species Interfering with Resource Management Goals in North American Natural Areas

Some characteristics of selected nonindigenous and indigenous plant species that may interfere with natural-resource management goals in North America (north of Mexico) and Hawaii are presented here. Indigenous ranges are based on published information; specific indigenous ranges are not well known for some species. Invaded ranges represent Canadian provinces and territories or American states from which the species have been reliably reported to occur as invaders; some of these species may not be established in all areas from where they have been reported (e.g., *Eichhornia crassipes* in Ontario). The references provided do not represent exhaustive surveys of literature. Instead, key references are cited that may serve as useful starting points should a researcher decide to initiate a more detailed literature search.

The species to be included in this list were suggested by Faith Thompson Campbell and Sarah Reichard. Geographical data were provided by John T. Kartesz. Bibliographic data were compiled by John W. Thieret.

Beach naupaka
(*Scaevola sericea*)

Abbreviations for Provinces, Territories, and States

Alberta	AB
British Columbia	BC
Manitoba	MB

255

New Brunswick	NB
Newfoundland	NF
Northwest Territories	NT
Nova Scotia	NS
Ontario	ON
Prince Edward Island	PE
Quebec	QU
Saskatchewan	SK
Yukon Territory	YT

Alabama	AL
Alaska	AK
Arizona	AZ
Arkansas	AR
California	CA
Colorado	CO
Connecticut	CT
Delaware	DE
Florida	FL
Georgia	GA
Hawaii	HI
Idaho	ID
Illinois	IL
Indiana	IN
Iowa	IA
Kansas	KS
Kentucky	KY
Louisiana	LA
Maine	ME
Maryland	MD
Massachusetts	MA
Michigan	MI
Minnesota	MN
Mississippi	MS
Missouri	MO
Montana	MT
Nebraska	NE
Nevada	NV
New Hampshire	NH
New Jersey	NJ
New Mexico	NM
New York	NY
North Carolina	NC
North Dakota	ND
Ohio	OH
Oklahoma	OK
Oregon	OR
Pennsylvania	PA
Rhode Island	RI
South Carolina	SC
South Dakota	SD
Tennessee	TN
Texas	TX
Utah	UT
Vermont	VT
Virginia	VA
Washington	WA
West Virginia	WV
Wisconsin	WI
Wyoming	WY

Species Information

Acacia confusa, Formosa koa
  Growth form: Shrub
  Indigenous range: Taiwan, Philippines
  Invaded range: HI
  Introduction data: Introduced about 1915 (Wagner et al. 1990)

Acacia farnesiana, sweet acacia
  Growth form: Shrub
  Indigenous range: Warm regions of New World, but now in both hemispheres
  Invaded range: AL AZ CA GA HI
  Introduction data: Mainland United States—Indigenous to Florida. Hawaii—First collected in 1860s but, because the species was present on all main islands prior to 1871, it was "probably introduced considerably earlier than the 1860s" (Wagner et al. 1990)
  References: Control, Bovey and Meyer (1989); Seed predation, Traveset (1991)

Acacia mearnsii, black wattle
  Growth form: Shrub
  Indigenous range: New South Wales, Tasmania
  Invaded range: CA HI
  Introduction data: Hawaii—Introduced from a California nursery in 1911 (Wagner et al. 1990)
  Reference: Cronk and Fuller (1995)

Aacia melanoxylon, blackwood
  Growth form: Shrub
  Indigenous range: Australia

Invaded range: CA HI
Introduced range: Hawaii—Planted by state foresters on at least four islands; by the 1980s it was naturalized on at least one island (Wagner et al. 1990)
References: Review, Cronk and Fuller (1995); Growth, Bi and Turvey (1994)

Agropyron repens, quack grass
Growth form: Perennial grass
Indigenous range: Eurasia
Invaded range: All provinces and territories; AK AR AZ CA CO CT DE IA ID IL IN KS KY MA MD ME MI MN MO MT NC ND NE NH NJ NM NV NY OH OK OR PA RI SD TN UT VA VT WA WI WV WY
Introduction data: Introduced probably during colonial times; "very troublesome" by 1822 (Eaton 1822)
References: Review, Holm et al. (1991), Werner and Rioux (1977); Control, Alcantara et al. (1989)

Ailanthus altissima, tree-of-heaven
Growth form: Tree
Indigenous range: Eastern Asia
Invaded range: BC ON; AL AR AZ CA CO CT DE FL GA IA ID IL IN KS KY LA MA MD ME MI MO MS NC NE NJ NM NY OH OK PA RI SC TN TX
Introduction data: Introduced to Europe in in 1751 (Hillier 1983). Introduced to United States in 1784 (Hu 1979)
References: Review, Anderson (1961), Cronk and Fuller (1995), Hu (1979), Randall and Marinelli (1996); Community effects, Heisey (1990), Lawrence et al. (1991)

Alliaria petiolata, garlic mustard
Growth form: Biennial or winter annual herb
Indigenous range: Europe
Invaded range: BC NB ON QU; AR CO CT IA IL IN KS KY MA MD ME MI MN MO NC ND NE NH NJ NY OH OK OR PA SC TN UT VA VT WI WV
Introduction data: First recorded in 1868, Long Island, New York (Nuzzo 1993)
References: Review, Cavers et al. (1979); Control, Nuzzo (1991, 1994), Nuzzo et al. (1991); Community impact, McCarthy (Chapter 10, this volume); Invasion, Nuzzo (1993)

Ammophila arenaria, European beach grass
Growth form: Perennial grass
Indigenous range: Western Europe
Invaded range: BC; CA HI MA OR WA
Introduction data: Mainland United States—Introduced in the "late 1800s" (Boyd 1992)
References: Review, Buell et al. (1995), Cronk and Fuller (1995), Huiskes (1979); Community effects, Boyd (1992)

Ampelopsis brevipedunculata, porcelain-berry
Growth form: Perennial woody vine
Indigenous range: Temperate eastern Asia
Invaded range: CT DE KY MA MD MI NC NH NJ NY OH PA RI VA
Introduction data: Introduced to Europe in 1870 (Hillier 1983)
References: Review, Randall and Marinelli (1996); Invasion, Robertson et al. (1994)

Andropogon virginicus, broom-sedge
Growth form: Perennial grass
Indigenous range: Eastern North America (ON; AL AR CT DE FL GA IA IL IN KS KY LA MA MD MI MO MS NC NJ NY OH OK PA RI SC TN TX VA WV), Mexico, Central America
Invaded range: CA HI
Introduction data: Hawaii—First collected in 1924 (Cuddihy and Stone 1990)
References: Review, Cronk and Fuller (1995), Cuddihy and Stone (1990); Community effects, Rice (1972); Control, Peters and Lowance (1974); Population differentiation, Nellessen and Ungar (1993)

Arundo donax, giant reed
Growth form: Perennial grass
Indigenous range: Mediterranean region
Invaded range: AL AR AZ CA FL GA HI IL KS KY LA MD MO MS NC NM NV OK SC TN TX UT VA WV
Introduction data: California—Introduced before 1820 (Hoshovsky (1993a). Hawaii—Naturalized prior to 1871 (Wagner et al. 1990)

References: Review, Hoshovsky (1993a), Jackson et al. (1994), Randall and Marinelli (1996); Community and ecosystem effects, Rieger and Kreager (1989)

Bromus inermis, smooth brome
Growth form: Perennial grass
Indigenous range: Eurasia
Invaded range: All provinces and territories; AK AR AZ CA CO CT DE IA ID IL IN KS KY LA MA MD ME MI MN MO MT NC ND NE NH NJ NM NV NY OH OK OR PA RI SD TN UT VA VT WA WI WV WY
Introduction data: Introduced to the United States in the 1880s (Blankespoor and Larson (1994)
References: Control, Blankespoor and Larson (1994), Grilz and Romo (1995)

Bromus tectorum, cheat grass
Growth form: Annual grass
Indigenous range: Europe and northern Asia
Invaded range: AB BC MB NB NS ON QU SK YT; all states
Introduction data: Introduced to North America before 1861 (Upadhyaya et al. 1986), to Pacific Northwest and Utah 1889–1894 (Mack 1981)
References: Review, Cronk and Fuller (1995), Upadhyaya et al. (1986); Community effects, Billings (1994); Invasion, Douglas et al. (1990), Mack (1981)

Cardaria draba, white-top
Growth form: Perennial herb
Indigenous range: Eurasia
Invaded range: AB BC MB NS ON QU SK; AZ CA CO CT IA ID IL IN KS KY MA MD ME MI MN MO MT ND NE NJ NM NV NY OH OK OR PA RI SD TX UT VA VT WA WI WV WY
Introduction data: First collected in the United States (New York) in 1862, in Canada (Ontario) 1878 (Mulligan and Findlay 1974)
Reference: Review, Mulligan and Findlay (1974)

Casuarina equisetifolia, Australian-pine
Growth form: Tree
Indigenous range: Australia
Invaded range: FL HI
Introduction data: (Some literature reports under this name may be based on misidentifications.) Mainland United States—Seeds available in the U.S. trade by 1825 (Mack 1991). Hawaii—First collected in 1895 but probably introduced in 1882 (Wagner et al. 1990)
References: Review, Cronk and Fuller (1995), Duever et al. (1986), Randall and Marinelli (1996); Community effects, Apfelbaum et al. (1983)

Celastrus orbiculatus, oriental bittersweet
Growth form: Perennial woody vine
Indigenous range: Eastern Asia
Invaded range: ON QU; AR CT GA IA IL IN KY MA MD ME MI NC NH NJ NY OH PA RI SC TN VA VT WV
Introduction data: Introduced to western horticulture around 1860 (Hillier 1983)
References: Review, Randall and Marinelli (1996); Invasion, Robertson et al. (1994)

Centaurea maculosa, spotted knapweed
Growth form: Biennial herb, short-lived perennial
Indigenous range: Europe
Invaded range: AB BC NB NS ON QU; AL AR AZ CA CO CT DE FL HI IA ID IL IN KS KY MA MD ME MI MN MO MT NC ND NH NJ NM NV NY OH OR PA RI SC SD TN UT VA VT WA WI WV WY
References: Review, Watson and Renney (1974); Control, Griffith and Lacey (1991), Muller-Scharer and Schroeder (1993); Reproduction, Harrod and Taylor (1995); Seed bank, Davis et al. (1993)

Centaurea repens, Russian knapweed
Growth form: Perennial herb
Indigenous range: Southern Russia and Asia
Invaded range: AB BC MB ON SK; AR AZ CA CO IA ID IL IN KS KY MI MN MO MT ND NE NM NV OH OK OR SD TX UT VA WA WI WY
References: Review, Watson 1980; Reproduction, Harrod and Taylor (1995)

Centaurea solstitialis, yellow starthistle
Growth form: annual herb
Indigenous range: Mediterranean region
Invaded range: AB MB ON SK; AZ CA CO CT DE FL IA ID IL IN KS KY MA MD MI

MN MO MT NC ND NE NH NJ NM NV NY OH OK OR PA RI SD TN TX UT VA WA WI WV WY
Introduction data: Introduced to the United States before 1814 (Pursh 1814)
References: Reproduction, Harrod and Taylor (1995); Seed germination, Callihan et al. (1993), Sheley et al. (1993)

Chondrilla juncea, rush-skeletonweed
Growth form: Biennial herb
Indigenous range: Eurasia
Invaded range: BC ON; CA DE GA ID IN MD MI NJ NY PA VA WA WV
References: Review, Panetta and Dodd (1987); Control, Grover and Cullen (1981), Supkoff et al. (1988), Zamora et al. (1989); Seed germination, Panetta (1989)

Clidemia hirta, Koster's curse
Growth form: Shrub
Indigenous range: Tropical America
Invaded range: HI
Introduction data: First reported in Hawaii in 1941 (Cuddihy and Stone 1990)
References: Review, Cuddihy and Stone (1990), Wester and Wood (1977)

Conicosia pugioniformis, narrow-leaf iceplant
Growth form: Short-lived perennial
Indigenous range: South Africa
Invaded range: CA

Cortaderia jubata, Andean pampas grass
Growth form: Perennial grass
Indigenous range: South America
Invaded range: CA
References: Review, Madison (1992), Randall and Marinelli (1996); Control, Costello (1986)

Crupina vulgaris, common crupina
Growth form: Winter annual herb
Indigenous range: Eurasia
Invaded range: CA ID MA OR WA
Introduction data: First reported in Idaho 1969 (Stickney 1972), in California 1975 (Davis and Sherman 1991)
References: Review, Prather et al. (1991); Control, Zamora et al. (1989); Invasion, Couderc-Levaillant and Talbot Roche (1993), Davis and Sherman (1991), Stickney (1972); Seed germination, Thill et al. 1986

Cynodon dactylon, Bermuda grass
Growth form: Perennial grass
Indigenous range: Africa
Invaded range: BC; AL AR AZ CA CO CT DE FL GA HI IA ID IL IN KS KY LA MA MD MI MO MS NC NE NH NJ NM NV NY OH OK OR PA SC TN TX UT VA WA WV
Introduction data: Mainland United States—Introduced before 1814 (Pursh 1814). Hawaii—Introduced in 1835 (Wagner et al. 1990)
Reference: Review, Holm et al. (1991)

Cytisus scoparius, Scotch broom
Growth form: Shrub
Indigenous range: Southern Europe, Asia minor, Russia
Invaded range: BC NS PE; AL CA CT DE GA HI ID MA MD ME MI MT NC NJ NY OH OR PA SC TN UT VA WA WV
Introduction data: Long cultivated. Mainland United States—Seeds available in U.S. trade by 1844–1845 (Mack 1991); introduced to California 1850s (Bossard and Rejmánek 1994). Hawaii—First collected 1909 (Wagner et al. 1990)
References: Review, Cronk and Fuller (1995), Hoshovsky (1993b), Randall and Marinelli (1996); Growth, Bossard and Rejmánek (1994); Seedling establishment, Bossard (1991)

Dioscorea batatas, Chinese yam
Growth form: Perennial herbaceous vine
Indigenous range: Asia
Invaded range: AL AR CT GA IL IN KS KY LA MD MO MS NC NJ NY OH PA SC TN VA VT WV

Dioscorea bulbifera, air-potato
Growth form: Perennial herbaceous vine
Indigenous range: Africa, Asia
Invaded range: FL HI LA MS TX
Introduced range: Hawaii—A Polynesian introduction (Wagner et al. 1990)
Reference: Review, Cronk and Fuller (1995)

Ehrharta calycina, veldt grass
Growth form: Perennial grass
Indigenous range: South Africa
Invaded range: CA
Reference: Control, Mubroy et al. (1992)

Ehrharta erecta
　Growth form: Perennial grass
　Indigenous range: South Africa
　Invaded range: CA
　Introduction data: "Became established as an adventive in northern California about 1930" (Stebbins 1985)
　Reference: Review, McIntyre and Ladiges (1985)

Eichhornia crassipes, water-hyacinth
　Growth form: Floating perennial aquatic
　Indigenous range: South America
　Invaded range: ON; AL AZ CA FL GA HI KY LA MO MS NC NY SC TN TX VA
　Introduction data: Florida—Introduced by the 1880s as an ornamental (Schmitz et al. 1993). Hawaii—Introduced probably in the latter part of the 1800s (Wagner et al. 1990)
　References: Review, Cronk and Fuller (1995), Duever et al. (1986), Holm et al. (1991), Madsen (Chapter 12, this volume), Randall and Marinelli (1996), Schmitz et al. (1993)

Elaeagnus angustifolia, Russian-olive
　Growth form: Shrub, small tree
　Indigenous range: Europe and western Asia
　Invaded range: AB BC MB NB ON; AZ CA CO CT IA ID IL KS KY MA MD ME MI MN MO MT ND NE NJ NM NV NY OH OK OR PA RI SD TN TX UT VA VT WI WY
　Introduction data: Cultivated in England since the sixteenth century (Hillier 1983). Introduced to the United States "prior to 1900" (Knopf and Olson 1984)
　References: Review, Randall and Marinelli (1996); Control, Bovey (1965); Community effects, Knopf and Olson (1984); Seed germination, Shafroth et al. (1995a)

Elaeagnus umbellata, autumn-olive
　Growth form: Shrub
　Indigenous range: Eastern Asia
　Invaded range: AL CT FL GA HI IA IL KY LA MA MD ME MI MO MS NC NH NJ NY OH PA RI SC TN VA VT WV
　Introduction data: Introduced to western horticulture 1830 (Hillier 1983). Hawaii—First collected 1963 (Wagner et al. 1990)
　References: Review, Eckardt (1993), Randall and Marinelli (1996); Community effects, Zimmerman et al. (1993); Invasion, Christensen (1963)

Euonymus fortunei, climbing euonymus
　Growth form: Perennial woody vine
　Indigenous range: Eastern Asia
　Invaded range: ON; AL IL IN KY MA MI MO NJ NY OH PA RI VA
　Introduction data: Introduced to western horticulture around 1865 (Hillier 1983)
　Reference: Reviews, Randall and Marinelli (1996)

Euphorbia esula, leafy spurge
　Growth form: Perennial herb
　Indigenous range: Europe
　Invaded range: AB BC MB NB NS ON PE QU SK YT; AZ CA CO CT DE IA ID IL IN KS MA MD ME MI MN MO MT ND NE NH NJ NM NV NY OH OR SD UT VA VT WA WI WV WY
　Introduction data: First U.S. record (Massachusetts) 1827 (Britton 1921)
　References: Review, Best et al. (1980), Britton (1921); Control, Lym and Messersmith (1994); Damage, Leistritz et al. (1992); Seed germination, Lacey et al. (1992)

Ficus microcarpa, laurel fig
　Growth form: Tree
　Indigenous range: Southern Asia, Australia, New Caledonia
　Invaded range: HI
　Introduction data: Cultivated in Hawaii at least since early 1900s but could not have become naturalized prior to 1938, when the pollinating wasp was introduced (Wagner et al. 1990)
　References: Review, Randall and Marinelli (1996); Invasion, McKey and Kaufmann (1991)

Genista monspessulana, French broom
　Growth form: Shrub
　Indigenous range: Mediterranean area, Azores
　Invaded range: BC; CA OR WA
　Reference: Review, Adams and Simmons (1991)

Hedera helix, English ivy
　Growth form: Perennial woody vine

Indigenous range: Europe to the Caucasus
Invaded range: BC; AL AZ CA FL GA HI IL KY LA MA MD MI MO MS NC NJ NY OR PA SC TN TX UT VA WA WV
Introduction data: Cultivated since ancient times. Mainland United States—Introduced probably in colonial times. Hawaii—Cultivated since at least the early 1900s, "sparingly naturalized" by 1990 (Wagner et al. 1990)
Reference: Review, Randall and Marinelli (1996); Control, Costello (1986)

Hedychium gardnerianum, Kahili-ginger
Growth form: Perennial herb
Indigenous range: Himalayas and adjacent regions
Invaded range: HI
Introduction data: Hawaii—First collected 1947 in Hawaii Volcanoes National Park (Cuddihy and Stone 1990)

Hydrilla verticillata, hydrilla
Growth form: Rooted submersed perennial aquatic
Indigenous range: Sri Lanka
Invaded range: AL AZ CA DE FL GA LA MD NC SC TN TX VA WA
Introduction data: Introduced to Florida in "the early 1950s" (Schmitz et al. 1993)
References: Review, Cronk and Fuller (1995), Duever et al. (1986), Madsen (Chapter 12, this volume), Schmitz et al. (1993)

Isatis tinctoria, woad
Growth form: Biennial herb
Indigenous range: Eurasia
Invaded range: BC NF ON QU; CA CO ID IL MT NJ NV UT VA WV WY
Introduction data: Long cultivated in Europe as a dye plant. Seeds available in U.S. trade by 1807 (Mack 1991). Apparently naturalized in the United States before 1822 (Eaton 1822). Reported as naturalized in northern Utah (where now locally common) by 1912 (Garrett 1912) but was not included in an eastern North American regional manual until 1950 (Fernald 1950: "se. Nfld.; w. Va.; doubtless elsewhere")
References: Review, Best et al. (1980), Callihan (1990), Evans (1991); Population biology, Farah et al. (1988)

Kochia scoparia, kochia
Growth form: Annual herb
Indigenous range: Temperate Asia
Invaded range: AB BC MB NB NS ON QU SK; AZ CA CO CT IA ID IL IN KS KY MA ME MI MN MO MS MT NC ND NE NH NJ NM NV NY OH OK OR PA RI SD TN TX UT VA WT WA WI WV WY
Introduction data: Early report, in a northeastern U.S. regional manual, 1889 as "sparingly naturalized" (Gray 1889)
Reference: Review, Eberlein and Fore (1984); Invasion, Forcella (1985)

Lantana camara, yellow-sage
Growth form: Shrub
Indigenous range: Tropical America
Invaded range: AL CA FL GA HI LA MS OK SC TX UT
Introduction data: Grown in Europe by 1687 (Morton 1994); many cultivars. Mainland United States—Introduced probably in eighteenth century (Morton 1994). Hawaii—Introduced 1858, naturalized before 1871 (Wagner et al. 1990)
References: Review, Cronk and Fuller (1995), Cuddihy and Stone (1990), Holm et al. (1991), Morton (1994), Sinha and Sharma (1984); Floral biology, Mathur and Mohan Ram (1986); Invasion, Fensham et al. (1994)

Lepidium latifolium, perennial pepperweed
Growth form: Perennial herb
Indigenous range: Eurasia
Invaded range: AB QU; AZ CA CO CT ID IL IN KS MA MO MT NE NM NV NY OR TX UT WA WY
Reference: Seed germination, Miller et al. (1986)

Lespedeza cuneata, silky bush-clover
Growth form: Perennial herb
Indigenous range: Eastern Asia
Invaded range: AL AR CT DE FL GA HI IA IL IN KS KY LA MA MD MI MO MS NC NE NJ NY OH OK PA SC TN TX VA WV
Reference: Control, Altom et al. (1992)

Ligustrum sinense, privet
  Growth form: Shrub
  Indigenous range: China
  Invaded range: AL AR CT FL GA KY LA MA MD MO MS NC NJ OK RI SC TN TX VA
  References: Review, Randall and Marinelli (1996); Control, Mowatt (1981)

Ligustrum vulgare, common privet
  Growth form: Shrub
  Indigenous range: Europe, northern Africa, Asia Minor
  Invaded range: BC NF NS ON; AL AR CT DE FL GA IL IN KY LA MA MD ME MI NC NH NJ NY OH PA RI SC TN TX UT VA VT WI WV
  Introduction data: Long cultivated. Introduced to North America probably in colonial times
  Reference: Review, Randall and Marinelli (1996)

Linaria dalmatica, Dalmatian toadflax
  Growth form: Perennial herb
  Indigenous range: Mediterranean region
  Invaded range: AB BC MB NB NS ON PE QU SK; AZ CA CO CT IA ID IL IN KS MA ME MI MN MT ND NE NH NJ NM NV NY OH OK OR PA SD UT VA VT WA WI WY
  Reference: Review, Robocker (1974)

Linaria vulgaris, butter-and-eggs
  Growth form: Perennial herb
  Indigenous range: Eurasia
  Introducton data: Introduced to New England before 1672 (Saner et al. 1995), perhaps as an ornamental.
  Invaded range: All provinces and territories; all states except HI
  References: Review, Mitich (1993), Saner et al. (1995)

Lonicera japonica, Japanese honeysuckle
  Growth form: Woody vine
  Indigenous range: Eastern Asia
  Invaded range: ON; AL AR AZ CA CT DE FL GA HI IL IN KS KY LA MA MD ME MI MO MS NC NE NH NJ NM NV NY OH OK PA RI SC TN TX UT VA WI WV
  Introduction data: Introduced to western horticulture 1806 (Hillier 1983). Seeds available in U.S. trade by 1823 (Mack 1991)
  References: Review, Randall and Marinelli (1996); Growth, Schierenbeck et al. (1994); Response to climate change, Sasek and Strain (1990)

Lonicera maackii, Amur honeysuckle
  Growth form: Shrub
  Indigenous range: Temperate eastern Asia
  Invaded range: ON; AR GA IL IN KS KY MA MD MI MO MS ND NJ OH PA SC TN TX VA WI WV
  Introduction data: Introduced to western horticulture around 1860. Earliest North American record, 1896, Ottawa, Ontario, in arboretum; introduced as an ornamental (Luken and Thieret 1995)
  References: Review, Luken and Thieret (1995), Randall and Marinelli (1996)

Lonicera tatarica, Tatarian honeysuckle
  Growth form: Shrub
  Indigenous range: Southern Russia to Altai and Turkestan
  Invaded range: AB MB NB NS ON QU SK; CA CO CT DE IA IL IN KS KY MA MD ME MI MN MT ND NE NH NJ NY OH PA RI SD UT VA VT WI WV WY
  Introduction data: Introduced to western horticulture 1752 (Hillier 1983)
  References: Review, Randall and Marinelli (1996); Community effects, Woods (1993)

Lygodium microphyllum, Old World climbing fern
  Growth form: Perennial herbaceous vine
  Indigenous range: Warm areas of the Old World
  Invaded range: FL
  Introduction data: A "recent introduction" to Florida (Beckner 1968)
  References: Review, Beckner (1968), Nauman and Austin (1978)

Lythrum salicaria, purple loosestrife
  Growth form: Perennial herb
  Indigenous range: Eurasia
  Invaded range: AB BC MB NB NF NS ON PE QU SK; AL AR CA CO CT DC DE IA ID IL IN KS KY MA MD ME MI MN MO MS MT NC ND NE NH NJ NV NY OH OR PA RI SD TN TX UT VA VT WA WI WV WY

Introduction data: Earliest North American record 1814 (Stuckey 1980)
References: Review, Mal et al. (1992), Randall and Marinelli (1996); Community effects, Anderson (1995); Invasion, Stuckey (1980)

Melaleuca quinquenervia, melaleuca
Growth form: Tree
Indigenous range: Eastern Australia, New Guinea, New Caledonia
Invaded range: FL HI LA
Introduction data: Florida—At least two dates are in literature for introduction of melaleuca to Florida, 1892 and 1920 (Duever et al. 1986), the former being more likely. Hawaii—Earliest collection 1920, from seed originating in Florida (Wagner et al. 1990)
References: Review, Bodle et al. (1994), Cronk and Fuller (1995), Duever et al. (1986), Randall and Marinelli (1996); Invasion, Laroche and Ferriter (1992); Seedling establishment, Myers (1983)

Melinis minutiflora, molasses grass
Growth form: Perennial grass
Indigenous range: Africa
Invaded range: FL HI
Introduction data: Hawaii—Introduced in the early 1900s as cattle fodder (Cuddihy and Stone 1990)
References: Review, Cronk and Fuller (1995), Cuddihy and Stone (1990)

Mesembryanthemum crystallinum, iceplant
Growth form: Annual herb
Indigenous range: Mediterranean region
Invaded range: AZ CA PA
Introduction data: Seeds available in U.S. trade by 1807 (Mack 1991)
References: Review, Cronk and Fuller (1995); Community effects, Kloot (1983), Vivrette and Muller (1977); Control, Moss (1994)

Miconia calvescens, miconia
Growth form: Tree
Indigenous range: Tropical America
Invaded range: HI
Introduction data: Introduced early 1980s
References: Review, Conant et al. (Chapter 18, this volume), Gillis (1992)

Microstegium vimineum, Nepal grass
Growth form: Annual grass
Indigenous range: Asia
Invaded range: AL AR CT DE FL GA IL IN KY LA MD MS NC NJ NY OH PA SC TN TX VA WV
Introduction data: Introduced to Tennessee from Asia 1919 (Hunt and Zaremba 1992)
References: Invasion, Barden (1987), Redman (1995), Hunt and Zaremba (1992)

Mimosa pigra, catclaw mimosa
Growth form: Perennial herb or shrub
Indigenous range: Pantropical
Invaded range: FL TX
Introduction data: "Apparently a newcomer to Florida" (Isely 1990)
References: Review, Cronk and Fuller (1995); Community effects, Braithwaite et al. (1989); Control, Creager (1992); Invasion, Lonsdale (1993)

Myrica faya, fayatree
Growth form: Tree
Indigenous range: Canary Islands, Azores, Madeira
Invaded range: HI
Introduction data: Hawaii—Introduced before 1900 (Cuddihy and Stone 1990)
References: Review, Cronk and Fuller (1995), Cuddihy and Stone (1990), Walker and Smith (Chapter 6, this volume); Community and ecosystem effects, Vitousek and Walker (1989)

Myriophyllum spicatum, Eurasian watermilfoil
Growth form: Rooted submersed perennial aquatic
Indigenous range: Eurasia
Invaded range: BC ON QU; AK AL AR CA CT DE FL GA IN KY LA MA MD MI MN MO MS NC NH NJ NY OH OK PA SC TN TX UT VA VT WA WI
Introduction data: Introduced to North America "in the late nineteenth century" (Aiken et al. 1979)
References: Review, Aiken et al. (1979); Control, Getsinger et al. (1982)

Neyraudia reynaudiana, Burma reed
Growth form: Perennial grass

Indigenous range: Southern Asia
Invaded range: FL
Introduction data: Introduced by U.S. Department of Agriculture to experimental garden, Miami, 1923 (Morton JF, pers. com.)

Nicotiana glauca, tree tobacco
Growth form: Shrub
Indigenous range: Argentina
Invaded range: AL AZ CA FL GA HI MD NJ NM NV TX
Introduction data: Hawaii—Introduced by 1864 (Wagner et al. 1990)
Reference: Review, Cronk and Fuller (1995)

Paederia foetida, skunkvine
Growth form: Vine
Indigenous range: Asia
Invaded range: FL HI LA SC TX
Introduction data: Hawaii—First recorded 1854 (Wagner et al. 1990)
Reference: Invasion, Brown (1992)

Panicum maximum, Guinea grass
Growth form: Perennial grass
Indigenous range: Africa
Invaded range: AL AZ FL GA HI LA TX
Introduction data: Hawaii—Probably naturalized prior to 1871 (Wagner et al. 1990)
Reference: Review, Holm et al. (1991)

Panicum repens, torpedo grass
Growth form: Perennial grass
Indigenous range: Warm areas of both hemispheres
Invaded range: AL FL HI LA MS NC TX
Reference: Review, Holm et al. (1991); Ecophysiology, Wilcut, Dute et al. (1988), Wilcut, Truelove et al. (1988)

Passiflora mollissima, banana poka
Growth form: Vine
Indigenous range: Andes
Invaded range: HI
Introduction data: Introduced to hide an outhouse in Hawaii early in the twentieth century; first collected in the wild 1926 (Wagner et al. 1990)
References: Review, Cronk and Fuller (1995), Cuddihy and Stone (1990)

Pennisetum clandestinum, Kikuyu grass
Growth form: Perennial grass
Indigenous range: Northern Africa and Middle East:
Invaded range: CA HI
Introduction data: Hawaii—First collected in 1938 (Wagner et al. 1990)
References: Review, Cronk and Fuller (1995), Cuddihy and Stone (l990), Holm et al. (1991)

Pennisetum setaceum, fountain grass
Growth form: Perennial grass
Indigenous range: Africa
Invaded range: ON; AZ CA CO FL HI LA TN
Introduction data: Mainland United States—Seeds available in U.S. trade by 1883 (Mack 1991). Hawaii—Introduced in early part of twentieth century (Cuddihy and Stone 1990)
References: Review, Cronk and Fuller (1995), Cuddihy and Stone (1990; Hawaii); Ecophysiology, Williams et al. (1995), Randall and Marinelli (1996)

Perilla frutescens, beefsteak mint
Growth form: Annual herb
Indigenous range: Southeastern Asia
Invaded range: ON; AL AR CT DE FL GA IA IL IN KS KY LA MA MD MN MO MS NC NE NJ NY OH OK PA SC TN TX VA WV
Introduction data: Early report as a wild plant, in a northeastern U.S. regional manual, 1889 (Gray 1889)

Polygonum cuspidatum, Japanese knotweed
Growth form: Perennial herb
Indigenous range: Eastern Asia
Invaded range: BC MB NB NF NS ON PE QU; AR CA CO CT DE GA IA ID IL IN KS KY LA MA MD ME MI MN MO MS NC NH NJ NY OH OR PA RI SC SD TN UT VA VT WA WI WV
Introduction data: Early report, in a northeastern U.S. regional manual, 1889 as "occasionally escaped from gardens" (Gray 1889)
References: Review, Cronk and Fuller (1995); Control, Pridham and Bing (1975), Scott and Marrs (1984); Growth form, Suzuki (1994)

Polygonum perfoliatum, mile-a-minute weed
  Growth form: Annual herbaceous vine
  Indigenous range: Eastern and southern Asia
  Invaded range: BC; MD MS NJ OR PA VA WV
  Introduction data: Appeared in a nursery in Maryland 1937 but was eradicated; a persistent and spreading population reported from Pennsylvania 1948 (Moul 1948)
  References: Control, McCormick and Hartwig (1995); Invasion, Hickman and Hickman (1978)

Psidium cattleianum, strawberry guava
  Growth form: Shrub
  Indigenous range: Brazil
  Invaded range: FL HI
  Introduction data: Mainland United States—Seeds available in U.S. trade by 1832 (Mack 1991). Hawaii—Present since 1825 (Cuddihy and Stone 1990)
  References: Review, Cronk and Fuller (1995), Cuddihy and Stone (1990), Tunison (1993); Seedling establishment, Huenneke and Vitousek (1990)

Psidium guajava, guava
  Growth form: Shrub, small tree
  Indigenous range: Tropical America
  Invaded range: FL HI
  Introduction data: Florida—Reportedly introduced in 1847, common over more than half of the state by 1886 (Morton 1987). Hawaii—Introduced in the early 1800s (Morton 1987)
  References: Review: Cronk and Fuller (1995), Cuddihy and Stone (1990; Hawaii), Morton (1987)

Pueraria montana, kudzu
  Growth form: Perennial herbaceous vine
  Indigenous range: Southeastern Asia
  Invaded range: AL AR CT DE FL GA HI IL IN KS KY LA MA MD MO MS NC NJ NY OH PA SC TN TX VA WV
  Introduction data: Mainland United States—Introduced 1876 from Japan for erosion control and forage (Frankel 1989). Hawaii—First collected 1915 but probably introduced by the Chinese in the late nineteenth century (Wagner et al. 1990)
  References: Review, Cronk and Fuller (1995); Invasion, Frankel (1989); Response to climate change, Sasek and Strain (1989, 1990)

Ranunculus ficaria, lesser celandine
  Growth form: Perennial herb
  Indigenous range: Europe
  Invaded range: BC NF ON QU; CT DC IL KY MA MD MI MO NH NJ NY OH OR PA RI TN VA WA WI WV
  Introduction data: Early report, in a northeastern U.S. regional manual, 1889 as an "escape from gardens about New York and Philadelphia" (Gray 1889)
  Reference: Review, Marsden-Jones (1935)

Rhamnus cathartica, common buckthorn
  Growth form: Shrub, small tree
  Indigenous range: Europe
  Invaded range: AB NB NS ON PE QU SK; CA CO CT IA IL IN KY MA ME MI MN MO MT ND NE NH NJ NY OH PA RI SD UT VT WI WY
  Introduction data: Long cultivated as an ornamental. Introduced to North America probably in colonial times; early mention 1822 (Eaton 1822)
  References: Review, Godwin (1943b), Randall and Marinelli (1996); Ecophysiology, Harrington et al. (1989); Invasion, Gill and Marks (1991)

Rhamnus frangula, glossy buckthorn
  Growth form: Shrub, small tree
  Indigenous range: Europe, western Asia, northern Africa:
  Invaded range: MB NB NS ON PE QU; CO CT IA IL IN KY MA MD ME MI MN NH NJ NY OH PA RI TN VT WI WV WY
  Introduction data: Long cultivated as an ornamental. Introduced to North America probably in colonial times.
  References: Review, Godwin (1943a), Randall and Marinelli (1996); Invasion, Catling and Porebski (1994)

Rosa multiflora, multiflora rose
  Growth form: Shrub
  Indigenous range: Eastern Asia

Invaded range: NB NS ON; AL AR CT DE FL GA IA IL IN KS KY LA MA MD ME MI MN MO MS NC NE NH NJ NY OH OK OR PA RI SC TN TX VA VT WA WI WV
Introduction data: Introduced into England from Japan 1804 (Hillier 1983). Present in the United States before 1822 (Eaton 1822); promoted by the U.S. Soil Conservation Service in 1930s and 1940s for erosion control and wildlife habitat
References: Review, Randall and Marinelli (1996); Community effects, Brothers and Spingarn (1992), Zimmerman et al. (1993); Dispersal, Stiles (1982)

Sapium sebiferum, Chinese tallowtree
Growth form: Tree
Indigenous range: China and Japan
Invaded range: AL AR FL GA LA NC SC TX
References: Review, Randall and Marinelli (1996); Community effects, Bruce et al. (1995); Ecophysiology, Jones and McLeod (1989)

Scaevola sericea, beach naupaka
Growth form: Shrub
Indigenous range: Indo-Pacific region, including HI
Invaded range: FL
Introduction data: Introduced to Florida as an ornamental but apparently not known to be naturalized until the 1970s (not included in Long and Lakela [1971] but appears in Wunderlin [1982])
Reference: Review, Randall and Marinelli (1996)

Schinus terebinthifolius, Brazilian peppertree
Growth form: Shrub, small tree
Indigenous range: South America
Invaded range: CA FL HI
Introduction data: Seeds available in U.S. trade by 1832 (Mack 1991). Florida—Seeds imported by U.S. Department of Agriculture as early as 1898; seeds or seedlings grown from them were forwarded to the Plant Introduction Station in Miami and distributed locally as ornamental (Morton 1978). Hawaii—First collected in 1911 (Wagner et al. 1990)

References: Review, Cronk and Fuller (1995), Randall and Marinelli (1996); Control, Doren and Whiteaker (1990a, 1990b, 1991)

Senecio mikanioides, German ivy
Growth form: Perennial herbaceous vine
Indigenous range: South Africa
Invaded range: CA HI
Introduction data: Hawaii—Naturalized by 1910 (Cuddihy and Stone 1990); cultivated as an ornamental
Reference: Review, Cuddihy and Stone (1990)

Solanum viarum, tropical soda-apple
Growth form: Perennial shrub
Indigenous range: Argentina and Brazil
Invaded range: FL GA MS
Introduction data: First observed in the United States 1987 (Mullahey et al. 1993b)
References: Review, Mullahey et al. (1993b); Control, Mullahey et al. (1993a); Invasion, Thompson Campbell (Chapter 16, this volume); Seed germination, Mullahey and Cornell (1994)

Taeniatherum caput-medusae, Medusa-head
Growth form: Annual grass
Indigenous range: Europe
Invaded range: CA CT ID NV NY OR UT WA
Introduction data: Seeds available in U.S. trade by 1866 (Mack 1991); first collected in the wild, Oregon 1887 (Howell 1903):
References: Review, Horton (1991); Control, Grey et al. (1995); Ecology and control, Young (1992); Invasion, Young and Evans (1970)

Tamarix spp., salt-cedar, tamarisk
Growth form: Shrub, tree
Indigenous range: Old World
Invaded range: Many North American reports of various species of *Tamarix* are open to question because of nomenclatural problems and difficulties of identification of some species.
T. africana: AZ CA LA SC TX
T. aphylla: AZ CA NV TX UT
T. aralensis: CA NC
T. canariensis: AZ GA LA NC SC

T. chinensis: BC MB ON QU; AR AZ CA CO LA MT NC NM NV OH OK TX WY

T. gallica: CA GA LA NC NM SC TX

T. parviflora: BC NS ON; AZ CA CO CT DE FL ID IL KS KY LA MA MI MO MS NC NJ NM NV OK OR PA TN TX UT WA VA

T. ramosissima: AR AZ CA CO GA KS LA MS NC ND NE NM NV OK SC SD TX UT VA

T. tetragyna: GA

Introduction data: Mainland United States—Nurserymen in the early 1800s were apparently the first to introduce tamarisk into the United States. Some early nursery and seed trade catalogs offered "French tamarisk" in 1823 and 1824, "Tamarix gallica" starting in 1825, and "Tamarix germanica" in 1825. A California nursery offered "tamarix" in 1856. The U.S. Department of Agriculture introduced some species of tamarisk starting in the late 1860s. Horton (1964) gave many additional records and a summary of tamarisk introduction into the mainland United States. Hawaii—Introduced for ornament into Hawaii before 1911 (Cuddihy and Stone 1990)

References: Review, Crins (1989), Cuddihy and Stone (1990), DeLoach (Chapter 13, this volume), Horton (1964), Walker and Smith (Chapter 6, this volume), Randall and Marinelli (1996); Control, Barrows (1993), Sudbrock (1993); Invasion, Everitt and DeLoach (1990); Seedling establishment, Shafroth et al. (1995b); Identification of U.S. *Tamarix*, Baum (1967) ("Collectors should be aware that most tamarisks are virtually unidentifiable in the vegetative state" [Crins 1989])

Ulex europaeus, gorse

Growth form: Shrub

Indigenous range: Western Europe

Invaded range: BC; CA HI MA NY OR PA VA WA WV

Introduction data: Long cultivated. Mainland United States—Present before 1822 (Eaton 1822), perhaps as early as 1807 (Mack 1991). Hawaii—Introduced before 1910 (Cuddihy and Stone 1990)

References: Review, Cronk and Fuller (1995), Cuddihy and Stone (1990), Hoshovsky (1993c)

References

Aarssen LW, Epp GA (1990) Neighbour manipulations in natural vegetation: a review. J Veg Sci 1:13–30

Aber JD (1987) Restored forests and the identification of critical factors in species-site interactions. In: Jordan WR III, Gilpin ME, Aber JD (eds) Restoration Ecology. Cambridge Univ Press, New York; pp. 241–250

Adams JM, Woodward FI (1992) The past as a key to the future: the use of paleoenvironmental understanding to predict the effects of man on the biosphere. Advances Ecol Res 22:257–314

Adams R, Simmons D (1991) The invasive potential of *Genista monspessulana* (Montpellier broom) in dry sclerophyll forest in Victoria. Victorian Naturalist 108:84–89

Aerts R, Berendse F (1988) The effect of increased nutrient availability on vegetation dynamics in wet heathlands. Vegetatio 76:63–69

Aiken SG, Newroth PR, Wile I (1979) The biology of Canadian weeds. 34. *Myriophyllum spicatum* L. Canad J Pl Sci 59:201–215

Alcantara EN, Wyse DL, Spitzmueller JM (1989) Quackgrass (*Agropyron repens*) biotype response to sethoxydim and haloxyfop. Weed Sci 37:107–111

Allaby M (ed) (1994) The Concise Oxford Dictionary of Ecology. Oxford Univ Press, New York

Allison TD, Moeller RK, Davis MB (1986) Pollen in laminated sediments provides evidence for a mid-Holocene forest pathogen outbreak. Ecology 67:1101–1105

Altom JV, Stritzke JF, Weeks DL (1992) Sericea lespedeza (*Lespedeza cuneata*) control with selected postemergence herbicides. Weed Technol 6:573–576

Alverson WS, Waller DM, Solheim SL (1988) Forests too deer: edge effects in northern Wisconsin. Conservation Biol 2:348–358

Amato M, Ritchie JT (1995) Small spatial scale soil water content measurement with time-domain reflectometry. J Soil Sci Soc Amer 59:325–329

Anable ME, McClaran MP, Ruyle GB (1992) Spread of introduced Lehmann lovegrass *Eragrostis lehmanniana* Nees. in southern Arizona, USA. Biol Conservation 61:181–188

Anderson B (1990) Identification and inventory of aquatic plant communities using remote sensing. Folia Geobot Phytotax 25:227–233

Anderson BW, Higgins A, Ohmart RD (1977) Avian use of saltcedar communities in the lower Colorado River Valley. U.S.D.A. Forest Serv Gen Techn Rep RM-43:128–136

Anderson BW, Ohmart RD (1977) Wildlife use and densities report of birds and mammals in the lower Colorado River Valley. Report to the U.S. Bureau of Reclamation, Boulder City, Nevada

Anderson BW, Ohmart RD (1984) Vegetation management study for the enhancement of wildlife along the lower Colorado River. Comprehensive Final Report to the U.S. Bureau of Reclamation, Boulder City, Nevada

Anderson E (1961) The tree of heaven, *Ailanthus altissima*. I. A blessing and a curse. Missouri Bot Gard Bull 49:105–107

Anderson JE (1982) Factors controlling transpiration and photosynthesis in *Tamarix chinensis* Lour. Ecology 63:48–56

References

Anderson MG (1995) Interactions between *Lythrum salicaria* and native organisms: a critical review. Environm Managem 19:225–231

Anderson RC (1990) The historic role of fire in the North American grassland. In: Collins SL, Wallace LL (eds) Fire in North American Tallgrass Prairies. Univ Oklahoma Press, Norman, Oklahoma; pp. 8–18

Anderson TW (1974) The chestnut pollen decline as a time horizon in lake sediments in eastern North America. Canad J Earth Sci 11:678–685

Andrascik RJ (1994) Process for developing a leafy spurge strategic management plan within Theodore Roosevelt National Park. Leafy Spurge News 16(3):5

Andres LA (1977) The economics of biological control of weeds. Aquatic Bot 3:111–123

Andres LA (1978) Biological control of puncturevine, *Tribulus terrestis* (Zygophyllaceae): post introduction collection records of *Microlarinus* spp. (Coleoptera: Curculionidae). In: Freeman TE (ed) Proceedings of the Fourth International Symposium on Biological Control of Weeds, 30 August–2 September 1976, Gainesville, Florida; pp. 132–136

Andres LA (1985) Interaction of *Chrysolina quadrigemina* and *Hypericum* spp. in California. In: Delfosse ES (ed) Proceedings of the Sixth International Symposium on Biological Control of Weeds, 19–25 August 1984, Vancouver, British Columbia, Canada; pp. 235–239

Andres LA, Bennett FD (1975) Biological control of aquatic weeds. Annual Rev Entomol 20:31–46

Andres LA, Coombs EM, McCaffrey JP (1995) Mediterranean sage. In: Nechols JR, Andres LA, Beardsley JW, Goeden RD, Jackson CG (eds) Biological Control in the Western United States. Accomplishments and Benefits of Regional Research Project W-84, 1964–1989. Univ Calif Div Agric Nat Resources Publ 3361; pp. 296–298

Andres LA, Coombs EM (1995) Scotch broom. In: Nechols JR, Andres LA, Beardsley JW, Goeden RD, Jackson CG (eds) Biological Control in the Western United States. Accomplishments and Benefits of Regional Research Project W-84, 1964–1989. Univ Calif Div Agric Nat Resources Publ 3361; pp. 303–305

Andres LA, Davis CJ, Harris P, Wapshere AJ (1976) Biological control of weeds. In: Huffaker CB, Messenger PS (eds) Theory and Practice of Biological Control. Academic Press, New York; pp. 481–499

Andres LA, Goeden RD (1995) Puncturevine. In: Nechols JR, Andres LA, Beardsley JW, Goeden RD, Jackson CG (eds) Biological Control in the Western United States. Accomplishments and Benefits of Regional Research Project W-84, 1964–1989. Univ Calif Div Agric Nat Resources Publ 3361; pp. 318–321

Andres LA, Rees NE (1995) Musk thistle. In: Nechols JR, Andres LA, Beardsley JW, Goeden RD, Jackson CG (eds) Biological Control in the Western United States. Accomplishments and Benefits of Regional Research Project W-84, 1964–1989. Univ Calif Div Agric Nat Resources Publ 3361; pp. 248–251

Andrews DS, Webb DH, Bates AL (1984) The use of aerial remote sensing in quantifying submersed aquatic macrophytes. In: Dennis WM, Isom BG (eds) Ecological Assessment of Macrophyton: Collection, Use, and Meaning of Data. ASTM STP 843. American Society for Testing and Materials, Washington, D.C.; pp. 92–99

Anonymous (1996) Biological Control of Weeds Regulatory Summit. 1 April 1996, Denver, CO. U.S. Dept. Agric., Animal and Plant Health Inspection Service, National Biological Control Institute, Riverdale, Maryland

Apfelbaum SI, Ludwig JP, Ludwig CE (1983) Ecological problems associated with disruption of dune vegetation dynamics by *Casuarina equisetifolia* L. at Sand Island, Midway Atoll. Atoll Res Bull 261:1–19

Aplet GH (1990) Alteration of earthworm community biomass by the alien *Myrica faya* in Hawaii. Oecologia 82:414–416

Aplet GH, Loh RK, Tunison JT, Vitousek PM (n.d.) Experimental restoration of a dense faya tree stand. Technical Report. Cooperative National Park Resources Studies Unit, Univ Hawaii, Manoa, Hawaii

Aplet GH, Vitousek PM (1994) An age-altitude matrix analysis of Hawaiian rainforest succession. J Ecol 82:137–147

Archer S, Scifres CJ, Bassham CR, Maggio R (1988) Autogenic succession in a subtropical savanna: conversion of grassland to thorn woodland. Ecol Monogr 58:111–127

Armesto JJ, Pickett STA, McDonnell MJ (1991) Spatial heterogeneity during succession: a cyclic model of invasion and exclusion. In: Kolasa J, Pickett STA (eds) Ecological Heterogeneity. Springer-Verlag, New York; pp. 256–269

Armour CL, Williamson SC (1988) Guidance for Modeling Causes and Effects in Environmental Problem-solving. U.S. Fish Wildlife Serv Biol Rep 89(4)

Aschmann H (1991) Human impacts on the biota of Mediterranean climate regions of Chile and California. In: Groves RH, di Castri F (eds) Biogeography of Mediterranean Invasions. Cambridge Univ Press, Cambridge; pp. 33–41

Ashby WC (1987) Forests. In: Jordan WR III, Gilpin ME, Aber JD (eds) Restoration Ecology. Cambridge Univ Press, New York; pp. 89–108

Ashton PJ, Mitchell DS (1989) Aquatic plants: patterns and modes of invasion, attributes of invading species and assessment of control programmes. In: Drake JA, Mooney HA, di Castri F, Groves RH, Kruger FJ, Rejmánek M, Williamson M (eds) Biological Invasions: A Global Perspective. John Wiley, New York; pp. 111–154

Atkinson IAE, Cameron EK (1993) Human influence on the terrestrial biota and biotic communities of New Zealand. Trends Ecol Evol 8:447–451

Auclair A (1976) Ecological factors in the development of intensive-management ecosystems in the midwestern United States. Ecology 57:431–444

Austin DF (1978) Exotic plants and their effects in southeastern Florida. Environm Conservation 5:25–34

Bailey LH, Bailey EZ (1941) Hortus the Second. Macmillan, New York

Baker HG (1965) Characteristics and modes of origin of weeds. In: Baker HG, Stebbins GL (eds) The Genetics of Colonizing Species. Academic Press, New York; pp. 147–172

Baker HG (1974) The evolution of weeds. Annual Rev Ecol Syst 5:1–24

Baker HG (1986) Patterns of plant invasion in North America. In: Mooney HA, Drake JA (eds) Ecology of Biological Invasions of North America and Hawaii. Springer-Verlag, New York; pp. 44–57

Baker RG, Bettis EA III, Schwert DP, Horton DG, Chumbley CA, Gonzalez LA, Reagan MK (1996) Holocene paleoenvironments of northeast Iowa. Ecol Monogr 66:203–234

Balciunas JK, Burrow DW, Purcell MF (1995) Australian insects for the biological control of the paperbark tree, *Melaleuca quinquenervia*, a serious pest of Florida, USA, wetlands. In: Delfosse ES, Scott RR (eds) Proceedings of the Eighth International Symposium on Biological Control of Weeds, 2–7 February 1992, Canterbury, New Zealand. DSIR/CSIRO, Melbourne, Australia; pp. 247–267

Bangsund DA, Leistritz FL (1991) Economic Impact of Leafy Spurge on Grazing Lands in the Northern Great Plains. Agricultural Economics Report No. 275-S, North Dakota State Univ, Fargo

Barbour MG, Johnson AF (1977) Beach and dune. In: Barbour MG, Major J (eds) Terrestrial Vegetation of California. John Wiley, New York; pp. 223–261

Barden LS (1987) Invasion of *Microstegium vimineum* (Poaceae), an exotic, shade-tolerant, C_4 grass, into a North Carolina floodplain. Am Midl Naturalist 118:40–45

Barnes PW, Flint SD, Caldwell MM (1990) Morphological responses of crop and weed species of different growth forms to ultraviolet-B radiation. Amer J Bot 77:1354–1360

Barrett SCH (1982) Genetic variation in weeds. In: Charudattan R, Walker HL (eds) Biological Control of Weeds with Plant Pathogens. John Wiley, New York; pp. 73–98

Barrett, SCH (1989) Waterplant invasions. Sci Amer 260(4):90–97

Barrows CW (1993) Tamarisk control. II. A success story. Restoration Managem Notes 11(1):35–38

Baskin JM, Baskin C (1992) Seed germination biology of the weedy biennial *Alliaria petiolata*. Nat Areas J 12:191–197

Batra SWT (1976) Some insects associated with hemp or marijuana (*Cannabis sativa* L.) in northern India. J Kansas Entomol Soc 49:385–388

Batra SWT (1981) Biological control of weeds in the northeast. Proc Northern Weed Sci Soc 35:21

Batra SWT, Coulson JR, Dunn PH, Boldt PE (1981) Insects and fungi associated with *Carduus* thistles. U.S.D.A. Techn Bull 1616

Batra SWT, Schroeder D, Boldt PE, Mendl W (1986) Insects associated with purple loosestrife (*Lythrum salicaria* L.) in Europe. Proc Entomol Soc Washington 88:748–759

Baum BR (1967) Introduced and naturalized tamarisks in the United States and Canada (Tamaricaceae). Baileya 15:19–25

Bazzaz FA (1986) Life history of colonizing plants: some demographic, genetic, and physiological features. In: Mooney HA, Drake JA (eds) Ecology of Biological Invasions of North America and Hawaii. Springer-Verlag, New York; pp. 96–110

Bazzaz FA, Carlson RW (1984) The response of plants to elevated CO_2. I. Competition among an assemblage of annuals at two levels of soil moisture. Oecologia 62:196–198

Beck G (1993) How do weeds affect us all? Proceedings of the Grazing Lands Forum. An Explosion in Slow Motion: Noxious Weeds and Invasive Alien Plants on Grazing Lands. Eighth

Forum. US Department of Agriculture Economic Research Service, Washington, D.C.; pp. 7–13

Beckner J (1968) *Lygodium microphyllum*, another fern escaped in Florida. Amer Fern J 58:93–94

Bedunah DJ (1992) The complex ecology of weeds, grazing and wildlife. Western Wildlands 18(2):6–11

Behn RD, Vaupel JW (1982) Quick Analysis for Busy Decision Makers. Basic Books, New York

Behre K-E (1981) The interpretation of anthropogenic indicators in pollen diagrams. Pollen & Spores 23:225–245

Behre K-E (1988) The rôle of man in European vegetation history. In: Huntley B, Webb T III (eds) Vegetation History. Kluwer, Dordrecht, Netherlands; pp. 633–672

Belcher JW, Wilson SD (1989) Leafy spurge and the species composition of a mixed-grass prairie. J Range Managem 42:172–175

Bell A (1983) Native plants facing extinction. Ecosystems 37:21–26

Bell GP (1994) Biology and growth habits of giant reed (*Arundo donax*). In: Jackson NE, Frandsen P, Duthoit S (compilers) Arundo donax Workshop Proceedings, November 19, 1993, Ontario, California; pp. 1–6

Bender B (1975) Farming in Prehistory: From Hunter-Gatherer to Food Producer. John Baker, London

Bennett FD, Crestana L, Habeck DH, Berti-Filho E (1990) Brazilian pepper tree—Prospects for biological control. In: Delfosse ES (ed) Proceedings of the Seventh International Symposium on Biological Control of Weeds, 6–11 March 1988, Rome, Italy. CSIRO, Melbourne, Australia; pp. 293–297

Bennett KD (1992) Holocene history of forest trees on the Bruce Peninsula, southern Ontario. Canad J Bot 70:6–18

Berglund BE (ed) (1991) The Cultural Landscape During 6000 Years in Southern Sweden—The Ystad Project. Ecol Bull 41

Bernabo JC (1981) Quantitative estimates of temperature changes over the last 2700 years in Michigan based on pollen data. Quatern Res 15:143–159

Bernstein BB, Zalinski J (1983) An optimum sampling design and power tests for environmental biologists. J Environm Managem 16:35–43

Best KF, Bowes GG, Thomas AG, Maw MG (1980) The biology of Canadian weeds. 39. *Euphorbia esula* L. Canad J Pl Sci 60:651–663

Betancourt JL, Davis OK (1984) Packrat middens from Canyon de Chelly, northeastern Arizona: paleoecological and archeological implications. Quatern Res 21:56–64

Betancourt JL, Long A, Donahue DJ, Jull AJT, Zabel TH (1984) Pre-Columbian age for North American *Corispermum* L. (Chenopodiaceae) confirmed by accelerator radiocarbon dating. Nature 311:653–655

Betancourt JL, Van Devender TR (1981) Holocene vegetation in Chaco Canyon, New Mexico. Science 214:656–658

Betancourt JL, Van Devender TR, Martin PS (eds) (1990) Packrat Middens: The Last 40,000 Years of Biotic Change. Univ Arizona Press, Tucson

Bhicy N, Filion L (1996) Mid-Holocene hemlock decline in eastern North America linked with phytophagous insect activity. Quatern Res 45:312–320

Bi H, Turvey ND (1994) Inter-specific competition between seedlings of *Pinus radiata*, *Eucalyptus regnans* and *Acacia melanoxylon*. Austral J Bot 42:61–70

Billings WD (1994) Ecological impacts of cheatgrass and resultant fire on ecosystems in the western Great Basin. In: Monsen SB, Kitchen SG (eds) Proceedings—Ecology and Management of Annual Rangelands. U.S.D.A. Forest Serv Gen Techn Rep INT-GTR 313. U.S.D.A. Forest Service, Intermountain Research Station, Ogden, Utah; pp. 22–30

Binford MW (1990) Calculation and uncertainty of ^{210}Pb dates for PIRLA project lake sediments. J Paleolimnol 3:253–267

Binford MW, Brenner M, Whitmore TJ, Higuera-Gundy A, Deevey ES, Leyden B (1987) Ecosystems, paleoecology and human disturbance in subtropical and tropical America. Quatern Sci Rev 6:115–128

Binggeli P (1993) Misuse of terminology and anthropomorphic concepts in the description of introduced species. Bull Brit Ecol Soc 25(1):10–13

Binkley D, Vitousek PM (1989) Soil nutrient availability. In: Pearcy RW, Ehleringer JR, Mooney HA, Rundel PW (eds) Plant Physiological Ecology. Chapman & Hall, London; pp. 75–96

Björkman L, Bradshaw R (n.d.) The immigration of *Fagus sylvatica* L. and *Picea abies* (L.) Karst. into a natural forest stand in southern Sweden during the last two thousand years. J Biogeogr

Blackburn WH, Knight RW, Schuster JL (1982) Saltcedar influence on sedimentation in the Brazos River. J Soil Water Conservation 37:298–301

Blackburn WH, Tueller PT (1970) Pinyon and juniper invasion in black sagebrush communities in east-central Nevada. Ecology 51:841–848

Blankespoor GW, Larson EA (1994) Response of smooth brome (*Bromus inermis* Leyss.) to burning under varying soil moisture conditions. Amer Midl Naturalist 131:266–272

Blatchley WS (1912) The Indiana Weed Book. Nature Publishing Company, Indianapolis

Block WM, With KA, Morrison ML (1987) On measuring bird habitat: influence of observer variability and sample size. Condor 89:241–251

Bobbink R (1991) Effects of nutrient enrichment in Dutch chalk grassland. J Appl Ecol 28:28–41

Bobbink R, Willems JH (1987) Increasing dominance of *Brachypodium pinnatum* in chalk grasslands: a threat to a species-rich ecosystem. Biol Conservation 40:301–314

Bock CE, Bock JH, Jepson KL, Ortega JC (1986) Ecological effects of planting African lovegrasses in Arizona. Natl Geogr Res 2:456–463

Bodle MJ, Ferriter AP, Thayer DD (1994) The biology, distribution, and ecological consequences of *Melaleuca quinquenervia* in the Everglades. In: Davis SM, Ogden JC (eds) Everglades: The Ecosystem and Its Restoration. St. Lucie Press, Delray Beach, Florida; pp. 341–355

Boeer WJ, Schmidly DJ (1977) Terrestrial mammals of the riparian corridor in Big Bend National Park. In: Johnson RR, Jones DA (eds) Proceedings of the Symposium on the Importance, Preservation and Management of Riparian Habitat, 9 July 1977, Tucson, Arizona. U.S.D.A. Forest Service Gen Techn Rep RM-43. Rocky Mountain Forest and Range Experiment Station, Fort Collins, Colorado; pp. 212–217

Boldt PE, Jackman JA (1993) Establishment of *Rhinocyllus conicus* Froelich on *Carduus macrocephalus* in Texas. Southwestern Entomol 18:173–181

Boldt PE, Robbins TO (1987) Phytophagous and pollinating insect fauna of *Baccharis neglecta* (Compositae) in Texas. Environm Entomol 16:887–895

Bonar SA, Vecht SA, Bennett CR, Pauley GB, Thomas GL (1993) Capture of grass carp from vegetated lakes. J Aquatic Pl Managem 31:168–174

Borchert R (1994a) Induction of rehydration and bud break by irrigation or rain in deciduous trees of a tropical dry forest in Costa Rica. Trees 8:198–204

Borchert R (1994b) Soil and stem water storage determine phenology and distribution of tropical dry forest trees. Ecology 75:1437–1449

Borgias D (1995) Preliminary site conservation and management plan for Euwauna Flat Preserve. Unpublished document on file at The Nature Conservancy, Portland, Oregon

Bossard CC (1991) The role of habitat disturbance, seed predation and ant dispersal on establishment of the exotic shrub *Cytisus scoparius* in California. Amer Midl Naturalist 126:1–13

Bossard CC, Rejmánek M (1994) Herbivory, growth, seed production, and resprouting of an exotic invasive shrub *Cytisus scoparius*. Biol Conservation 67:193–200

Bottema S (1995) Ancient palynology. Amer J Archaeol 99(1):93–96

Bovey RW (1965) Control of Russian olive by aerial application of herbicides. J Range Managem 18:194–195

Bovey RW, Meyer RE (1989) Control of huisache [*Acacia farnesiana*] and honey mesquite with a carpeted herbicide applicator. J Range Managem 42:407–411

Bowler PA (1992) Shrublands: in defense of disturbed land. Restoration Managem Notes 10:144–149

Boyd RS (1992) Influence of *Amophila arenaria* on foredune plant microdistributions at Point Reyes National Seashore, California. Madroño 39:67–76

Boyd WA, Stewart RM (1994) Status of *Hydrellia pakistanae* modeling efforts and approach for future development of the insect simulation. In: Proceedings, 28th Annual Meeting of the Aquatic Plant Control Research Program, 15–18 November 1993, Baltimore, Maryland. Misc Paper A-94-2, U.S. Army Engineer Waterways Experiment Station, Vicksburg, Mississippi; pp. 53–57

Bradley J (1988) Bringing Back the Bush: The Bradley Method of Bush Regeneration. Lansdowne Press, Sydney, Australia

Bradshaw R, Hannon G (1992) Climatic change, human influence and disturbance regime in the control of vegetation dynamics within Fiby Forest, Sweden. J Ecol 80:625–632

Bradshaw RHW, Zackrisson O (1990) A two thousand year history of a northern Swedish boreal forest stand. J Veg Sci 1:519–528

Braithwaite RW, Lonsdale WM, Estbergs JA (1989) Alien vegetation and native biota in tropical Australia: the impact of *Mimosa pigra*. Biol Conservation 48:189–210

Bramble WC, Byrnes WR (1976) Development of a stable, low plant cover on a utility right-of-way. In: Tillman R (ed) Proceedings of the First National Symposium on Environmental Concerns in Rights-of-Way Management. Mississippi State Univ, Starkville; pp. 167–176

Brandt CA, Rickard WH (1994) Alien taxa in the North American shrub-steppe four decades after cessation of livestock grazing and cultivation agriculture. Biol Conservation 68:95–105

Bratton SP (1982) The effects of exotic plant and animal species on nature preserves. Nat Areas J 2:3–13

Brenchley WE (1920) Weeds of Farm Land. Longmans, Green, London

Breternitz DA, Robinson CK, Gross GT (1986) Dolores Archaeological Program: Final Synthetic Report. U.S. Bureau of Reclamation, Denver

Breytenbach GJ (1986) Impacts of alien organisms on terrestrial communities with emphasis on communities of the south-western Cape. In: MacDonald IAW, Kruger FJ, Ferrar AA (eds) The Ecology and Management of Biological Invasions in Southern Africa. Oxford Univ Press, Cape Town; pp. 239–246

Brezonik PL, Fox JL (eds) (1975) The Proceedings of a Symposium on Water Quality Management through Biological Control, 23–30 January 1975. Report No. ENV 07-75-1. Univ Florida, Gainesville, Florida

Bridges DC (ed) (1992) Crop Losses due to Weeds in the United States—1992. Weed Science Society of America, Champaign, Ilinois

Bridgewater PB, Backshall DJ (1981) Dynamics of some Western Australian ligneous formations with special reference to the invasion of exotic species. Vegetatio 46:141–148

Brieman L, Friedman JA, Olshen RA, Stone CJ (1984) Classification and Regression Trees. Wadsworth, Belmont, California

Bright C (1995) Biological invasions: the spread of the world's most aggressive species. World Watch 8:10–19

Brightman HJ (1988) Group Problem-solving: An Improved Managerial Approach. Georgia State Univ, Atlanta, Georgia

Britton NL (1921) The leafy spurge becoming a pest. J New York Bot Gard 22:73–75

Brock JH (1994) *Tamarix* spp. (salt cedar), an invasive exotic woody plant in arid and semi-arid riparian habitats of western USA. In: de Waal LC, Child LE, Wade PM, Brock JH (eds) Ecology and Management of Invasive Riverside Plants. John Wiley, New York; pp. 27–44

Brockie RE, Loope LL, Usher MB, Hamann O (1988) Biological invasions of island nature reserves. Biol Conservation 44:9–36

Brokaw NVL, Walker LR (1991) Summary of the effects of Caribbean hurricanes on vegetation. Biotropica 23:442–447

Brookreson B (n.d.) Purple Loosestrife. Washington State Department of Agriculture, Olympia

Brothers TS, Spingarn A (1992) Forest fragmentation and alien plant invasion in central Indiana old-growth forests. Conservation Biol 6:91–100

Brotherson JD, Field D (1987) *Tamarix*: impacts of a successful weed. Rangelands 9:110–112

Brown DE, Minnich RA (1986) Fire and changes in creosote bush scrub of the western Sonoran Desert, California. Amer Midl Naturalist 116: 411–422

Brown JH (1989) Patterns, modes and extents of invasions by vertebrates. In: Drake JH, Mooney HA, di Castri F, Groves RH, Kruger FJ, Rejmánek M, Williamson M (eds) Biological Invasions: A Global Perspective. John Wiley, New York; pp. 85–109

Brown JH (1995) Macroecology. Univ Chicago Press, Chicago

Brown LE (1992) *Cayratia japonica* (Vitaceae) and *Paederia foetida* (Rubiaceae) adventive in Texas. Phytologia 72:45–47

Bruce KA, Cameron GN, Harcombe PA (1995) Initiation of a new woodland type on the Texas Coastal Prairie by the Chinese tallow tree (*Sapium sebiferum* (L.) Roxb.). Bull Torrey Bot Club 122:215–225

Bruckart WL (1989) Host range determination of *Puccinia jaceae* from yellow starthistle. Pl Dis 73:155–160

Bruckart WL, Dowler WM (1986) Evaluation of exotic rust fungi in the United States for classical biological control of weeds. Weed Sci 34(Suppl 1):11–14

Bruckart WL, Peterson GL (1990) Comparison of *Puccinia* spp. from *Carduus* thistles using isozyme analysis. In: Delfosse ES (ed) Proceedings of the Seventh International Symposium on Biological Control of Weeds, 6–11 March 1988, Rome, Italy. CSIRO, Melbourne, Australia; pp. 445–448

Brugam RB (1978) Pollen indicators of land-use change in southern Connecticut. Quatern Res 9:349–362

Brush GS (1989) Rates and patterns of estuarine sediment accumulation. Limnol Oceanogr 34:1235–1246

Buckingham GR (1994) Biological control of aquatic weeds. In: Rosen D, Bennett FD, Capinera JL (eds) Pest Management in the Subtropics: Biological Control—A Florida Perspective. Intercept, Lavoisier, Paris, pp. 413–480

Buckingham GR, Okrah EA (1993) Biological and Host Range Studies with Two Species of *Hydrellia* (Diptera: Ephydridae) that Feed on Hydrilla. Techn Rep A-93-7. U.S. Army Engineer Waterways Experiment Station, Vicksburg, Mississippi

Buell AC, Pickart AJ, Stuart JD (1995) Introduction history and invasion patterns of *Ammophila arenaria* on the north coast of California. Conservation Biol 9:1587–1593

Buffington LC, Herbel CH (1965) Vegetation changes on a semi-desert grassland range from 1858 to 1963. Ecol Monogr 35:139–164

Bunting AH (1960) Some reflections on the ecology of weeds. In: Harper JL (ed) The Biology of Weeds. Blackwell Scientific Publications, Oxford, U.K.

Burden ET, McAndrews JH, Norris G (1986) Palynology of Indian and European forest clearance and farming in lake sediment cores from Awenda Provincial Park, Ontario. Canad J Earth Sci 23:43–54

Burdon JJ, Chilvers JA (1977) Preliminary studies on a native eucalypt forest invaded by exotic pines. Oecologia 31:1–12

Burns C, Sauer J (1992) Resistance by natural vegetation in the San Gabriel Mountains of California to invasion by introduced conifers. Global Ecol Biogeogr Lett 2:46–51

Busch DE (n.d.) Effects of fire on southwestern riparian plant community structure. Southw Naturalist

Busch DE, Ingraham NL, Smith SD (1992) Water uptake in woody riparian phreatophytes of the southwestern United States: a stable isotope study. Ecol Applic 2:450–459

Busch DE, Smith SD (1993) Effects of fire on water and salinity relations of riparian woody taxa. Oecologia 94:186–194

Busch DE, Smith SD (1995) Mechanisms associated with the decline of woody species in riparian ecosystems of the southwestern U.S. Ecol Monogr 65:347–370

Bush MB, Piperno DR, Colinvaux PA, de Oliveira PE, Krissek LA, Miller MC, Rowe WE (1992) A 14,300-yr paleoecological profile of a lowland tropical lake in Panama. Ecol Monogr 62:251–275

Butler RS, Moyer EJ, Hulon MW, Williams VP (1992) Littoral zone invertebrate communities as affected by a habitat restoration project on Lake Tohopekaliga, Florida. J Freshwater Ecol 7:317–328

Byers D (1988) Life history variation of *Alliaria petiolata* in a range of habitats in New Jersey. M.S. thesis. Rutgers Univ, Piscataway, New Jersey

Byl TD, Sutton HD, Klaine SJ (1994) Evaluation of peroxidase as a biochemical indicator of toxic chemical exposure in the aquatic plant *Hydrilla verticillata*, Royle. Environm Toxicol Chem 13:509–515

Byrne R, McAndrews JH (1975) Pre-Columbian purslane (*Portulaca oleracea*) in the New World. Nature 253:726–727

CAB (1994) Using Biodiversity to Protect Biodiversity. Commonwealth Agricultural Bureaux International, Wallingford, Oxon, U.K.

Cagne BH, Loope LL, Medeiros AC, Anderson SJ (1992) *Miconia calvescens*: a threat to native forests in the Hawaiian Islands (Abstr). Pacific Sci 46:390–391

Calcote R (1995) Pollen source area and pollen productivity: evidence from forest hollows. J Ecol 83:591–602

Caldwell MM, Richards JH, Manwaring JH, Eissenstat DM (1987) Rapid shifts in phosphate acquisition show direct competition between neighbouring plants. Nature 327:615–616

Caldwell MM, Richards JH (1986) Competing root systems: morphology and models of absorption. In: Givnish T (ed) On the Economy of Plant Form and Function. Cambridge Univ Press, Cambridge, U.K.; pp. 251–273

Caldwell MM, Richards JH (1989) Hydraulic lift: water efflux from upper roots improves effectiveness of water uptake by deep roots. Oecologia 79:1–5

Caldwell MM, Teramura AH, Tevini M (1989) The changing solar ultraviolet climate and the ecological consequences for higher plants. Trends Ecol Evol 4:363–367

Callihan RH (1990) Dyers woad [*Isatis tinctoria*]: biology, distribution and control. Idaho Agric Exp Sta Bull 857

Callihan RH, Prather TS, Northam FE (1993) Longevity of yellow starthistle (*Centaurea solstitialis*) achenes in soil. Weed Technol 7:33–35

Callison J, Brotherson JD, Downs JE (1985) The effects of fire on the blackbrush (*Coleogyne ramosissima*) community of southwestern Utah. J Range Managem 38:535–538

Cameron G, Spencer SR (1989) Rapid leaf decay and nutrient release in a Chinese tallow forest. Oecologia 80:222–228

Cameron PJ, Hill RJ, Bain J, Thomas WP (eds) (1989) A Review of Biological Control of Invertebrate Pests and Weeds in New Zealand 1874 to 1987. Techn Commun 10. CAB International Institute of Biological Control, DSIR Entomology Division, Wallingford, Oxon, U.K.

Carlquist S (1980) Hawaii: A Natural History. SB Printers, Honolulu, Hawaii

Carlson RW, Bazzaz FA (1982) Photosynthetic and growth response to fumigation with SO_2 at elevated CO_2 for C_3 and C_4 plants. Oecologia 54:50–54

Carpenter SR, Adams MS (1976) The macrophyte tissue nutrient pool of a hardwater eutrophic lake: implications for macrophyte harvesting. Aquatic Bot 3:239–255

Carpenter SR, Adams MS (1978) Macrophyte control by harvesting and herbicides: implications for phosphorus cycling in Lake Wingra, Wisconsin. J Aquatic Pl Managem 16:20–23

Carpenter SR, Gasith A (1978) Mechanical cutting of submersed macrophytes: immediate effects on littoral chemistry and metabolism. Water Res 12:55–57

Carter DR, Carter S, Allen JL (1994) Submerged macrophyte control using plastic blankets. Water Sci Technol 29:119–126

Carter DR, Peterson KM (1983) Effects of a CO_2-enriched atmosphere on the growth and competitive interaction of a C_3 and a C_4 grass. Oecologia 58:188–193

Carter RN, Prince SD (1988) Epidemic models used to explain biogeographical distribution limits. Nature 293:644–645

Catling PM, Porebski ZS (1994) The history of invasion and current status of glossy buckthorn, *Rhamnus frangula*, in southern Ontario. Canad Field-Naturalist 108:305–310

Cavers PB, Heagy MI, Kokron RF (1979) The biology of Canadian weeds. 35. *Alliaria petiolata* (M. Bieb.) Cavara and Grande. Canad J Pl Sci 59:217–229

Center TD, Frank JH, Dray FA Jr (1994) Scientific Methods for Managing Invasive Non-Indigenous Species—Biological Control. In: Schmitz DC, Brown TC (1994) An Assessment of Invasive Non-Indigenous Species in Florida's Public Lands. Florida Department of Environmental Protection, Tallahassee, Florida; pp. 209–247

Center TD, Frank JH, Dray FA Jr (1995) Biological invasions: stemming the tide in Florida. Florida Entomol 78:45–55

Center TD, Van TK (1989) Alteration of waterhyacinth (*Eichhornia crassipes* (Mart.) Solms) leaf dynamics and phytochemistry by insect damage and plant density. Aquatic Bot 35:181–195

Ceska O, Ceska A (1986) *Myriophyllum* Haloragaceae species in British Columbia: problems with identification. In: Proceedings, First International Symposium on Watermilfoil (*Myriophyllum spicatum*) and Related Haloragaceae Species, 23–24 July 1985, Vancouver, British Columbia, Canada. Aquatic Plant Management Society, Vicksburg, Mississippi; pp. 39–50

Chaloupka MY, Domm SB (1986) Role of anthropochory in the invasion of coral cays by alien flora. Ecology 67:1536–1547

Chandler JM, Hamill AS, Thomas AG (1984) Crop losses due to weeds in Canada and the United States. Special Report. Weed Science Society of America, Champaign, Illinois

Chapin FS III, Vitousek PM, Van Cleve K (1986) The nature of nutrient limitation in plant communities. Amer Naturalist 127:48–58

Chapin FS III, Walker LR, Fastie CL, Sharman LC (1994) Mechanisms of primary succession following deglaciation at Glacier Bay, Alaska. Ecol Monogr 64:149–175

Chapman J, Stewart RB, Yarnell RA (1974) Archaeological evidence for pre-Columbian introduction of *Portulaca oleracea* and *Mollugo verticillata* into eastern North America. Econ Bot 28:411–412

Charudattan R, DeLoach CJ (1988) Management of pathogens and insects for weed control in agrosystems. In: Altieri MA, Liebman M (eds) Weed Management in Agroecosystems: Ecological Approaches. CRC Press, Boca Raton, Florida; pp. 245–264

Charudattan R, Walker HL (1982) Biological Control of Weeds with Plant Pathogens. John Wiley, New York

Chilvers GA, Burdon JJ (1983) Further studies on a native Australian eucalypt forest invaded by exotic pines. Oecologia 59:239–245

Christensen EM (1963) Naturalization of Russian olive (*Elaeagnus angustifolia* L.) in Utah. Amer Midl Naturalist 70:133–137

Clark JD, Brandt SA (1984) From Hunters to Farmers: The Causes and Consequences of Food Production in Africa. Univ California Press, Berkeley

Clark JS (1988) Stratigraphic charcoal analysis on petrographic thin sections: recent fire history in northwestern Minnesota. Quatern Res 30:81–91

Clark JS, Patterson WA III (1984) Pollen, PB-210, and opaque spherules: an integrated approach to dating and sedimentation in the intertidal environment. J Sediment Petrol 54:1251–1265

Cleland CE (1983) Indians in a changing environment. In: Flader SL (ed) The Great Lakes Forest:

An Environmental History. Univ Minnesota Press, Minneapolis; pp. 83–95

Clement SL, Cristofaro M (1995) Open-field tests in host-specificity determinations of insects for biological control of weeds. Biocontrol Sci Technol 5:395–406

Clements FE (1928) Plant Succession and Indicators. Wilson, New York

Clifford HT (1959) Seed dispersal by motor vehicles. J Ecol 47:311–315

Coblentz BE (1980) Effects of feral goats on the Santa Catalina Island ecosystem. In: Power DM (ed) The California Islands: Proceedings of a Multidisciplinary Symposium, Santa Barbara Museum of Natural History, Santa Barbara, California; pp. 167–170

Coblentz BE (1991) A response to Temple and Lugo. Conservation Biol 5:5–6

Coblentz BE (1993a) Invasive ecological dominants: environments boar-ed to tears and living on burro-ed time. In: McKnight BN (ed) Biological Pollution: The Control and Impact of Invasive Exotic Species. Indiana Academy of Science, Indianapolis; pp. 223–224

Coblentz BE (1993b) Letter to the editor. Nat Areas J 13:3

Cochran WG (1977) Sampling Techniques. 3rd ed. John Wiley, New York

Coffey BT, McNabb CD (1974) Eurasian watermilfoil in Michigan. Michigan Bot 13:159–165

Cofrancesco AF Jr (1994) Biological control technology overview—fiscal year 1993. In: Proceedings, 28th Annual Meeting of the Aquatic Plant Control Research Program, 15–18 November 1993, Baltimore, Maryland. Misc Paper A-94-2, U.S. Army Engineer Waterways Experiment Station, Vicksburg, Mississippi; pp. 169–171

Cofrancesco AF Jr (1995) Review of biological control research for 1994. In: Proceedings, 29th Annual Meeting of the Aquatic Plant Control Research Program, 14–17 November 1994, Vicksburg, Mississippi. Misc Paper A-95-3, U.S. Army Engineer Waterways Experiment Station, Vicksburg, Mississippi; pp. 101–105

COHMAP Members (1988) Climatic changes of the last 18,000 years: observations and model simulations. Science 241:1043–1052

Comrack L (1986) Exotic plant control—Tamarisk removal, Anza Borrego Desert State Park, 1985/86, 1984/85, 1983/84. In: Statewide Resource Management Program, Project Status Report. State of California, Sacramento, California

Conant P (1996) New Hawaiian Pest Plant Records for 1995. In: Evenhuis NL, Scott EM (eds) Records of the Hawaii Biological Survey for 1995. Part 2. Occas Papers Bernice P. Bishop Mus 46

Connell JH, Slatyer RO (1977) Mechanisms of succession in natural communities and their role in community stability and organization. Amer Naturalist 3:1119–1144

Conyers DL, Cooke GD (1983) A comparison of the costs of harvesting and herbicides and their effectiveness in nutrient removal and control of macrophyte biomass In: Lake Restoration, Protection and Management. EPA 440/5-83-001. U.S. Environmental Protection Agency, Washington, D.C.; pp. 317–321

Cooke GD (1980a) Covering bottom sediments as a lake restoration technique. Water Resources Bull 16:921–926

Cooke GD (1980b) Lake level drawdown as a macrophyte control technique. Water Resources Bull 16:317–322

Cooke GD, Martin AB, Carlson RE (1990) The effect of harvesting on macrophyte regrowth and water quality in LaDue Reservoir, Ohio. J Iowa Acad Sci 97:127–132

Cordo HA (1985) Host specificity studies of the Argentine weevil, *Heilipodus ventralis*, for biological control of snakeweeds (*Gutierrezia* spp.) in the U.S. In: Delfosse ES (ed) Proceedings of the Sixth International Symposium on Biological Control of Weeds, 19–25 August 1984, Vancouver, British Columbia, Canada; pp. 709–720

Costello LR (1986) Control of ornamentals gone wild: pampas grass [*Cortaderia jubata*], bamboo, English and Algerian ivy [*Hedera helix*]. Proceedings of the 38th California Weed Conference. California Weed Conference Office, Sacramento, California; pp. 162–170

Couderc-LeVaillant M, Talbott Roche C (1993) Evidence of multiple introduction of *Crupina vulgaris* in infestations in the western United States. Madroño 40:63–65

Coughlan BAK, Armour CL (1992) Group Decision-making Techniques for Natural Resource Management Applications. U.S. Fish Wildlife Serv Resource Publ 185

Coulson JR, Soper RS (1989) Protocols for the introduction of biological control agents in the U.S. In: Kahn RP (ed) Plant Protection and Quarantine. Vol. III Special Topics. CRC Press, Boca Raton, Florida; pp. 1–35

Coulson JR, Vail PV, Dix ME, Nordlund DA, Kauffmann WC (eds) (n.d.) 110 Years of Biological Control Research and Development in the

United States Department of Agriculture 1883–1993. U.S. Dept Agric Misc Publ

Cowles HC (1901) The physiographic ecology of Chicago and vicinity. Bot Gaz 31:73–108, 145–182

Crance JH (1987) Guidelines for Using the Delphi Technique to Develop Habitat Suitability Index Curves. U.S. Fish Wildlife Serv Biol Rep 82(10.134)

Crawley MJ (1987) What makes a community invasible? In: Gray AJ, Crawley MJ, Edwards PJ (eds) Colonization, Succession and Stability. Blackwell Scientific Publications, London; pp. 429–453

Creager RA (1992) Seed germination, physical and chemical control of catclaw mimosa (*Mimosa pigra* var. *pigra*). Weed Technol 6:884–891

Creed RP Jr, Sheldon SP (1992) Further investigations into the effect of herbivores on Eurasian watermilfoil (*Myriophyllum spicatum*). In: Proceedings, 26th Annual Meeting of the Aquatic Plant Control Research Program, 18–22 November 1991, Baltimore Maryland. Misc Paper A-92-2, U.S. Army Engineer Waterways Experiment Station, Vicksburg, Mississippi; pp. 244–252

Creed RP Jr, Sheldon SP (1993) The effect of the weevil *Euhrychiopsis lecontei* on Eurasian watermilfoil: results from Brownington Pond and North Brook Pond. In: Proceedings, 27th Annual Meeting of the Aquatic Plant Control Research Program, 16–19 November 1992, Bellevue, Washington. Misc Paper A-93-2, U.S. Army Engineer Waterways Experiment Station, Vicksburg, Mississippi; pp. 99–117

Creed RP Jr, Sheldon SP, Cheek DM (1992) The effect of herbivore feeding on the buoyancy of Eurasian watermilfoil. J Aquatic Pl Managem 30:75–76

Creed RP Jr, Sheldon SP (1994) The effect of two herbivorous insect larvae on Eurasian watermilfoil. J Aquatic Pl Managem 32:21–26

Crins WJ (1989) The Tamaricaceae in the southwestern United States. J Arnold Arbor 70:403–425

Cronk CB, Fuller JL (1995) Plant Invaders: The Threat to Natural Ecosystems. Chapman and Hall, London, U.K.

Cross JR (1982) The invasion and impact of *Rhododendron ponticum* in native Irish vegetation. J Life Sci Roy Dublin Soc 3:209–220

Crosson H (1987) Vermont Eurasian Watermilfoil Control Program. Vermont Department of Environmental Conservation, Waterbury, Vermont

Crowder LB, Cooper WE (1982) Habitat structural complexity and the interaction between bluegills and their prey. Ecology 63:1802–1813

Cuddihy LW, Stone CP (1990) Alteration of Native Hawaiian Vegetation: Effects of Humans, Their Activities and Introductions. Cooperative National Park Resources Studies Unit, Univ Hawaii, Honolulu, Hawaii

Cullen JM (1978) Evaluating the success of the programme for the biological control of *Chondrilla juncea* L. In: Freeman TE (ed) Proceedings of the Fourth International Symposium on Biological Control of Weeds, 30 August–2 September 1976, Gainesville, Florida. Institute of Food and Agricultural Sciences, Univ Florida, Gainesville, Florida; pp. 117–121

Cullen JM (1990) Current problems in host-specificity screening. In: Delfosse ES (ed) Proceedings of the Seventh International Symposium on Biological Control of Weeds, 6–11 March 1988, Rome, Italy. CSIRO, Melbourne, Australia; pp. 27–36

Cullen JM, Delfosse ES (1985) *Echium plantagineum*: catalyst for conflict and change in Australia. In: Delfosse ES (ed) Proceedings of the Sixth International Symposium on Biological Control of Weeds, 19–25 August 1984, Vancouver, British Columbia, Canada; pp. 249–292

Culler RC (1970) Water conservation by removal of phreatophytes. Trans Amer Geophys Union 51:684–689

Cyr H, Downing JA (1988) Empirical relationships of phytomacrofaunal abundance to plant biomass and macrophyte bed characteristics. Canad J Fish Aquatic Sci 45:976–984

D'Antonio CM (1990a) Invasion of coastal plant communities by the introduced succulent, *Carpobrotus edulis* (Aizoaceae). Ph.D. dissertation. Univ California, Santa Barbara, California

D'Antonio CM (1990b) Seed production and dispersal in the non-native, invasive succulent *Carpobrotus edulis* (Aizoaceae) in coastal strand communities of central California. J Appl Ecol 27:693–702

D'Antonio CM, Mahall BE (1991) Root profiles and competition between the invasive, exotic perennial, *Carpobrotus edulis*, and two native shrub species in California coastal scrub. Amer J Bot 78:885–894

D'Antonio CM, Vitousek PM (1992) Biological invasions by exotic grasses, the grass/fire cycle, and global change. Annual Rev Ecol Syst 23:63–87

Daehler CC, Strong DR (1993) Prediction and biological invasions. Trends Ecol Evol 8:380–381

Dalton FN, Herkelrath WN, Rawlins DS, Rhoades JD (1984) Time domain reflectometry: simultaneous assessment of the soil water content and electrical conductivity with a single probe. Science 224:989–990

Darwin C (1909) The Voyage of the Beagle. PF Collier and Son, New York

Davenport DC, Hagan RM, Gay LW, Bonde EK, Kreith F, Anderson JE (1978) Factors influencing usefulness of antitranspirants applied on phreatophytes to increase water supplies. California Water Resources Center, Davis, California

Davenport DC, Martin PE, Hagan RM (1982) Evapotranspiration from riparian vegetation: water relations and irrecoverable losses for saltcedar. J Soil Water Conservation 37:233–236

Davis AM, McAndrews JH, Wallace BL (1988) Paleoenvironment and the archaeological record at the L'Anse aux Meadows site, Newfoundland. Geoarchaeology 3:53–64

Davis ES, Fay PK, Chicoine TK, Lacey CA (1993) Persistence of spotted knapweed (*Centaurea maculosa*) seed in soil. Weed Sci 41:57–61

Davis LH, Sherman RJ (1991) *Crupina vulgaris* Cass. (Asteraceae: Cynareae), established in Sonoma County, California at Annadel State Park. Madroño 38:296

Davis MB (1976) Pleistocene biogeography of temperate deciduous forests. Geosci & Man 13:13–26

Davis MB (1981a) Outbreaks of forest pathogens in Quaternary history. Proceedings, Fourth International Palynological Conference, Lucknow (1976–1977). Birbal Sahni Institute of Palaeobotany, Lucknow, India; volume 3, pp. 216–228

Davis MB (1981b) Quaternary history and stability of forest communities. In: West DC, Shugart HH (eds) Forest Succession. Springer-Verlag, New York; pp. 132–153

Davis MB (1987) Invasions of forest communities during the Holocene: beech and hemlock in the Great Lakes region. In: Gray AJ, Crawley MJ, Edwards PJ (eds) Colonization, Succession, and Stability. Blackwell, London, U.K.; pp. 373–394

Davis MB (1989a) Insights from paleoecology on global change. Bull Ecol Soc Amer 70:222–228

Davis MB (1989b) Lags in vegetation response to greenhouse warming. Climate Change 15:75–82

Davis MB (1989c) Research questions posed by the paleoecological record of global change. In: Bradley RS (ed) Global Changes of the Past. UCAR/Office for Interdisciplinary Studies, Boulder, Colorado; pp. 385–395

Davis MB, Schwartz MW, Woods KD (1991a) Detecting a species limit from pollen in sediments. J Biogeogr 18:653–668

Davis MB, Sugita S, Calcote RR, Frelich L (1991b) Effects of invasion by *Tsuga canadensis* on a North American forest ecosystem. Proceedings of the First European Ecosystem Conference, Firenze, Italy. Blackwell, Oxford, U.K.; pp. 1–11

Davis MB, Sugita S, Calcote RR, Ferrari JB, Frelich LE (1994) Historical development of alternate communities in a hemlock-hardwood forest in northern Michigan, USA. In: Edwards PJ, May R, Webb NR (eds) Large-scale Ecology and Conservation Biology. Blackwell, Oxford, U.K.; pp. 19–39

Davis OK (1987) Spores of the dung fungus *Sporormiella*: increased abundance in historic sediments and before Pleistocene megafaunal extinction. Quatern Res 28:290–294

Davis OK (1992) Rapid climatic change in coastal southern California inferred from pollen analysis of San Joaquin Marsh. Quatern Res 37:89–100

Davis OK, Turner RM (1986) Palynological evidence for the historic expansion of juniper and desert shrubs in Arizona, USA. Rev Palaeobot Palynol 49:177–193

Davis S, Buckingham G (1988) Biocontrol mistake, Part II: Rebuttal. Aquaphyte 8:1–2, 14–15.

Dawson FH (1981) The reduction of light as a technique for the control of aquatic plants—an assessment. In: Proceedings of the Conference on Aquatic Weeds and Their Control, 7th and 8th April, 1981, Christ Church, Oxford, England. AAB Office, Wellesbourne, Warwick, U.K.; pp. 157–164

Dawson FH (1986) Light reduction techniques for aquatic plant control. Lake Reservoir Managem 2:258–262

Dawson FH, Hallows HB (1983) Practical applications of a shading material for macrophyte control in watercourses. Aquatic Bot 17:301–308

Dawson FH, Kern-Hansen U (1978) Aquatic weed management in natural streams: the effect of shade by the marginal vegetation. Verh Int Vereinigung Theor Angew Limnol 20:1451–1456

Dawson TE, Ehleringer JR (1991) Streamside trees that do not use stream water. Nature 350:335–337

Dayton WA (1950) Glossary of botanical terms commonly used in range research. U.S.D.A. Misc Publ 110

References

de Kozlowski SJ (1994) South Carolina grass carp policy. In: Proceedings of the Grass Carp Symposium, 7–9 March 1994, Gainesville, Florida. U.S. Army Engineer Waterways Experiment Station, Vicksburg, Mississippi; pp. 11–15

De Pietri DE (1992) Alien shrubs in a national park: can they help in the recovery of natural degraded forest? Biol Conservation 62:127–130

De Steno FM (1992) Controlling Eurasian water milfoil infestations using an integrated approach. Aquatics 14(2):4–8

Deam CC (1940) Flora of Indiana. Department of Conservation, Division of Forestry, Indianapolis

DeBach P (1974) Biological Control by Natural Enemies. Cambridge Univ Press, London, U.K.

Decker JP (1961) Salt secretion by *Tamarix pentandra* Pall. Forest Sci 7:214–217

Défago G, Bruckart WL, Sedlar L (1985) Evaluation of plant pathogens in Europe for the biological control of introduced weed species in North America. In: Delfosse ES (ed) Proceedings of the Sixth International Symposium on Biological Control of Weeds, 19–25 August 1984, Vancouver, British Columbia, Canada. Agriculture Canada, Ottawa, Canada; pp. 601–607

DeFerrari CM, Naiman RJ (1994) A multi-scale assessment of the occurrence of exotic plants on the Olympic Peninsula, Washington. J Veg Sci 5:247–258

Delbecq AL, Van de Ven AH, Gustafsen DH (1975) Group Techniques for Program Planning: A Guide to Nominal Group and Delphi Processes. Scott, Foreman, Glenview, Illinois

Delcourt PA (1980) Goshen Springs: late Quaternary vegetation record for southern Alabama. Ecology 61:371–386

Delcourt PA, Delcourt HR (1977) The Tunica Hills, Louisiana-Mississippi: late glacial locality for spruce and deciduous forest species. Quatern Res 7:218–237

Delcourt PA, Delcourt HR (1987) Long-term Forest Dynamics of the Temperate Zone. Springer-Verlag, New York

Delcourt PA, Delcourt HR, Cridlebaugh PA, Chapman J (1986) Holocene ethnobotanical and paleoecological record of human impact on vegetation in the Little Tennessee River Valley, Tennessee. Quatern Res 25:330–349

Delfosse ES (ed) (1981) Proceedings of the Fifth International Symposium on Biological Control of Weeds, 22–29 July 1980, Brisbane, Queensland, Australia. CSIRO, Melbourne, Australia

Delfosse ES (ed) (1985) Proceedings of the Sixth International Symposium on Biological Control of Weeds, 19–25 August 1984, Vancouver, British Columbia, Canada. Canadian Government Publishing Centre, Ottawa, Canada

Delfosse ES (ed) (1990) Proceedings of the Seventh International Symposium on Biological Control of Weeds, 6–11 March 1988, Rome, Italy. CSIRO, Melbourne, Australia

Delfosse ES, Scott RR (eds) (1995) Proceedings of the Eighth International Symposium on Biological Control of Weeds, 2–7 February 1992, Lincoln University, Canterbury, New Zealand. DSIR/CSIRO, Melbourne, Australia

DeLoach CJ (1981) Prognosis for biological control of weeds of southwestern U.S. rangelands. In: Delfosse ES (ed) Proceedings of the Fifth International Symposium on Biological Control of Weeds, 22–29 July 1980, Brisbane, Australia. CSIRO, Melbourne, Australia; pp. 175–199

DeLoach CJ (1990a) Prospects for biological control of saltcedar (*Tamarix* spp.) in riparian habitats of the southwestern United States. In: Delfosse ES (ed), Proceedings of the Seventh International Symposium on the Biological Control of Weeds, 6–11 March 1994, Rome, Italy. CSIRO Publications, Melbourne, Australia; pp. 307–314

DeLoach CJ (1990b) Terrestrial weeds. In: Habeck DH, Bennett FD, Frank JH (eds) Classical Biological Control in the Southern United States. Southern Cooperative Series Bulletin 355, Institute of Food and Agricultural Sciences, Univ Florida, Gainesville, Florida; pp. 157–164

DeLoach CJ (1991) Past successes and current prospects in biological control of weeds in the United States and Canada. Nat Areas J 11:129–142

DeLoach CJ (1995) Progress and problems in introductory biological control of native weeds in the U.S. In: Delfosse ES, Scott RR (eds) Proceedings of the Eighth International Symposium on Biological Control of Weeds, 2–7 February 1992, Lincoln University, Canterbury, New Zealand. CSIRO, Melbourne, Australia; pp. 111–112

DeLoach CJ, Boldt PE, Cordo HA, Johnson HB, Cuda JP (1986) Weeds common to Mexican and U.S. rangelands: proposals for biological control and ecological studies. In: Patton DR, Gonzales V CE, Medina AL, Segura T LA, Hamre RH (eds) Management and Utilization of Arid Land Plants, February 18–22, 1985, Saltillo, Mexico. U.S.D.A. Forest Service, Rocky Mountain Forest and Range Experiment Station, Fort Collins, Colorado; pp. 49–67

DeLoach CJ, Gerling D, Fornasari L, Sobhian R, Myartseva S, Mityaev ID, Lu QG, Tracy JL, Wang R, Wang JF, Kirk A, Pemberton RW, Chikatunov V, Jashenko RV, Johnson JE, Zheng H, Jiang SL, Liu MT, Liu AP, Cisneroz J (1996) Biological control programme against saltcedar (*Tamarix* spp.) in the United States of America: progress and problems. In: Moran VC, Hoffmann JH (eds), Proceedings of the Ninth International Symposium on Biological Control of Weeds, 19–26 January 1996, Stellenbosch, South Africa. University of Cape Town, Cape Town, South Africa; pp. 253–260

Deschenes P, Ludlow J (1993) Maintenance control of hydrilla in the Winter Park Chain of Lakes, Florida. Aquatics 15(2):13–15

Dewey MR, Jennings CA (1992) Habitat use by larval fishes in a backwater lake of the upper Mississippi River. J Freshwater Ecol 7:363–372

Dewey SA, Price KP, Ramsey D (1991) Satellite remote sensing to predict potential distribution of dyers woad (*Isatis tinctoria*). Weed Technol 5:479–484

Dial KP, Czaplewski NJ (1990) Do woodrat middens accurately represent the animal's environments and diets? The Woodhouse Mesa study. In: Betancourt JL, Van Devender TR, Martin PS (eds) Packrat Middens: The Last 40,000 Years of Biotic Change. Univ Arizona Press, Tucson, Arizona; pp. 43–58

di Castri F, Hansen AJ, Debussche M (eds) (1990) Biological Invasions in Europe and the Mediterranean Basin. Kluwer, Dordrecht, Netherlands

DiTommaso A, Aarssen LW (1989) Resource manipulations in natural vegetation: a review. Vegetatio 84:9–29

Dodd AP (1940) The Biological Campaign against Prickly-pear. Commonwealth Prickly Pear Board, Brisbane, Australia

Donald WW (1990) Management and control of Canada thistle. Rev Weed Sci 5:193–250

Doren RF, Whiteaker LD (1990a) Comparison of economic feasibility of chemical control strategies on differing age and density classes of *Schinus terebinthifolius*. Nat Areas J 10:28–34

Doren RF, Whiteaker LD (1990b) Effects of fire on different size individuals of *Schinus terebinthifolius*. Nat Areas J 10:107–113

Doren RF, Whiteaker LD, LaRosa AM (1991) Evaluation of fire as a management tool for controlling *Schinus terebinthifolius* as secondary successional growth on abandoned agricultural land. Environm Managem 15:121–129

Douglas BJ, Thomas AG, Derksen DA (1990) Downy brome (*Bromus tectorum*) invasion into southwestern Saskatchewan. Canad J Pl Sci 70:1143–1151

Drake JA, Mooney HA, di Castri F, Groves RH, Kruger FJ, Rejmánek M, Williamson M (eds) (1989) Biological Invasions: A Global Perspective. John Wiley, New York

Dudley T, Collins B (1995) Biological Invasions in California Wetlands. Pacific Institute for Studies in Development, Environment, and Security, Oakland, California

Dudley TL, Grimm NB (1994) Modification of macrophyte resistance to disturbance by an exotic grass, and implications for desert stream succession. Verh Int Verein Limnol 25:1456–1460

Duever MJ, Carlson JE, Meeder JF, Duever LC, Gunderson LH, Riopelle LA, Alexander TR, Myers RL, Spangler DP (1986) The Big Cypress National Preserve. Res Rep 6. National Audubon Society, New York

Duncan K (1994) Saltcedar: establishment effects and management. Wetland J (summer) 6:10–13

Dunn P, Campobasso G (1990) Host damage by *Pteronlonche inspersa* (Lepidoptera: Pterolonchidae) and *Bangasternus fausti* (Coleoptera: Curculionidae) on diffuse knapweed (*Centaurea diffusa*) (Abstr). In: Delfosse ES (ed) Proceedings of the Seventh International Symposium on Biological Control of Weeds, 6–11 March 1988, Rome, Italy. CSIRO, Melbourne, Australia; p. 171

Dunwiddie PW (1987) Macrofossil and pollen representation of coniferous trees in modern sediments from Washington. Ecology 68:1–11

Dunwiddie PW (1989) Forest and heath: the shaping of the vegetation on Nantucket Island. J Forest Hist (July):126–133

Durocher PP (1994) Status of the grass carp program in Texas. In: Proceedings of the Grass Carp Symposium, 7–9 March 1994, Gainesville, Florida. U.S. Army Engineer Waterways Experiment Station, Vicksburg, Mississippi; pp. 16–17

Dye T, Steadman DW (1990) Polynesian ancestors and their animal world. Amer Sci 78:207–215

Eamus D, Jarvis PG (1989) The direct effects of increase in the global atmospheric CO_2 concentration on natural and commercial temperate trees and forests. Advances Ecol Res 19:1–55

Eaton A (1822) A Manual of Botany, for the Northern and Middle States of America. Websters and Skinners, Albany, New York

Eberlein CV, Fore ZQ (1984) Kochia [*Kochia scoparia*] biology. Weeds Today 15(3):5–7

Eckardt N (1993) Element Stewardship Abstract: *Elaeagnus umbellata*. Computer printout. The Nature Conservancy, Minneapolis, Minnesota

Eddleman LE, Miller PM (1992) Potential impacts of western juniper on the hydrological cycle. In: Clary WP, McArthur ED, Bedunch D, Wambolt CL (compilers) Symposium on Ecology and Management of Riparian Shrub Communities. U.S.D.A. Forest Serv Gen Techn Rep INT-289. Intermountain Research Station, Ogden, Utah; pp. 176–180

Edwards ME (1986) Disturbance histories of four Snowdonian woodlands and their relation to Atlantic bryophyte distributions. Biol Conservation 37:301–320

Eggeman D (1994) Integrated hydrilla management plan utilizing herbicides and triploid grass carp in Lake Istokpoga. In: Proceedings of the Grass Carp Symposium, 7–9 March 1994, Gainesville, Florida. U.S. Army Engineer Waterways Experiment Station, Vicksburg, Mississippi; pp. 164–165

Egler FE (1942) Indigene versus alien in the development of arid Hawaiian vegetation. Ecology 23:14–23

Ehleringer JR, Dawson TE (1992) Water uptake by plants: perspectives from stable isotope composition. Pl Cell Environm 15:1073–1082

Ehleringer JR, Phillips SL, Schuster WSF, Sandquist DR (1991) Differential utilization of summer rains by desert plants. Oecologia 88:430–434

Ehrlich PR (1989) Attributes of invaders and the invading process: vertebrates. In: Drake JA, Mooney HA, di Castri F, Groves RH, Kruger FJ, Rejmánek M, Williamson M (eds) Biological Invasions: A Global Perspective. John Wiley, London, U.K.; pp. 315–328

Ehrlich PR, Dobkin DS, Wheye D (1988) The Birders Handook: A Field Guide to the Natural History of North American Birds. Simon and Schuster, New York

Ehrlich PR, Mooney HA (1983) Extinction, substitution, and ecosystem services. BioScience 33:248–254

Eichler LW, Bombard RT Jr, Sutherland JW, Boylen CW (1993) Suction harvesting of Eurasian watermilfoil and its effect on native plant communities. J Aquatic Pl Managem 31:144–148

Eichler LW, Bombard RT Jr, Sutherland JW, Boylen CW (1995) Recolonization of the littoral zone by macrophytes following the removal of benthic barrier material. J Aquatic Pl Managem 33:51–54

Elliott KJ, White AS (1989) Competition effects of various grasses and forbs on ponderosa pine seedlings. Forest Sci 33:356–366

Ellison CA, Evans HC (1990) Preliminary assessment of fungal pathogens as biological control agents for *Rottboellia cochinchinensis* (Gramineae). In: Delfosse ES (ed) Proceedings of the Seventh International Symposium on Biological Control of Weeds, 6–11 March 1988, Rome, Italy. CSIRO, Melbourne, Australia; pp. 477–482

Ellstrand NC, Marshall DL (1985) Interpopulation gene flow by pollen in wild radish, *Raphanus sativus*. Amer Naturalist 126:606–616

Elton CS (1958) The Ecology of Invasions by Animals and Plants. Methuen, London, U.K.

Emerson RW (1878) Fortune of the Republic. Houghton and Osgood, Boston, Massachusetts

Engel S (1984) Evaluating stationary blankets and removable screens for macrophyte control in lakes. J Aquatic Pl Managem 22:43–48

Engel S (1990) Ecological impacts of harvesting macrophytes in Halverson Lake, Wisconsin. J Aquatic Pl Managem 28:41–45

Engel S, Nichols SA (1984) Lake sediment alteration for macrophyte control. J Aquatic Pl Managem 22:38–41

Engel-Wilson RW, Ohmart RD (1978) Floral and attendant faunal changes on the lower Rio Grande between Fort Quitman and Presidio, Texas. In: Johnson RR, McCormick FJ (eds) Proceedings Symposium Strategies for Protection and Management of Floodplain Wetlands and Other Riparian Ecosystems, 11–13 December 1978, Callaway Gardens, Georgia. U.S.D.A. Forest Serv Gen Techn Rep WO-12. U.S.D.A. Forest Service, Washington, D.C.; pp. 139–147

[EPA] Environmental Protection Agency (1986) Effects of ozone on natural ecosystems and their components. In: Air Quality Criteria for Ozone and Other Photochemical Oxidants, EPA-600-8-84-020cF; pp. 7-1 to 7-66

Erickson SM (1981) Management Tools for Everyone: Twenty Techniques. Petrocelli Books, New York

Esler D (1990) Avian community responses to *Hydrilla* invasion. Wilson Bull 102:427–440

Evans JO (1991) The importance, distribution, and control of dyers woad (*Isatis tinctoria*). In: James LF, et al. (eds) Noxious Range Weeds. Westview Press, Boulder, Colorado; pp. 387–393

Evans JR (1986) Education: creative thinking and innovative education in the decision sciences. Decision Sci 17:250–262

Everitt BL (1980) Ecology of saltcedar—a plea for research. Environm Geol 3:77–84

Everitt JH, Anderson GL, Escobar DE, Davis MR, Spencer NR, Andrascik RJ (1995) Use of remote sensing for detecting and mapping leafy spurge (*Euphorbia esula*). Weed Technol 9:599–609

Everitt JH, DeLoach CJ (1990) Remote sensing of Chinese tamarisk (*Tamarix chinensis*) and associated vegetation. Weed Sci 38:273–278

Ewel JJ (1986) Invasibility: lessons from south Florida. In: Mooney HA, Drake JA (eds) Ecology of Biological Invasions of North America and Hawaii. Springer-Verlag, New York; pp. 214–230

Faber-Langendoen D, Davis MA (1995) Effects of fire frequency on tree canopy cover at Allison Savanna, eastcentral Minnesota, USA. Nat Areas J 15:319–328

Fahey TJ, Young DR (1984) Soil and xylem water potential and soil water content in contrasting *Pinus contorta* ecosystems, southeastern Wyoming, USA. Oecologia 61:346–351

Falk DA (1990) The theory of integrated conservation strategies for biological diversity. In: Mitchell RS, Sheviak CJ, Leopold DJ (eds) Ecosystem Management: Rare Species and Significant Habitats. New York State Mus Bull 471; pp. 5–10

Farah KO, Tanaka AF, West NE (1988) Autecology and population biology of dyers woad (*Isatis tinctoria*). Weed Sci 36:186–193

Farone SM, McNabb TM (1993) Changes in nontarget wetland vegetation following a large-scale fluridone application. J Aquatic Pl Managem 31:185–189

Fægri K, Kaland PE, Krzywinski K (1989) Textbook of Pollen Analysis. 4th ed. John Wiley, New York

Federal Register (1995) Part III Department of the Interior, Fish and Wildlife Service. 50 CFR Part 17. Endangered and threatened species: Southwestern willow flycatcher; Final Rule. Volume 60:10694–10715

Fenner P, Brady WW, Patton DR (1985) Effects of regulated water flows on regeneration of Fremont cottonwood. J Range Managem 38:135–139

Fensham RJ, Fairfax RJ, Cannell RJ (1994) The invasion of *Lantana camara* L. in Forty Mile Scrub National Park, north Queensland. Austral J Ecol 19:297–305

Fenwood J (1992) Using silviculture to achieve a desired future condition for biological diversity in the southern region. In: Murphy D (compiler), Proceedings of the Workshop, Getting to the Future Through Silviculture. U.S.D.A. Forest Serv Gen Techn Rep INT-291; pp. 40–41

Fernald ML (1950) Gray's Manual of Botany. 8th ed. American Book Co., New York

Field RP (1990) Progress towards biological control of ragwort in Australia. In: Delfosse ED (ed) Proceedings of the Seventh International Symposium on Biological Control of Weeds, 6–11 March 1988, Rome, Italy. CSIRO, Melbourne, Australia; pp. 315–322

Finley RB Jr (1990) Woodrat ecology and behavior and the interpretation of paleomiddens. In: Betancourt JL, Van Devender TR, Martin PS (eds) Packrat Middens: The Last 40,000 Years of Biotic Change. Univ Arizona Press, Tucson, Arizona; pp. 28–42

Flanagan LB, Ehleringer JR, Marshall JD (1992) Differential uptake of summer precipitation among co-occurring trees and shrubs in a pinyon-juniper woodland. Pl Cell Environm 15:831–836

Flenley JR, King ASM, Jackson J, Chew C, Teller JT, Prentice ME (1991) The Late Quaternary vegetational and climatic history of Easter Island. J Quatern Sci 6:85–115

Fogg JM Jr (1945) Weeds of Lawn and Garden. Univ Pennsylvania Press, Philadelphia, Pennsylvania

Fonteyn PJ, Mahall BE (1978) Competition among desert perennials. Nature 275:544–545

Fonteyn PJ, Schlesinger WH, Marion GM (1987) Accuracy of soil thermocouple hygrometer measurements in desert ecosystems. Ecology 68:1121–1124

Forcella F (1985a) Final distribution is related to rate of spread in alien weeds. Weed Res 25:181–191

Forcella F (1985b) Spread of kochia [*Kochia scoparia*] in the northwestern United States. Weeds Today 16(4):4–6

Forcella F, Harvey SJ (1988) Patterns of weed migration in northwestern USA. Weed Sci 36:194–201

Forcella F, Wood JT, Dillon SP (1986) Characteristics distinguishing invasive weeds within *Echium* (bugloss). Weed Res 26:351–364

Forest Resources Management (1992) Evaluation of Effects of Melaleuca Control Methods Contract No. C-3150. 6 October 1992. South Florida Water Management District, West Palm Beach, Florida

Foster DR, Zebryk TM (1993) Long-term vegetation dynamics and disturbance history of a *Tsuga*-dominated forest in New England. Ecology 74:982–998

Foster DR, Zebryk T, Schoonmaker P, Lezberg A (1992) Post-settlement history of human land-use and vegetation dynamics of a *Tsuga canadensis* (hemlock) woodlot in central New England. J Ecol 80:773–786

Fowler MC, Robson TO (1978) The effects of food preferences and stocking rates of grass carp (*Ctenopharyngodon idella* Val.) on mixed plant communities. Aquatic Bot 5:261–272

Fox MD, Fox BJ (1986) The susceptibility of natural communities to invasion. In: Groves RH, Burdon JJ (eds) Ecology of Biological Invasions: An Australian Perspective. Cambridge Univ Press, New York; pp. 57–66

Foy CL, Forney DR, Cooley WE (1983) History of weed introductions. In: Wilson CL, Graham CL (eds) Exotic Plant Pests and North American Agriculture. Academic Press, New York; pp. 65–92

Frankel E (1989) Distribution of *Pueraria lobata* in and around New York City. Bull Torrey Bot Club 116:390–394

Franklin JF, Swanson FJ, Harmon ME, Perry DA, Spies TA, Dale VH, McKee A, Ferrell WK, Means JE, Gregory SV, Lattin JD, Schowalter TD, Larsen D (1992) Effects of Global Climatic Change on Forests in Northwestern North America. In: Peters RL, Lovejoy TE (eds) Global Warming and Biological Diversity. Yale Univ Press, New Haven, Connecticut; pp. 244–257

Frazier BE, Moore BC (1993) Some tests of film types for remote sensing of purple loosestrife, *Lythrum salicaria*, at low densities. Wetlands 13:145–152

Fredskild B (1978) Palaeobotanical investigations of some peat deposits of Norse age at Quagssiarssuk, South Greenland. Meddel Grønland 204(5):1–41

Fredskild B (1988) Agriculture in a marginal area—South Greenland from the Norse landnam (985 A.D.) to the present (1985 A.D.). In: Birks HH, Birks HJB, Kaland PE, Moe D (eds) The Cultural Landscape—Past, Present and Future. Cambridge Univ Press, Cambridge, U.K.; pp. 381–393

Freeman DC, Klikoff LG, Harper KT (1976) Differential resource utilization by the sexes of dioecious plants. Science 193:597–599

Freeman TE (ed) (1978) Proceedings of the IV International Symposium on Biological Control of Weeds, Gainesville, Florida, USA, August 30-September 2, 1976. Center for Environmental Programs, Univ Florida, Gainesville, Florida

Frenkel RE (1970) Ruderal Vegetation along some California Roadsides. Univ Calif Publ Geogr 20

Frick KE (1974) Biological control of weeds: introduction, history, theoretical and practical applications. In: Maxwell FG, Harris FA (eds) Proceedings of the Summer Institute on Biological Control of Plant Insects and Diseases, Univ Press of Mississippi, Jackson, Mississippi; pp. 204–223

Frick KE (ed) (1978) Biological control of thistles in the genus *Carduus* in the United States. A progress report. U.S.D.A. Science & Education Administration, New Orleans, Louisiana

Frick KE, Chandler JM (1978) Augmenting the moth *Bactra verutana* in field plots for early-season suppression of purple nutsedge (*Cyperus rotundus*). Weed Sci 26:703–710

Frick KE, Johnson GR (1973) *Longitarsus jacobaeae* (Coleoptera: Chrysomelidae), a flea beetle for the biological control of tansy ragwort. 4. Life history and adult aestivation of an Italian biotype. Ann Entomol Soc Amer 66:358–367

Fritz GL (1994) Precolumbian *Cucurbita argyrosperma* ssp. *argyrosperma* (Cucurbitaceae) in the eastern woodlands of North America. Econ Bot 48:280–292

Funasaki GY, Lai P-Y, Nakahara LM, Beardsley JW, Ota AK (1988) A review of biological control introductions in Hawaii: 1890 to 1985. Proc Hawaiian Entomol Soc 28:105–160

Gagne BH, Loope LL, Medeiros AC, Anderson SJ (1992) *Miconia calvescens*: A threat to native forest of the Hawaiian islands. Pacif Sci 46:390–391

Gaillard M-J, Berglund BE (1988) Land-use history during the last 2700 years in the area of Bjäresjö, Southern Sweden. In: Birks HH, Birks HJB, Kaland PE, Moe D (eds) The Cultural Landscape—Past, Present and Future. Cambridge Univ Press, Cambridge, U.K.; pp. 409–428

Gajewski K (1988) Late Holocene climate changes in eastern North America estimated from pollen data. Quatern Res 29:255–262

Gallagher JE, Haller WT (1990) History and development of aquatic weed control in the United States. Rev Weed Sci 5:115–192

Galloway JN, Schlesinger WH, Levy H II, Michaels A, Schnoor JL (1995) Nitrogen fixation: anthropogenic enhancement—environmental response. Global Biogeochem Cycles 9:235–252

Garbutt K, Bazzaz FA (1984) The effects of elevated CO_2 on plants. II. Flower, fruit and seed production and abortion. New Phytol 98:433–446

Garrett AO (1912) Spring Flora of the Wasatch Region. 2nd ed. New Era Printing Co., Lancaster, Pennsylvania

Gatewood JS, Robinson TW, Colby BR, Hem JD, Halpenny LC (1950) Use of Water by Bottomland Vegetation in Lower Safford Valley, Arizona. U.S. Geol Surv Water Supply Paper 1103

Gaubert H (1992) Les invasions biologiques en milieu insulare: le cas de *Miconia calvescens* a Tahiti. Unpublished report at Library of Centre ORSTOM de Tahiti, Papeete, French Polynesia

Gaudreau DC, Jackson ST, Webb T III (1989) Spatial scale and sampling strategy in paleoecological studies of vegetation patterns in mountainous terrain. Acta Bot Neerl 38:369–390

Gavin TA (1989) What's wrong with the questions we ask in wildlife research? Wildlife Soc Bull 17:345–350

Gay LW, Fritschen LJ (1979) An energy budget analysis of water use by saltcedar. Water Resource Res 15:1589–1592

Geber MA, Dawson TE (1993) Evolutionary responses of plants to global change. In: Kareiva PM, Kingsolver JG, Huey RB (eds) Biotic Interactions and Global Change. Sinauer Associates, Sunderland, Massachusetts; pp. 179–197

Gehring JL, Linhart YB (1992) Population structure and genetic differentiation in native and introduced populations of *Deschampsia caespitosa* (Poaceae) in the Colorado alpine. Amer J Bot 97:1337–1343

Geiger NS (1983) Winter drawdown for the control of Eurasian water milfoil in an Oregon oxbow lake (Blue Lake, Multnomah County). In: Lake Restoration, Protection and Management. EPA 440/5-83-001. U.S. Environmental Protection Agency, Washington, D.C.; pp. 193–197

Gerling D, Kugler J (1973) Evaluation of Enemies of Noxious Plants in Israel as Potential Agents for the Biological Control of Weeds. Final Technical Report, 1 September 1970–31 August 1973, Department of Zoology, Tel Aviv University, Tel Aviv, Israel

Getsinger KD (1991) Chemical control technology: history and overview. In: Proceedings, 25th Annual Meeting, Aquatic Plant Control Research Program, 26–30 November 1990, Orlando, Florida. Misc Paper A-91-3, U.S. Army Engineer Waterways Experiment Station, Vicksburg, Mississippi; pp. 197–200

Getsinger KD (1993) Long Lake Project: chemical control technology transfer. In: Proceedings, 27th Annual Meeting, Aquatic Plant Control Research Program, Bellevue, Washington. Misc Paper A-93-2, U.S. Army Engineer Waterways Experiment Station, Vicksburg, Mississippi; pp. 10–16

Getsinger KD, Davis GJ, Brinson MM (1982) Changes in a *Myriophyllum spicatum* L. community following 2,4-D treatment. J Aquatic Pl Managem 20:4–8

Getsinger KD, Dick GO, Crouch RM, Nelson LS (1994) Mesocosm evaluation of bensulfuron methyl activity on Eurasian watermilfoil, vallisneria, and American pondweed. J Aquatic Pl Managem 32:1–6

Getsinger KD, Hanlon C, Joyce JC, Fox AM, Haller WT (1991) Herbicide application technique development for flowing water: Relationship of water exchange and submersed application methods. In: Proceedings, 25th Annual Meeting, Aquatic Plant Control Research Program, 26–30 November 1990, Orlando, Florida. Misc Paper A-91-3, U.S. Army Engineer Waterways Experiment Station, Vicksburg, Mississippi; pp. 210–218

Gibbons MV, Gibbons HL Jr (1988) Efficacy of rotovation in controlling Eurasian watermilfoil in the Pend Oreille River, Washington. Lake Reservoir Managem 4:153–160

Gibbons MV, Gibbons HL Jr, Sytsma MD (1994) A Citizen's Manual for Developing Integrated Aquatic Vegetation Management Plans. 1st ed. Washington Department of Ecology, Olympia, Washington

Gibson DJ, Seastedt TR, Briggs JM (1993) Management practices in tallgrass prairie: large- and small-scale experimental effects on species composition. J Appl Ecol 30:247–255

Gill DS, Marks PL (1991) Tree and shrub seedling colonization of old fields in central New York. Ecol Monogr 61:183–205

Gillis AM (1992) Keeping aliens out of paradise. BioScience 42:482–485

Gish JW (1991) Current perceptions, recent discoveries, and future directions in Hohokam palynology. Kiva 56:237–254

Given DR (1992) An Overview of the Terrestrial Biodiversity of Pacific Islands. Report to South Pacific Regional Environmental Programme, Apia, Western Samoa.

Givens CR, Givens FM (1987) Age and significance of fossil white spruce (*Picea glauca*), Tunica Hills, Louisiana-Mississippi. Quatern Res 27:283–296

Gleason HA, Cronquist A (1991) Manual of Vascular Plants of Northeastern United States and Adjacent Canada. 2nd ed. The New York Botanical Garden, Bronx, New York

Glyphis JP, Milton SJ, Siegfried WR (1981) Dispersal of *Acacia cyclops* by birds. Oecologia 48:138–141

Godfrey RK (1988) Trees, Shrubs, and Woody Vines of Northern Florida and Adjacent Georgia and Alabama. Univ Georgia Press, Athens, Georgia

Godwin H (1943a) *Frangula alnus* Miller (*Rhamnus frangula* L.). J Ecol 31:77–92

Godwin H (1943b) *Rhamnus cathartica* L. J Ecol 31:69–76

Goeden RD (1977) Biological control of weeds. In: Truelove B (ed) Research Methods in Weed Science. Auburn Printing, Auburn, Alabama; pp. 45–47

Goeden RD (1978) Part II. Biological control of weeds. In: Clausen CP (ed) Introduced Parasites and Predators of Arthropod Pests and Weeds. U.S.D.A. Agric Handb 480:357–414

Goeden RD (1983) Critique and revision of Harris' scoring system for selection of insect agents in biological control of weeds. Protect Ecol 5:287–301

Goeden RD (1995a) Italian thistle. In: Nechols JR, Andres LA, Beardsley JW, Goeden RD, Jackson CG (eds) Biological Control in the Western United States. Accomplishments and Benefits of Regional Research Project W-84, 1964–1989. Univ Calif Div Agric Nat Resources Publ 3361; pp. 242–244

Goeden RD (1995b) Milk thistle. In: Nechols JR, Andres LA, Beardsley JW, Goeden RD, Jackson CG (eds) Biological Control in the Western United States. Accomplishments and Benefits of Regional Research Project W-84, 1964–1989. Univ Calif Div Agric Nat Resources Publ 3361; pp. 245–247

Goeden RD, Louda SM (1976) Biotic interference with insects imported for weed control. Annual Rev Entomol 21:325–342

Goeden RD, Pemberton RW (1995) Russian thistle. In: Nechols JR, Andres LA, Beardsley JW, Goeden RD, Jackson CG (eds) Biological Control in the Western United States. Accomplishments and Benefits of Regional Research Project W-84, 1964–1989. Univ Calif Div Agric Nat Resources Publ 3361; pp. 276–280

Goeden RD, Ricker DW (1981) Santa Cruz Island—revisited, sequential photography records the causation, rates of progress, and lasting benefits of successful biological weed control. In: Delfosse ES (ed) Proceedings of the Fifth International Symposium on Biological Control of Weeds, 22–29 July 1980, Brisbane, Australia. CSIRO, Melbourne, Australia; pp. 355–365

Gove PB (ed) (1986) Webster's Third New International Dictionary of the English Language Unabridged. Merriam-Webster, Springfield, Massachusetts

Gowanloch JN (1944) The economic status of the waterhyacinth in Louisiana. Louisiana Conservationist 2:3–8

Grace JB, Wetzel RG (1978) The production biology of Eurasian watermilfoil (*Myriophyllum spicatum* L.): a review. J Aquatic Pl Managem 16:1–11

Graf WL (1978) Fluvial adjustments to the spread of tamarisk in the Colorado Plateau region. Geol Soc Amer Bull 89:1491–1501

Graf WL (1982) Tamarisk and river-channel management. Environm Managem 6:283–296

Graham RL, Hunsaker CT, O'Neill RV (1991) Ecological risk assessment at the regional scale. Ecol Applic 1:196–206

Graumlich LJ, Brubaker LB (1995) Long-term records of growth and distribution of conifers: integration of paleoecology and physiological ecology. In: Smith WK, Hinckley TM (eds) Ecophysiology of Coniferous Forests. Academic Press, San Diego, California; pp. 36–62

Graumlich LJ, Davis MB (1993) Holocene variation in spatial scales of vegetation pattern in the Upper Great Lakes. Ecology 74:826–839

Gray A (1889) Manual of the Botany of the Northern United States. 6th ed; revised by S. Watson and J.M. Coulter. American Book Co., New York

Greathead DJ (1971) A Review of Biological Control in the Ethiopian Region. Commonwealth Inst Biol Control Techn Commun 5. Commonwealth Agricultural Bureaux, Farnham Royal, Slough, U.K.

Greathead DJ (1973) Progress in the biological control of *Lantana camara* in East Africa and discussion of problems raised by the unexpected reaction of some of the more promising insects to *Sesamum indicum*. In: Dunn PH (ed) Proceedings of the Second International Symposium on Biological Control Of Weeds, 4–7 October 1971, Rome, Italy. Commonwealth Agricultural Bureaux, Farnham Royal, Slough, U.K.; pp. 89–92

Great Plains Flora Association (1986) Flora of the Great Plains. Univ Press of Kansas, Lawrence, Kansas

Green H, Hunter C, Moore B (1990) Application of the Delphi technique in tourism. Ann Tourism Res 17:270–279

Green RH (1979) Sampling Design and Statistical Methods for Environmental Biologists. John Wiley, New York

Green RH, Young RC (1993) Sampling to detect rare species. Ecol Applic 3:351–356

Green WR, Westerdahl HE (1990) Response of Eurasian watermilfoil to 2,4-D concentration and exposure times. J Aquatic Pl Managem 28:27–32

Grey WE, Quimby PC Jr, Mathre DE, Young JA (1995) Potential for biological control of downy brome (*Bromus tectorum*) and medusahead (*Taeniatherum caput-medusae*) with crown and root rot fungi. Weed Technol 9:362–365

Griffin GF, Stafford Smith DM, Morton SR, Allan GE, Masters KA (1989) Status and implications of the invasion of tamarisk (*Tamarix aphylla*) on the Finke River, Northern Territory, Australia. J Environm Managem 29:297–315

Griffith D, Lacey JR (1991) Economic evaluation of spotted knapweed [*Centaurea maculosa*] control using picloram. J Range Managem 44:43–47

Grilz PL, Romo JT (1995) Management considerations for controlling smooth brome in fescue prairie. Nat Areas J 15:148–156

Grime JP (1977) Evidence for the existence of three primary strategies in plants and its relevance to ecological and evolutionary theory. Amer Naturalist 111:1169–1194

Grime JP, Hodgson JG, Hunt R (1988) Comparative Plant Ecology: A Functional Approach to Common British Species. Unwin Hyman, London, U.K.

Grimm EC (1983) Chronology and dynamics of vegetation change in the prairie-woodland section of southern Minnesota, U.S.A. New Phytol 93:311–350

Grimm MP, Backx JJGM (1990) The restoration of shallow eutrophic lakes, and the role of northern pike, aquatic vegetation and nutrient concentration. Hydrobiologia 200/201:557–566

Grinnell J, Storer TI (1916) Animal life as an asset of national parks. Science 44:375–380

Grodowitz MJ, Center TD, Snoddy E, Dray FA Jr (1994) Release and establishment of insect biocontrol agents for the management of hydrilla. In: Proceedings, 28th Annual Meeting of the Aquatic Plant Control Research Program, 15–18 November 1993, Baltimore, Maryland. Misc Paper A-94-2, U.S. Army Engineer Waterways Experiment Station, Vicksburg, Mississippi; pp. 181–201

Grodowitz MJ, Center TD, Snoddy E (1995) Current status on the use of insect biocontrol agents for the management of hydrilla. In: Proceedings, 29th Annual Meeting of the Aquatic Plant Control Research Program, 14–17 November 1994, Vicksburg, Mississippi. Misc Paper A-95-3, U.S. Army Engineer Waterways Experiment Station, Vicksburg, Mississippi; pp. 134–141

Groves RH (1986) Plant invasions of Australia: an overview. In: Groves RH, Burdon JJ (eds) Ecology of Biological Invasions. Cambridge Univ Press, Cambridge, U.K.; pp. 137–149

Groves RH, Burdon JJ (1986) Ecology of Biological Invasions: Some Australian Case Histories. John Wiley, Brisbane, Australia

Groves RH, Cullen JM (1981) *Chondrila juncea*: the ecological control of a weed in Australia. In: Ketching RL, Jones RE (eds) The Ecology of Pests: Some Australian Case Histories. CSIRO, Melbourne, Australia; pp. 6–17

Groves RH, di Castri F (eds) (1991) Biogeography of Mediterranean Invasions. Cambridge Univ Press, Cambridge, U.K.

Grumbine RE (1994) What is ecosystem management? Conservation Biol 8:27–38

Gunner HB, Limpa-Amara Y, Bouchard BS, Weilerstein PJ, Taylor ME (1990) Microbiological Control of Eurasian Watermilfoil: Final Report. Techn Rep A-90-2. U.S. Army Engineer Waterways Experiment Station, Vicksburg, Mississippi

Gunnison G, Barko JW (1992) Factors influencing gas evolution beneath a benthic barrier. J Aquatic Pl Managem 30:23–28

Haag KH, Glenn MS, Jordan JC (1988) Selective patterns of herbicide application for improved biological control of waterhyacinth. J Aquatic Pl Managem 26:17–19

Haag KH, Habeck DH (1991) Enhanced biological control of waterhyacinth following limited herbicide application. J Aquatic Pl Managem 29:24–28

Habeck DH (1990) Insects associated with poison ivy as potential biocontrol agents. In: Delfosse ES (ed) Proceedings of the Seventh International Symposium on Biological Control of Weeds, 6–11 March 1988, Rome, Italy. CSIRO, Melbourne, Australia; pp. 329–337

Habib R, Hasan SA (1982) Insect enemies attacking tamarisk, *Tamarix* spp., in Pakistan. Final Report, June 1975–June 1980. Pakistan Station, Commonwealth Institute of Biological Control, Rawalpindi, Pakistan

Hairston NG Sr (1989) Ecological Experiments: Purpose, Design, and Execution. Cambridge Univ Press, Cambridge, U.K.

Haller WT (1994) Probable grass carp stocking scenarios. In: Proceedings of the Grass Carp Symposium, 7–9 March 1994, Gainesville, Florida. U.S. Army Engineer Waterways Experiment Station, Vicksburg, Mississippi; pp. 236–238

Haller WT, Fox AM, Hanlon CA (1992) Inhibition of hydrilla tuber formation by bensulfuron methyl. J Aquatic Pl Managem 30:48–49

Halvorson WL, Meander GJ (eds) The Fourth California Island Symposium: Update on the Status of Resources. Santa Barbara Museum of Natural History, Santa Barbara, California; pp. 351–364

Hammond D (1994) Florida Game and Fresh Water Fish Commission concerns regarding the use of grass carp. In: Proceedings of the Grass Carp Symposium, 7–9 March 1994, Gainesville, Florida. U.S. Army Engineer Waterways Experiment Station, Vicksburg, Mississippi; pp. 7–8

Harlan JR, deWet JMJ (1965) Some thoughts about weeds. Econ Bot 19:16–24

Harley KLS, Forno IW (1992) Biological Control of Weeds. A Handbook for Practitioners and Students. CSIRO Division of Entomology, Brisbane, Australia

Harper JL (1977) Population Biology of Plants. Academic Press, London, U.K.

Harrington RA, Brown BJ, Reich PB (1989) Ecophysiology of exotic and native shrubs in southern Wisconsin I. Relationship of leaf characteristics, resource availability, and phenology to seasonal patterns of carbon gain. Oecologia 80:356–367

Harris GA (1967) Some competitive relationships between *Agropyron spicatum* and *Bromus tectorum*. Ecol Monogr 37:89–111

Harris GA, Wilson AM (1970) Competition for moisture among seedlings of annual and perennial grasses as influenced by root elongation at low temperature. Ecology 51:530–534

Harris P (1973) The selection of effective agents for the biological control of weeds. Canad Entomol 105:1495–1503

Harris P (1979) The cost of biological control of weeds by insects in Canada. Weed Sci 27:242–250

Harris P (1988) Environmental impact of weed-control insects. BioScience 38:542–548

Harris P (1990) Environmental impact of introduced biological control agents. In: Mackauer M, Ehler LE, Roland J (eds) Critical Issues in Biological Control. Intercept, Andover, Hants, U.K.; pp. 289–300

Harris P (1993) Effects, constraints and the future of weed biocontrol. Agric Eco-syst Environm 46:289–303

Harris P, Myers JH (1984) *Centaurea diffusa* Lam. and *C. maculosa* Lam. s. lat., diffuse and spotted knapweeds (Compositae). In: Kelleher JS, Hulme MA (eds) Biological Control Programmes Against Insects and Weeds in Canada 1969–1980. Commonwealth Agricultural Bureaux, Farnham Royal, Slough, U.K.; pp. 127–137

Harris P, Wilkinson ATS (1984) *Cirsium vulgare* (Savi) Ten., bull thistle (Compositae). In: Kelleher JS, Hulme MA (eds) Biological Control Programmes Against Insects and Weeds in Canada 1969–1980. Commonwealth Agricultural Bureaux, Farnham Royal, Slough, U.K.; pp. 147–153

Harris P, Wilkinson ATS, Myers JH (1984) *Senecio jacobaea* L., tansy ragwort (Compositae). In: Kelleher JS, Hulme MA (eds) Biological Control Programmes against Insects and Weeds in Canada 1969–1980. Commonwealth Agricultural Bureaux, Farnham Royal, Slough, U.K.; pp. 195–201

Harris P, Zwölfer H (1968) Screening of phytophagous insects for biological control of weeds. Canad Entomol 100:295–303

Harrod RJ, Taylor RJ (1995) Reproduction and pollination biology of *Centaurea* and *Acroptilon* species, with emphasis on *C. diffusa*. Northwest Sci 69:97–105

Harte J, Shaw R (1995) Shifting dominance within a montane vegetation community: results of a climate-warming experiment. Science 267:876–880

Harty FM (1993) How Illinois kicked the exotic habit. In: McKnight BN (ed) Biological Pollution: The Control and Impact of Invasive Exotic Species. Indiana Academy of Science, Indianapolis, Indiana; pp. 195–209

Harvey JL, Varley DR, Evans HC (1995) European surveys for pathogens of Eurasian watermilfoil. In: Proceedings, 29th Annual Meeting of the Aquatic Plant Control Research Program, 14–17 November 1994, Vicksburg, Mississippi. Misc Paper A-95-3, U.S. Army Engineer Waterways Experiment Station, Vicksburg, Mississippi; pp. 130–133

Hastings JR, Turner RM (1965) The Changing Mile. Univ Arizona Press, Tucson, Arizona

Havera SP, Suloway LB (1994) Wetlands. In: The Changing Illinois Environment: Critical Trends. Techn Rep 3, Ecological Resources. Illinois Department of Energy and Natural Resources, Springfield, Illinois; pp. 87–152

Hawkes RB (1981) Biological control of tansy ragwort in the state of Oregon, U.S.A. In: Delfosse ES (ed) Proceedings of the Fifth International Symposium on the Biological Control of Weeds, 22–29 July 1980, Brisbane, Australia. CSIRO, Melbourne, Australia; pp. 623–626

Heil GW, Diemont WH (1983) Raised nutrient levels change heathland into grassland. Vegetatio 53:113–120

Heisey RM (1990) Evidence for allelopathy by tree-of-heaven (*Ailanthus altissima*). J Chem Ecol 16:2039–2055

Hellquist CB (1993) Taxonomic considerations in aquatic vegetation assessments. Lake Reservoir Managem 7:175–183

Henderson JE (1993) Economic evaluations of aquatic plant control. In: Proceedings, 27th Annual Meeting, Aquatic Plant Control Research Program, 16–19 November 1992, Bellevue, Washington. Misc Paper A-93-2, U.S. Army Engineer Waterways Experiment Station, Vicksburg, Mississippi; pp. 23–29

Henderson JE (1994) Valuation of aquatic plant alternatives at Lake Guntersville: preliminary results from the recreation study. In: Proceedings, 28th Annual Meeting of the Aquatic Plant Control Research Program, 15–18 November 1993, Baltimore, Maryland. Misc Paper A-94-2, U.S. Army Engineer Waterways Experiment Station, Vicksburg, Mississippi; pp. 21–27

Henderson JE (1995) Use of economic information in the evaluation of aquatic plant control programs: the Lake Guntersville Recreation Study. In: Proceedings, 29th Annual Meeting of the Aquatic Plant Control Research Program, 14–17 November 1994, Vicksburg, Mississippi. Misc Paper A-95-3, U.S. Army Engineer Waterways Experiment Station, Vicksburg, Mississippi; pp. 8–18

Hengeveld R (1989) Dynamics of Biological Invasions. Chapman and Hall, London, U.K.

Henry RD (1992) Letter to the editor. Nat Areas J 12:126

Henry RD (1993) Letter to the editor. Nat Areas J 13:71

Henry RD (1994) Letter to the editor. Nat Areas J 14:87

Herkert JR (ed)(1991) Endangered and Threatened Species of Illinois: Status and Distribution. Volume 1—Plants. Illinois Endangered Species Board, Springfield, Illinois

Herrera CM (1995) Plant-vertebrate seed dispersal systems in the Mediterranean: ecological, evolutionary, and historical determinants. Annual Rev Ecol Syst 26:705–727

Hester AJ, Hobbs RJ (1992) Influence of fire and soil nutrients on native and non-native annuals at remnant vegetation edges in the Western Australian wheatbelt. J Veg Sci 3:101–108

Heyerdahl T (1971) The Ra Expeditions. Doubleday, New York

Heywood VH (1989) Patterns, extents and modes of invasions by terrestrial plants. In: Drake JH, Mooney HA, di Castri F, Groves RH, Kruger FJ, Rejmánek M, Williamson M (eds) Biological Invasions: A Global Perspective. John Wiley, New York; pp. 31–60

Heywood VH (1993) Flowering Plants of the World. Oxford Univ Press, New York

Hickman JC (ed) (1993) The Jepson Manual: Higher Plants of California. Univ California Press, Berkeley, California

Hickman JC, Hickman CS (1978) *Polygonum perfoliatum*: a recent Asiatic adventive. Bartonia 45:18–23

Hiebert R (1990) Managing alien plants. Trends 27:8–16

Hiebert RD, Stubbendieck J (1993) Handbook for Ranking Exotic Plants for Management and Control. U.S. Natl Park Serv Nat Resources Rep NPS/NRMWRO/NRR-93/08

Hight SD (1993) Control of the ornamental purple loosestrife (*Lythrum salicaria*) by exotic organisms. In: McKnight BN (ed) Biological Pollution: The Control and Impact of Invasive Exotic Species. Indiana Academy of Sciences, Indianapolis, Indiana; pp. 147–148

Hight SD, Drea JJ (1991) Prospects for a classical biological control project against purple loosestrife (*Lythrum salicaria* L.). Nat Areas J 11:151–157

Hill JD, Canham CD, Wood DM (1995) Patterns and causes of resistance to tree invasion in rights-of-way. Ecol Applic 5:459–470

Hillier HG (1983) Hillier's Manual of Trees & Shrubs. Van Nostrand Reinhold, New York

Hils MH, Vankat JL (1982) Species removals from a first-year old-field plant community. Ecology 63:705–711

Hink VC, Ohmart RD (1984) Middle Rio Grande Biological Survey. Final report. Center for Environmental Studies, Arizona State Univ, Tempe, Arizona

Hintze J (1990) Number Cruncher Statistical System (v. 5.8), Reference Manual for Advanced Table Analysis. NCSS, Kaysville, Utah

Hirose Y (ed) (1992) Biological Control in South and East Asia. IOBC/SEARS, Kyushu Univ Press, Fukuoka, Japan

Hobbs RJ (1989) The nature and effects of disturbance relative to invasions. In: Drake JH, Mooney HA, di Castri F, Groves RH, Kruger FJ, Rejmánek M, Williamson M (eds) Biological Invasions: A Global Perspective. John Wiley, New York; pp. 389–405

Hobbs RJ, Huenneke LF (1992) Disturbance, diversity, and invasion: implications for conservation. Conservation Biol 6:324–337

Hobbs RJ, Humphries SE (1995) An integrated approach to the ecology and management of plant invasions. Conservation Biol 9:761–770

Hobbs RJ, Mooney HA (1986) Community changes following shrub invasion of grassland. Oecologia 70:508–513

Hocking PJ, Meyer CP (1985) Responses of Noogoora burr (*Xanthium occidentale*) to nitrogen supply and carbon dioxide enrichment. Ann Bot (Oxford) 55:835–844

Hocutt GE, Dimmick RW (1971) Summer food habits of juvenile wood ducks in east Tennessee. J Wildlife Managem 35:286–292

Hoffman JR (1991) Agriculture Ecosystems and Environment. Biological Control of Weeds in South Africa. Volume 37, Special Issue. Elsevier, Amsterdam, Netherlands

Hoffmann AJ, Armesto JJ (1995) Modes of seed dispersal in the Mediterranean regions in Chile, California, and Australia. In: Kalin Arroyo MT, Zedler PH, Fox MD (eds) Ecology and Biogeography of Mediterranean ecosystems in Chile, California and Australia. Springer-Verlag, New York: pp. 289–310

Hogue CL (1993) Insects of the Los Angeles Basin. Natural History Museum of Los Angeles County, Los Angeles, California

Holloway JK (1964) Projects in biological control of weeds. In: DeBach PE (ed) Biological Control of Insect Pests and Weeds. Chapman and Hall, London, U.K.; pp. 650–670

Holloway JK, Huffaker CB (1951) The role of *Chrysolina quadrigemina* in the biological control of Klamath weed. J Econ Entomol 44:244–247

Holm LG, Plucknett DL, Pancho JV, Herberger JP (1991) The World's Worst Weeds. Krieger, Malabar, Florida

Holm LG, Plucknett PL, Pancho JV, Herberger JP (1979) A Geographical Atlas of World Weeds. John Wiley, New York

Honnell DR, Madsen JD, Smart RM (1993) Effects of Selected Exotic and Native Aquatic Plant Communities on Water Temperature and Dissolved Oxygen. Inform Exch Bull A-93-2. U.S. Army Engineer Waterways Experiment Station, Vicksburg, Mississippi

Hooghiemstra H (1984) Vegetational and Climatic History of the High Plain of Bogotá, Colombia: A Continuous Record of the Last 3.5 Million Years. J. Cramer, Vaduz, Liechtenstein

Hooghiemstra H, Sarmiento G (1991) Long continental pollen record from a tropical intermontane basin: late Pliocene and Pleistocene history from a 540-meter core. Episodes 14:107–115

Horsley SB (1977a) Allelopathic inhibition of black cherry by fern, grass, goldenrod, and aster. Canad J Forest Res 7:205–216

Horsley SB (1977b) Allelopathic inhibition of black cherry. II. Inhibition by woodland grass, ferns, and club moss. Canad J Forest Res 7:515–519

Horton JS (1964) Notes on the Introduction of Deciduous Tamarisk. U.S.D.A. Forest Serv Res Note RM-16

Horton JS, Campbell CJ (1974) Management of Phreatophyte and Riparian Vegetation for Maximum Multiple Use Values. U.S.D.A. Forest Serv Res Paper RM-117

Horton WH (1991) Medusahead: importance, distribution, and control. In: James LF et al. (eds) Noxious Range Weeds. Westview Press, Boulder, Colorado; pp. 394–398

Hoshovsky M (1993a) Element Stewardship Abstract: *Arundo donax*. Computer printout. The Nature Conservancy, San Francisco, California

Hoshovsky M (1993b) Element Stewardship Abstract: *Cytisus scoparius*. Computer printout. The Nature Conservancy, San Francisco, California

Hoshovsky M (1993c) Element Stewardship Abstract: *Ulex europaeus*. Computer printout. The Nature Conservancy, San Francisco, California

Howarth FG (1985) Impacts of alien land arthropods and mollusks on native plants and animals in Hawaii. In: Stone CP, Scott JM (eds) Hawaii's Terrestrial Ecosystems: Preservation and Management, Proceedings of a Symposium, June 5–6, 1984, Hawaii Volcanoes National Park. Cooperative National Park Research Studies Unit, Univ Hawaii, Honolulu, Hawaii; pp. 149–179

Howe HF (1986) Seed dispersal by fruit-eating birds and mammals. In: Murray DR (ed) Seed Dispersal. Academic Press, New York; pp. 123–189

Howe HF, Smallwood J (1982) Ecology of seed dispersal. Annual Rev Ecol Syst 13:201–228

Howell T (1903) A Flora of Northwest America. Publisher not indicated, Portland, Oregon

Hoy JM (1961) *Eriococcus orariensis* Hoy and other Coccoidea associated with *Leptospermum* Forst. species in New Zealand. New Zealand Dep Sci Industr Res Bull 141

Hu SY (1979) *Ailanthus*. Arnoldia 39:29–50

Hubbs CL, Pope TEB (1937) The spread of the sea lamprey through the Great Lakes. Trans Amer Fish Soc 66:172–176

Huenneke LF, Hamburg SP, Koide R, Mooney HA, Vitousek PM (1990) Effects of soil resources on plant invasion and community structure in Californian serpentine grassland. Ecology 71:478–491

Huenneke LF, Mooney HA (1989) The California annual grassland: an overview. In: Huenneke LF, Mooney HA (eds) Grassland Structure and Function. Kluwer, Boston, Massachusetts; pp. 213–218

Huenneke LF, Vitousek PM (1990) Seedling and clonal recruitment of the invasive tree *Psidium cattleianum*: implications for the management of native Hawaiian forests. Biol Conservation 53:199–211

Huffaker CB (1957) Fundamentals of biological control of weeds. Hilgardia 27:101–157

Huffaker CB (1959) Biological control of weeds with insects. Annual Rev Entomol 4:251–276

Huffaker CB (1964) Fundamentals of biological weed control. In: DeBach P (ed) Biological Control of Insect Pests and Weeds. Chapman and Hall, London, U.K.; pp. 631–649

Huffaker CB, Kennett CE (1959) A ten-year study of vegetational changes associated with biological control of Klamath weed. J Range Managem 12:69–82

Hughes F, Vitousek PM, Tunison T (1991) Alien grass invasion and fire in the seasonal submontane zone of Hawaii. Ecology 72:743–746

Hughes WC (1972) Simulation of saltcedar evapotranspiration. Amer Soc Civil Engin 98:533–542

Huiskes AHL (1979) *Ammophila arenaria* (L.) Link (*Psamma arenaria* (L.) Roem. et Schult.; *Calamagrostis arenaria* (L.) Roth. J Ecol 67:363–382

Hulsman RB (1985) The neutron probe and the microcomputer. Soil Sci 140:153–157

Humphries SE, Groves RH, Mitchell DS (1991) Plant invasions of Australian ecosystems: a status review and management directions. Kowari 2:1–134

Hunt DM, Zaremba RE (1992) The northeastward spread of *Microstegium vimineum* (Poaceae) into New York and adjacent states. Rhodora 94:167–170

Hunt GS, Lutz RW (1959) Seed production by curly-leaved pondweed and its significance to waterfowl. J Wildlife Managem 23:405–408

Hunt R, Hand DW, Hannah MA, Neal AM (1995) Temporal and nutritional influences on the response to elevated CO_2 in selected British grasses. Ann Bot (Oxford) 75:207–216

Hunter WC (1984) Status of Nine Bird Species of Special Concern along the Colorado River. State of California, Resources Agency, Department of Fish and Game, Sacramento, California

Hunter WC, Ohmart RD, Anderson BW (1988) Use of exotic saltcedar (*Tamarix chinensis*) by birds in arid riparian systems. Condor 90:113–123

Huntley B (1990) European post-glacial forests: compositional changes in response to climatic change. J Veg Sci 1:507–518

Huntley B, Prentice IC (1993) Holocene vegetation and climates of Europe. In: Wright HE Jr, Kutzbach JE, Webb T III, Ruddiman WF, Street-Perrott FA, Bartlein PJ (eds) Global Climates Since the Last Glacial Maximum. Univ Minnesota Press, Minneapolis, Minnesota; pp. 136–168

Huntley B, Webb T III (eds) (1988) Vegetation History. Kluwer, Dordrecht, Netherlands

Hurlbert SJ (1984) Pseudoreplication and the design of ecological field experiments. Ecol Monogr 54:187–211

Ingold JL, Craycraft MJ (1983) Avian frugivory on honeysuckle (*Lonicera*) in southwesern Ohio in fall. Ohio J Sci 83:256–258

Isely D (1990) Vascular Flora of the Southeastern United States. Volume 3, Part 2. Leguminosae (Fabaceae). Univ North Carolina Press, Chapel Hill, North Carolina

Iversen J (1941) Landnam i Danmarks Stenalder (Land occupation in Denmark's Stone Age). Danmarks Geol Undersøg II. Raekke, Nr. 66

Iversen J (1973) The development of Denmark's nature since the last glacial. Danmarks Geol Undersøg V. Raekke, Nr. 7-C

Iverson LR (1992) Illinois Plant Information Network (ILPIN). A Data Base on the Ecology, Biology, Distribution, Taxonomy and Literature of the 3200 Plant Species Found in Illinois. Illinois Natural History Survey, Champaign, Illinois

Iverson ME (1994) Effects of *Arundo donax* on water resources. In: Jackson NE, Frandsen P, Duthoit S (compilers) *Arundo donax* Workshop Proceedings, November 19, 1994, Ontario, California; pp. 19–25

Jackson LE (1985) Ecological origins of California's Mediterranean grasses. J Biogeogr 12:349–361

Jackson NE, Frandsen P, Duthoit S (compilers) (1994) *Arundo donax* Workshop Proceedings, November 19, 1994, Ontario, California

Jackson ST (1989) Postglacial Vegetational Changes along an Elevational Gradient in the Adirondack Mountains (New York): A Macrofossil Study. New York State Mus Bull 465

Jackson ST (1990) Pollen source area and representation in small lakes of the northeastern United States. Rev Palaeobot Palynol 63:53–76

Jackson ST (1994) Pollen and spores in Quaternary lake sediments as sensors of vegetation composition: theoretical models and empirical evidence. In: Traverse A (ed) Sedimentation of Organic Particles. Cambridge Univ Press, Cambridge, U.K.; pp. 253–286

Jackson ST, Futyma RP, Wilcox DA (1988) A paleoecological test of a classical hydrosere in the Lake Michigan Dunes. Ecology 69:928–936

Jackson ST, Givens CR (1994) Late Wisconsinan vegetation and environment of the Tunica Hills region, Louisiana/Mississippi. Quatern Res 41: 316–325

Jackson ST, Overpeck JT, Webb T III, Keattch SE, Anderson KH (n.d.) Mapped plant macrofossil and pollen records of Late Quaternary vegetation change in eastern North America. Quatern Sci Rev (submitted)

Jackson ST, Webb T III, Prentice IC, Hansen JE (1995) Exploration and calibration of pollen/vegetation relationships: a PC program for the extended R-value models. Rev Palaeobot Palynol 84:365–374

Jackson ST, Whitehead DR (1991) Holocene vegetation patterns in the Adirondack Mountains. Ecology 72:641–653

Jacobi JD, Scott JM (1985) An assessment of the current status of native upland habitats and associated endangered species on the island of Hawaii. In: Stone CP, Scott JM (eds) Hawaii's Terrestrial Ecosystems: Preservation and Management. Cooperative National Park Resources Studies Unit, Univ Hawaii, Honolulu, Hawaii; pp. 3–22

Jacobson GL Jr (1979) The palaeoecology of white pine (*Pinus strobus*) in Minnesota. J Ecol 67:697–726

Jaggers BV (1994) Economic considerations of integrated hydrilla management: a case history of Johns Lake, Florida. In: Proceedings of the Grass Carp Symposium, 7–9 March 1994, Gainesville, Florida. U.S. Army Engineer Waterways Experiment Station, Vicksburg, Mississippi; pp. 151–163

James D (1994) Exotic plants... environmental weeds? Pl Press (Arizona Native Plant Society) 18(3):1, 12–13

Janzen DH (1983) Dispersal of seeds by vertebrate guts. In: Futuyma DJ, Slatkin M (eds) Coevolution. Sinauer Associates, Sunderland, Massachusetts; pp. 232–263

Janzen DH (1984) Dispersal of small seeds by big herbivores: foliage is the fruit. Am Naturalist 123:338–353

Johnson AP (1994) Coastal impacts of non-indigenous species. In: Schmitz DC, Brown TC (1994) An Assessment of Invasive Non-Indigenous Species in Florida's Public Lands. Florida Department of Environmental Protection, Tallahassee, Florida; pp. 119–126

Johnson DH (1981) How to measure habitat—a statistical perspective. In: Capen DE (ed) The Use of Multivariate Statistics in Studies of Wildlife Habitat. U.S.D.A. Forest Serv Gen Techn Rep RM-87; pp. 53–57

Johnson HB (1985) Consequences of species introductions and removals on ecosystem functions—implications for applied ecology. In: Delfosse ES (ed) Proceedings of the Sixth International Symposium on the Biological Control of Weeds, 19–25 August 1984, Vancouver, Canada; pp. 27–56

Johnson HB, Mayeux HS (1992) Viewpoint: a view on species additions and deletions and the balance of nature. J Range Managem 45:322–333

Johnson HB, Polley HW, Mayeux HS (1993) Increasing CO_2 and plant-plant interactions: effects on natural revegetation. Vegetatio 104/105:157–170

Jones JJ, Drobney RD (1986) Winter feeding ecology of scaup and common goldeneye in Michigan. J Wildlife Managem 50:446–452

Jones RC, Walti K, Adams MS (1983) Phytoplankton as a factor in the decline of the submersed macrophyte *Myriophyllum spicatum* L. in Lake Wingra, Wisconsin, U.S.A. Hydrobiologia 107: 213–219

Jones RH, McLeod KW (1989) Shade tolerance in seedlings of Chinese tallow tree, American sycamore and cherrybark oak. Bull Torrey Bot Club 116:371–377

Jordan WR III, Gilpin ME, Aber JD (1987) Restoration Ecology. Cambridge Univ Press, Cambridge, U.K.

Jorga W, Heym W-D, Weise G (1982) Shading as a measure to prevent mass development of submersed macrophytes. Int Rev Gesamten Hydrobiol 67:271–281

Joyce JC (1991) Future of chemical technology in aquatic plant management operations. In: Proceedings, 25th Annual Meeting, Aquatic Plant Control Research Program, 26–30 November 1990, Orlando, Florida. Misc Paper A-91-3, U.S. Army Engineer Waterways Experiment Station, Vicksburg, Mississippi; pp. 240–244

Joye GF (1990) Biological control of hydrilla with an endemic plant pathogen. In: Proceedings, 24th Annual Meeting, Aquatic Plant Control Research Program, 13–16 November 1989, Huntsville, Alabama. Misc Paper A-90-3, U.S. Army Engineer Waterways Experiment Station, Vicksburg, Mississippi; pp. 75–78

Jubinsky G (1993) A Review of the Literature: *Sapium sebiferum*. Florida Dep Nat Res Bur Aquatic Pl Managem Techn Serv Sect Publ TSS 93-0

Julien MH (1992) Biological Control of Weeds: A World Catalogue of Agents and Their Target Weeds. 3rd ed. Commonwealth Agricultural Bureaux, Wallingford, Oxon, U.K.

Kangas J (1994) An approach to public participation in strategic forest management planning. Forest Ecol Managem 70:75–88

Kangasniemi BJ (1983) Observations on herbivorous insects that feed on *Myriophyllum spicatum* in British Columbia. In: Proceedings, Second Annual Conference of the North American Lake Management Society, 26–29 October 1982, Vancouver, British Columbia, Canada. U.S. Environmental Protection Agency, EPA 440/5-83-001, Washington, D.C.; pp. 214–218

Kareiva P, Groom M, Parker I, Ruesink J (1991) Risk analysis as tool for making decisions about the introduction of non-indigenous species in the United States. Contractor report prepared for the U.S. Office of Technology Assessment

Karr JR, Dudley DR (1981) Ecological perspective on water quality goals. Environm Managem 5:55–68

Keeley JE (1990) The California grassland. In: Schoenherr AA (ed) Endangered Plant Communities of Southern California. Southern Calif Botanists Spec Publ 3:2–23

Keeley JE (1995) Future of California floristics and systematics: wildfire threats to the California flora. Madroño 42:175–179

Kelleher JS, Hulme MA (eds) (1984) Biological Control Programmes Against Insects and Weeds in Canada 1969–1980. Commonwealth Agricultural Bureaux, Farnham Royal, Slough, U.K.

Kelley T, Anderson RC (1990) Examination of the allelopathic properties of garlic mustard (*Alliaria petiolata*). Trans Illinois State Acad Sci 83 (Suppl):31–32 (Abstr)

Kelly DL (1981) The native forest vegetation of Killarney, south-west Ireland: an ecological account. J Ecol 69:437–472

Kimbel JC, Carpenter SR (1981) Effects of mechanical harvesting on *Myriophyllum spicatum* L. regrowth and carbohydrate allocation to roots and shoots. Aquatic Bot 11:121–127

Kincaid DR, Hold GA, Dalton PD, Tixier JS (1959) The spread of Lehmann lovegrass as affected by mesquite and native perennial grasses. Ecology 40:738–742

Kingsbury JM (1964) Poisonous Plants of the United States and Canada. Prentice-Hall, Englewood Cliffs, New Jersey

Klick K, O'Brien S, Lobik-Klick L (1989) Exotic plants of Indiana Dunes National Lakeshore: a management review of their extent and implications. Report to the U.S. National Park Service

Kline VM, Howell EA (1987) Prairies. In: Jordan WR III, Gilpin ME, Aber JD (eds) Restoration Ecology. Cambridge Univ Press, New York; pp. 75–83

Kloot PM (1983) The role of common iceplant (*Mesembryanthemum crystallinum*) in the deterioration of medic pastures. Austral J Ecol 8:301–306

Knight RS (1986) A comparative analysis of fleshy fruit displays in alien and indigenous plants. In: MacDonald IA, Kruger FJ, Ferrar AA (eds) The Ecology and Management of Biological Invasions in Southern Africa. Oxford Univ Press, Cape Town, South Africa; pp. 171–178

Knopf FL, Olson TE (1984) Naturalization of Russian-olive: implications to Rocky Mountain wildlife. Wildlife Soc Bull 12:289–298

Knops JMH, Griffin JR, Royalty AC (1995) Introduced and native plants of the Hastings Reservation, central coastal California: a comparison. Biol Conservation 71:115–123

Kok LT, McAvoy TJ, Malecki RA, Hight SD, Drea JJ, Coulson JR (1992) Host specificity tests of *Galerucella calmariensis* (L.) and *G. pusilla* (Duft.) (Coleoptera: Chrysomelidae), potential biological control agents of purple loosestrife, *Lythrum salicaria* L. (Lythraceae). Biol Control 2:282–290

Kok LT, Surles WW (1975) Successful biocontrol of musk thistle by an introduced weevil, *Rhinocyllus conicus*. Environm Entomol 4:1025–1027

Kovach WL (1987) MultiVariate Statistical Package software (v. 2.1e). Pentraeth, Anglesey, Wales, U.K.

Kovalev OV (1995) Co-evolution of the tamarisks (Tamaricaceae) and pest arthropods (Insecta: Arachnida: Acarina), with special reference to biological control prospects. Proc Zool Inst Russ Acad Sci 259

Krebs CJ (1984) Ecological Methodology. Harper & Row, New York

Kress R, Morgan D (1995) Application of New Technologies for Aquatic Plant Management. Misc Paper A-95-1, U.S. Army Engineer Waterways Experiment Station, Vicksburg, Mississippi

Kruger FJ (1977) Invasive woody plants in the Cape fynbos with special reference to the biology and control of *Pinus pinaster*. Proceedings of the Second National Weeds Conference of South Africa, Stellenbosch, 2–4 February 1977. Balkema, Cape Town, South Africa; pp. 57–74

Kruger FJ, Breytenback GJ, Macdonald IAW, Richardson DM (1986) The characteristics of invaded Mediteranean climate regions. In: Drake JA, Mooney HA, di Castri F, Groves RH, Kruger FJ, Rejmánek M, Williamson M (eds) Biological Invasions, A Global Perspective. John Wiley, London, U.K.; pp. 181–213

Kummerow M (1992) Weeds in wilderness: a threat to biodiversity. Western Wildlands 18(2):12–17

Kuno E (1972) Some notes on population estimation by sequential sampling. Res Populat Ecol 14:58–73

Kutzbach JE, Guetter P (1986) The influence of changing orbital parameters and surface boundary conditions on climate simulations for the past 18,000 years. J Atmos Sci 43:1726–1759

Kutzbach JE, Webb T III (1991) Late-Quaternary climatic and vegetational change in eastern North America: concepts, models, and data. In: Shane LCK, Cushing EJ (eds) Quaternary Landscapes. Univ Minnesota Press, Minneapolis, Minnesota; pp. 175–217

Lacey JR, Wallander R, Olson-Rutz K (1992) Recovery, germinability, and viability of leafy spurge (*Euphorbia esula*) seeds ingested by sheep and goats. Weed Technol 6:599–602

Laing JE, Hamai J (1976) Biological control of insect pests and weeds by imported parasites, predators, and pathogens. In: Huffaker CB, Messenger PS (eds) Theory and Practice of Biological Control. Academic Press, New York; pp. 685–743

Laird CA, Page LM (1996) Non-native fishes inhabiting the streams and lakes of Illinois. Illinois Nat Hist Surv Bull 35:1–51

Langdon KR, Johnson KD (1994) Additional notes on invasion of *Paulownia tomentosa* in natural areas. Nat Areas J 14:139–140

Langeland KA (1989) Karyotypes of *Hydrilla* (Hydrocharitaceae) populations in the United States. J Aquatic Pl Managem 27:111–115

Langeland KA (1994) Bensulfuron methyl residues in Florida lakes. J Aquatic Pl Managem 32:80–81

LaRoche FB (ed) (1994) Melaleuca Management Plan for Florida. 2nd ed. Exotic Pest Plant Council and South Florida Water Management District, West Palm Beach, Florida

LaRoche FB, Ferriter AP (1992) The rate of expansion of melaleuca in South Florida. J Aquatic Pl Managem 30:62–65

LaRosa AM, Smith CW, Gardner DE (1985) Role of alien and native birds in the dissemination of firetree (*Myrica faya* Ait.—Myricaceae) and associated plants in Hawaii. Pacific Sci 39:372–378

Larsen LM, Olsen O, Ploger A, Sorensen H (1983) Sinapine-O-β-D-glucopyranoside in seeds of *Alliaria officinalis*. Phytochemistry 22:219–222

Lawrence JG, Colwell A, Sexton OJ (1991) The ecological impact of allelopathy in *Ailanthus altissima* (Simaroubaceae). Amer J Bot 78:948–958

Lawton JH (1985) Ecological theory and choice of biological control agents. In: Delfosse ES (ed) Proceedings of the Sixth International Symposium on Biological Control of Weeds, 19–25 August 1984, Vancouver, British Columbia, Canada; pp. 13–26

Leishman MR, Westoby M, Jurado E (1995) Correlates of seed size variation: a comparison among five temperate floras. J Ecol 83:517–530

Leistritz FL, Thompson F, Leitch JA (1992) Economic impact of leafy spurge (*Euphorbia esula*) in North Dakota. Weed Sci 40:275–280

Lembi CA, Chand T (1992) Response of hydrilla and Eurasian watermilfoil to flurprimidol concentrations and exposure times. J Aquatic Pl Managem 30:6–9

Levin DA (1980) Polyploidy and novelty in flowering plants. Amer Naturalist 122:11–25

Lewis R (1990) Wetlands restoration/creation/enhancement terminology: suggestions for standardization. In: Kusler JA, Kentula ME (eds) Wetland Creation and Restoration: The Status of the Science. Island Press, Washington, D.C.; pp. 417–422

Leyden BW (1987) Man and climate in the Maya lowlands. Quatern Res 28:407–414

Lidicker WZ (1989) Impacts of non-domesticated vertebrates on California grasslands. In: Huenneke LF, Mooney H (eds) Grassland Structure and Function: California Annual Grassland. Kluwer, Dordrecht, Netherlands; pp. 135–150

Lillie RA, Budd J (1992) Habitat architecture of *Myriophyllum spicatum* as an index to habitat quality for fish and macroinvertebrates. J Freshwater Ecol 7:113–125

Lime DW, Koth BA, Vlaming JC (1993) Effects of restoring wolves on Yellowstone area big game and grizzly bears: opinions of scientists. In: Cook RS (ed) Ecological issues on reintroducing wolves into Yellowstone National Park. U.S. Natl Park Serv Sci Monogr NPS/NRYELL/NRSM-93/22:306–328

Lipp CC (1994) Physiological and Community-level Constraints to the Invasion of *Myrica faya*, an

Alien Tree in Hawaii Volcanoes National Park. Ph.D. dissertation. Univ Hawaii, Manoa, Hawaii

Little EL Jr (1971) Atlas of United States Trees. Volume 1. Conifers and Important Hardwoods. U.S.D.A. Misc Publ 1146

Lodge DJ, McDowell WH, McSwiney CP (1994) The importance of nutrient pulses in tropical forests. Trends Ecol Evol 9:384–387

Lodge DM (1993a) Biological invasions: lessons for ecology. Trends Ecol Evol 8:133–137

Lodge DM (1993b) Species invasions and deletions: community effects and responses to climate and habitat change. In: Kareiva PM, Kingsolver JG, Huey RB (eds) Biotic Interactions and Global Change. Sinauer Associates, Sunderland, Massachusetts; pp. 367–387

Loh RL, Tunison JT, Walker LR, Vitousek PM (1995) Population growth and nitrogen fixation of faya tree, 1986–1992. Technical Report. Cooperative National Park Resources Studies Unit, Univ Hawaii, Manoa, Hawaii

Long RW, Lakela O (1971) A Flora of Tropical Florida. Univ Miami Press, Coral Gables, Florida

Lonsdale WM (1993) Rates of spread of an invading species—*Mimosa pigra* in northern Australia. J Ecol 81:513–521

Lonsdale WM (1994) Inviting trouble: introduced pasture species in northern Australia. Austral J Ecol 19:345–354

Lonsdale WM, Lane AM (1994) Tourist vehicles as vectors of weed seeds in Kakadu National Park, Northern Australia. Biol Conservation 69:277–283

Loope LL, Hamann O, Stone CP (1988) Comparative conservation biology of oceanic archipelagos: Hawaii and the Galápagos. BioScience 38:272–282

Loope LL, Mueller-Dombois D (1989) Characteristics of invaded islands, with special reference to Hawaii. In: Drake JA, Mooney HA, di Castri F, Groves RH, Kruger FJ, Rejmánek M, Williamson M (eds) Biological Invasions: A Global Perspective. John Wiley, New York; pp. 257–280

Loope LL, Sanchez PG, Tarr PW, Loope WL, Anderson RL (1988) Biological invasions of arid land nature reserves. Biol Conservation 44:95–118

Lopez EG (1993) Effect of glyphosate on different densities of waterhyacinth. J Aquatic Pl Managem 31:255–257

Lovejoy TE (1985) Rehabilitation of degraded tropical forest lands. Environmentalist 5:13–20

Lovich JF (1994) Tamarisk control on public lands in the desert of southern California: two case stuides. In: Proceedings, 46th Annual California Weed Science Conference, California Weed Science Society, San Jose, California; pp. 166–177

Lubchenco J, Olson AM, Brubaker LB, Carpenter SR, Holland MJ, Hubbell SP, Levin SA, MacMahon JA, Matson PA, Melillo JM, Mooney HA, Peterson CH, Pullman HR, Real LA, Regal PJ, Risser PG (1991) The sustainable biosphere initiative: an ecological research agenda. Ecology 72:371–412

Luckmann WH, Metcalf RL (1975) The pest-management concept. In: Ludlow J (1995) Management of aquatic plant communities in Rodman Reservoir from 1969–1994. Aquatics 17(3):11–15

Lugo AE (1988) The future of the forest: ecosystem rehabilitation in the tropics. Environment 30:16–20, 41–45

Lugo AE (1990) Removal of exotic organisms. Conservation Biol 4:345

Lugo AE (1992) More on exotic species. Conservation Biol 6:6

Luken JO (1990) Directing Ecological Succession. Chapman and Hall, London

Luken JO (1994) Valuing plants in natural areas. Nat Areas J 14:295–299

Luken JO, Goessling N (1995) Seedling distribution and potential persistence of the exotic shrub *Lonicera maackii* in fragmented forests. Amer Midl Naturalist 133:124–130

Luken JO, Mattimiro DT (1991) Habitat-specific resilience of the invasive shrub Amur honeysuckle (*Lonicera maackii*) during repeated clipping. Ecol Applic 1:104–109

Luken JO, Thieret JW (1993) *Erucastrum gallicum* (Brassicaceae): invasion and spread in North America. Sida 15:569–582

Luken JO, Thieret JW (1995) Amur honeysuckle (*Lonicera maackii*) (Caprifoliaceae): its ascent, decline, and fall. Sida 16:479–503

Lutzow-Felling CJ, Gardner DE, Markin GP, Smith CW (eds) (1995) *Myrica faya*: Review of the Biology, Ecology, Distribution and Control, Including an Annotated Bibliography. Techn Rep 94. Cooperative National Park Resources Studies Unit, Univ Hawaii, Manoa, Hawaii

Lym RG, Messersmith CG (1994) Leafy spurge (*Euphorbia esula*) control, forage production, and economic return with fall-applied herbicides. Weed Technol 8:824–829

MacArthur RH, Wilson EO (1967) The Theory of Island Biogeography. Princeton Univ Press, Princeton, New Jersey

MacDonald GM (1993) Fossil pollen analysis and the reconstruction of plant invasions. Advances Ecol Res 24:67–110

Macdonald IAW (1990) Strategies for limiting the invasion of protected areas by introduced organisms. Monogr Syst Bot 32:189–199

Macdonald IAW, Frame GW (1988) The invasion of introduced species into nature reserves in tropical savannas and dry woodlands. Biol Conservation 44:67–93

Macdonald IAW, Kruger FJ, Ferrar AA (eds) (1986) The Ecology and Management of Biological Invasions in Southern Africa. Oxford Univ Press, Cape Town, South Africa

Macdonald IAW, Loope LL, Usher MB, Hamann O (1989) Wildlife conservation and the invasion of nature reserves by introduced species: a global perspective. In: Drake JA, Mooney HA, di Castri F, Groves RH, Kruger FJ, Rejmánek M, Williamson M (eds) Biological Invasions: A Global Perspective. John Wiley, New York; pp. 215–256

Macdonald IAW, Richardson DM (1986) Alien species in terrestrial ecosystems of the fynbos biome. In: Macdonald IAW, Kruger FJ, Ferrar AA (eds) The Ecology and Management of Biological Invasions in Southern Africa. Oxford Univ Press, Cape Town, South Africa; pp. 77–91

Maceina MJ, Cichra MF, Betsill RK, Bettoli PW (1992) Limnological changes in a large reservoir following vegetation removal by grass carp. J Freshwater Ecol 7:81–95

Mack RN (1981) Invasion of *Bromus tectorum* L. into western North America: an ecological chronicle. Agro-Ecosystems 7:145–165

Mack RN (1986) Alien plant invasion into the intermountain west: a case history. In: Mooney HA, Drake JA (eds) Ecology of Biological Invasions of North America and Hawaii. Springer-Verlag, New York; pp. 191–213

Mack RN (1989) Temperate grasslands vulnerable to plant invasion: characteristics and consequences. In: Drake JA, Mooney HA, di Castri F, Groves RH, Kruger FJ, Rejmánek M, Williamson M (eds) Biological Invasions: A Global Perspective. John Wiley, New York; pp. 155–179

Mack RN (1991) The commercial seed trade: an early disperser of weeds in the United States. Econ Bot 45:257–273

Maddox DM (1981) Seed and stem weevils of puncturevine: a comparative study of impact, interaction, and insect strategy. In: Delfosse ES (ed) Proceedings of the Fifth International Symposium on Biological Control of Weeds, 22–29 July 1980, Brisbane, Australia. CSIRO, Melbourne, Australia; pp. 447–467

Madison J (1992) Pampas grasses: one a weed [*Cortaderia jubata*] and one a garden queen. Pacif Hortic 53(1):48–53

Madsen JD (1993) Biomass techniques for monitoring and assessing control of aquatic vegetation. Lake Reservoir Managem 7:141–154

Madsen JD, Adams MS, Ruffier P (1988) Harvest as a control for sago pondweed (*Potamogeton pectinatus* L.) in Badfish Creek, Wisconsin: frequency, efficiency and its impact on stream community oxygen metabolism. J Aquatic Pl Managem 26:20–25

Madsen JD, Bloomfield JA (1992) Aquatic Vegetation Monitoring and Assessment Protocol Manual: A Report to the Finger Lakes Water Resources Board. New York Fresh Water Institute, New York State Department of Environmental Conservation, Albany, New York

Madsen JD, Bloomfield JA (1993) Aquatic vegetation quantification symposium: an overview. Lake Reservoir Managem 7:137–140

Madsen JD, Dick GO, Honnell D, Shearer J, Smart RM (1994a) Ecological Assessment of Kirk Pond. Misc Paper A-94-1, U.S. Army Engineer Waterways Experiment Station, Vicksburg, Mississippi

Madsen JD, Getsinger KD, Turner EG (1994b) Response of native vegetation to an application of triclopyr. In: Proceedings, 28th Annual Meeting, Aquatic Plant Control Research Program, 15–18 November 1993, Baltimore, Maryland. Misc Paper A-94-2. U.S. Army Engineer Waterways Experiment Station, Vicksburg, Mississippi; pp. 271–274

Madsen JD, Hartleb CF, Boylen CW (1991a) Photosynthetic characteristics of *Myriophyllum spicatum* and six submersed aquatic macrophyte species native to Lake George, New York. Freshwater Biol 26:233–240

Madsen JD, Sutherland JW, Bloomfield JA, Eichler LW, Boylen CW (1991b) The decline of native vegetation under dense Eurasian watermilfoil canopies. J Aquatic Pl Managem 29:94–99

Maffei MD (1994) Invasive non-indigenous species on national wildlife refuges in Florida. In: Schmitz DC, Brown TC (eds) An Assessment of Invasive Non-Indigenous Species in Florida's Public Lands. Florida Department of Environmental Protection, Tallahassee, Florida; pp. 179–184

Magnuson JJ (1990) Long-term ecological research and the invisible present. BioScience 40:495–501

Maguire LA (1986) Using decision analysis to manage endangered species populations. J Environm Managem 22:345–360

Maguire LA (1991) Risk analysis for conservation biology. Conservation Biol 5:123–125

Maguire LA, Seal US, Brussard PF (1987) Managing critically endangered species: the Sumatran rhino as a case study. In: Soulé ME (ed) Viable Populations for Conservation. Cambridge Univ Press, Cambridge, Massachusetts; pp. 141–158

Magurran AE (1988) Ecological Diversity and its Measurement. Princeton Univ Press, Princeton, New Jersey

Mahall BE, Callaway RM (1992) Root communication mechanisms and intracommunity distributions of two Mojave Desert shrubs. Ecology 73:2145–2151

Mahat TBS, Griffin DM, Shepherd KR (1987) Human impacts on some forests of the middle hills of Nepal. Part 3. Forests in the subsistence economy of Sindhu Palchok and Kabhre Palanchok. Mountain Res Developm 7:53–70

Mal TK, Lovett-Doust J, Lovett-Doust L, Mulligan GA (1992) The biology of Canadian weeds. 100. *Lythrum salicaria.* Canad J Pl Sci 72:1305–1330

Malecki RA, Blossey B, Hight SD, Schroeder D, Kok LT, Coulson JR (1993) Biological control of purple loosestrife. BioScience 43:680–686

Malo JE, Suárez F (1995a) Establishment of pasture species on cattle dung: the role of endozoochorous seeds. J Veg Sci 6:169–174

Malo JE, Suárez F (1995b) Herbivorous mammals as seed dispersers in a Mediterranean dehesa. Oecologia 104:246–255

Mann J (1969) Cactus-Feeding Insects and Mites. U.S. Natl Mus Bull 256

Markin GP, Burkhart RM (1995) Clidemia (Koster's curse). In: Nechols JR, Andres LA, Beardsley JW, Goeden RD, Jackson CG (eds) Biological Control in the Western United States. Accomplishments and Benefits of Regional Research Project W-84, 1964–1989. Univ Calif Div Agric Nat Res Publ 3361; pp. 306–308

Markin GP, Pemberton RW (1995) Banana poka. In: Nechols JR, Andres LA, Beardsley JW, Goeden RD, Jackson CG (eds) Biological Control in the Western United States. Accomplishments and Benefits of Regional Research Project W-84, 1964–1989. Univ Calif Div Agric Nat Resources Publ 3361; pp. 309–311

Markin GP, Yoshioka ER, Brown RE (1995) Gorse. In: Nechols JR, Andres LA, Beardsley JW, Goeden RD, Jackson CG (eds) Biological Control in the Western United States. Accomplishments and Benefits of Regional Research Project W-84, 1964–1989. Univ Calif Div Agric Nat Resources Publ 3361; pp. 299–302

Markin GP, Yoshioka ER (1990) Present status of biological control of the weed gorse (*Ulex europaeus* L.) in Hawaii. In: Delfosse ED (ed) Proceedings of the 7th International Symposium on the Biological Control of Weeds, 6–11 March 1988, Rome, Italy. CSIRO, Melbourne, Australia; pp. 357–362

Marks M, Lapin B, Randall J (1994) *Phragmites australis* (*P. communis*): threats, management, and monitoring. Nat Areas J 14:285–294

Marrs RH (1993) Soil fertility and nature conservation in Europe: theoretical considerations and practical management solutions. Advances Ecol Res 24:241–300

Marrs RH, Lowday JE (1992) Control of bracken and the restoration of heathland. II. Regeneration of the heathland community. J Appl Ecol 29:204–211

Marsden-Jones EM (1935) *Ranunculus ficaria* Linn.: life-history and pollination. J Linn Soc Bot 50: 39–55

Martin DF, Martin BB (1992) Aquashade, an annotated bibliography. Florida Sci 55:264–266

Martin PS, Byers W (1965) Pollen and archaeology at Wetherill Mesa. Mem Soc Amer Archaeol 19:122–135

Mathur G, Mohan Ram HY (1986) Floral biology and pollination of *Lantana camara.* Phytomorphology 36:79–100

Mayewski PA, Legrand MR (1990) Recent increase in nitrate concentration of antarctic snow. Nature 346:258–260

McAndrews JH (1988) Human disturbance of North American forests and grasslands: the fossil pollen record. In: Huntley B, Webb T III (eds) Vegetation History. Kluwer, Dordrecht, Netherlands; pp. 673–697

McAndrews JH, Boyko-Diakonow M (1989) Pollen analysis of varved sediment at Crawford Lake, Ontario: evidence of Indian and European farming. In: Fulton RJ (ed) Quaternary Geology of Canada and Greenland. Geological Society of America, Boulder, Colorado; pp. 528–530

McCaffrey JP, Campbell CL, Andres LA (1995) St. Johnswort. In: Nechols JR, Andres LA, Beardsley JW, Goeden RD, Jackson CG (eds) Biological Control in the Western United States. Accomplishments and Benefits of Regional Research Project W-84, 1964–1989. Univ Calif Div Agric Nat Resources Publ 3361; pp. 281–288

McCain AH, Noviello C (1985) Biological control of *Cannabis sativa*. In: Delfosse ES (ed) Proceedings of the Sixth International Symposium on Biological Control of Weeds, 19–25 August 1984, Vancouver, Canada. Agriculture Canada, Ottawa, Canada; pp. 635–642

McCarthy BC (1994) Understory responses to fire in a mixed oak forest of western Maryland. Amer J Bot 81(Suppl):59(Abstr)

McClaren MP, Anable ME (1992) Spread of introduced Lehmann lovegrass along a grazing intensity gradient. J Appl Ecol 29:92–98

McClellan AJ, Fitter AH, Law R (1995) On decaying roots, mycorrhizal colonization and the design of removal experiments. J Ecol 83:225–230

McComas S (1993) LakeSmarts: The First Lake Maintenance Handbook. Terrene Institute, Washington, D.C.

McCormick LH, Hartwig NL (1995) Control of the noxious weed mile-a-minute (*Polygonum perfoliatum*) in reforestation. Northern J Appl For 12:127–132

McDonnell MJ, Pickett STA (1990) Ecosystem structure and function along urban-rural gradients: an unexploited opportunity for ecology. Ecology 71:1232–1237

McDonnell MJ, Pickett STA, Pouyat RV (1993) The application of the ecological gradient paradigm to the study of urban effects. In: McDonnell MJ, Pickett STA (eds) Humans as Components of Ecosystems. Springer-Verlag, New York; pp. 175–189

McGlone MS (1983) Polynesian deforestation of New Zealand: a preliminary synthesis. Archaeol Oceania 18:11–25

McGlone MS (1989) The Polynesian settlement of New Zealand in relation to environmental and biotic changes. New Zealand J Ecol 12(Suppl):115–129

McIntyre S, Ladiges PY (1985) Aspects of the biology of *Ehrharta erecta* Lam. Weed Res 25(1):21–22

McIntyre S, Lavorel S (1994) Predicting richness of native, rare and exotic plants in response to habitat and disturbance variables across a variegated landscape. Conservation Biol 8:521–531

McKey DB, Kaufmann SC (1991) Naturalization of exotic *Ficus* species (Moraceae) in South Florida. In: Center TD, Doren RF, Hofstetter RL, Meyers RL, Whiteaker LD (eds) Proceedings of the Symposium on Exotic Pest Plants, Miami, Florida (Nov 1988). Techn Rep NPS/NREVER/NRTR-91/06. U.S. National Park Service; pp. 221–235

McKnight BN (ed) (1993) Biological Pollution: The Control and Impact of Invasive Exotic Species. Indiana Academy of Sciences, Indianapolis, Indiana

McNabb CD, Batterson TR (1991) Occurrence of the common reed, *Phragmites australis*, along roadsides in lower Michigan. Michigan Acad 23:211–220

[MDA] Montana Department of Agriculture (1992) Final Noxious Weed Trust Fund Programmatic Environmental Impact Statement. Montana Department of Agriculture, Agricultural and Biological Sciences Division, Helena, Montana

[MDNR] Michigan Department of Natural Resources (1990) State of Michigan Inland Lake Self-help Program: 1989 Annual Report. Inland Lakes Management Unit, Lake and Water Division, Michigan Department of Natural Resources, Lansing, Michigan

[MDNR] Minnesota Department of Natural Resources (1991) Report and Recommendations of the Minnesota Interagency Exotic Species Task Force. Minnesota Department of National Resources, St. Paul, Minnesota

Melgoza G, Nowak RS (1991) Competition between cheatgrass and two native species after fire: implications from observations and measurements of root distribution. J Range Managem 44:27–33

Melgoza G, Nowak RS, Tausch RJ (1990) Soil water exploitation after fire: competition between *Bromus tectorum* (cheatgrass) and two native species. Oecologia 83:7–13

Mesner N, Narf R (1987) Alum injection into sediments for phosphorus inactivation and macrophyte control. Lake Reservoir Managem 3:256–265

Metcalf RL, Luckmann WH (eds) (1994) Introduction to Insect Pest Management. John Wiley, New York

Metzler K, Rozsa R (1987) Additional notes on the tidal wetlands of the Connecticut River. Connecticut Bot Soc Newslett 15:1–6

Meyer J-Y (1994) Mécanismes d'invasion de *Miconia calvescens* en Polynesia Française. Ph.D. thesis. Univ Montpellier II Sciences et Techniques du Languedoc, Montpellier, France

Meyer J-Y (1996) Status of *Miconia calvescens* (Melastomataceae), a dominant invasive tree in the Society Islands (French Polynesia). Pacific Sci 50:66–76

Michaud JP (1991) A Citizen's Guide to Understanding and Monitoring Lakes and Streams. Publ #94-149. Washington Department of Ecology, Olympia, Washington

Mikol GF (1984) Effects of mechanical control of aquatic vegetation on biomass, regrowth rates, and juvenile fish populations at Saratoga Lake, New York. In: Lake and Reservoir Management. EPA 440/5-84-001. U.S. Environmental Protection Agency, Washington, D.C.; pp. 456–462

Milberg P, Lamont BB (1995) Fire enhances weed invasion of roadside vegetation in southwestern Australia. Conservation Biol 73:45–50

Miller GK, Young JA, Evans RA (1986) Germination of seeds of perennial pepperweed (*Lepidium latifolium*). Weed Sci 34:252–255

Miller M, Aplet G (1993) Biological control: a little knowledge is a dangerous thing. Rutgers Law Rev 45:285–334

Miller NF (1995) Archaeology: macroremains. Amer J Archaeol 99:91–93

Milton SJ, Siegfried WR, Dean WJ (1990) The distribution of epizoochoric plant species: a clue to the prehistoric use of arid Karoo rangelands by large herbivores. J Biogeogr 17:25–34

Mishra BK, Ramakrishnan PS (1983) Secondary succession subsequent to slash and burn agriculture at higher elevations of north-east India. I. Species diversity, biomass, and litter production. Acta Oecol Oecol Applic 4:95–107

Mitchell JFB, Manabe S, Meleshko V, Tokoika T (1990) Equlibrium climate change and its implications for the future. In: Houghton JT, Jenkins GJ, Ephraums JJ (eds) Climatic Change: The IPCC Scientific Assessment. Cambridge Univ Press, Cambridge, U.K.; pp. 131–172

Mitich LW (1993) Yellow toadflax. Weed Technol 7:791–793

Mityaev ID (1958) A review of insect pests of *Tamarix* in the Balkhash-Alakul depression. (In Russian) Trudy Inst Zool Akad Nauk Kazakh 8:74–97

Mohlenbrock RH (1986) Guide to the Vascular Flora of Illinois. Southern Illinois Univ Press, Carbondale, Illinois

Mohyuddin AI (1981) Investigations on the Insect Enemies of *Abutilon, Amaranthus, Rumex* and *Sorghum* in Pakistan. September 1975–September 1980. Final Report. Commonwealth Institute of Biological Control, Pakistan Station, Rawalpindi, Pakistan

Mooney HA (1991) Biological response to climate change: an agenda for research. Ecol Applic 1:112–117

Mooney HA, Drake JA (eds) (1986) Ecology of Biological Invasions of North America and Hawaii. Springer-Verlag, New York

Mooney HA, Hamburg SP, Drake JA (1986) The invasions of plants and animals into California. In: Mooney HA, Drake JA (eds) Ecology of Biological Invasions of North America and Hawaii. Springer-Verlag, New York; pp. 250–272

Moore AD, Noble IR (1990) An individualistic model of vegetation stand dynamics. J Environm Managem 31:61–81

Moore ML (1989) NALMS Management Guide for Lakes and Reservoirs. North American Lake Management Society, Washington, D.C.

Moore PD, Webb JA, Collinson ME (1991) Pollen Analysis. 2nd ed. Blackwell, Oxford, U.K.

Moran VC, Hoffmann JH (eds) (1996) Proceedings of the Ninth International Symposium on Biological Control of Weeds, 19–26 January 1992, Stellenbosch, South Africa. University of Cape Town, Cape Town, South Africa

Moran VC, Zimmerman HG (1991) Biological control of cactus weeds of minor importance in South Africa. Agric Eco-syst Environm 37:37–55

Morrison ML (1984) Influence of sample size on discriminant function analyses of habitat use by birds. J Field Ornithol 55:330–335

Morrison ML (1988) On sample sizes and reliable information. Condor 90:275–278

Morrison ML, Marcot BG (1994) An evaluation of resource inventory and monitoring program used in national forest planning. Environm Managem 19:147–156

Morton JF (1978) Brazilian pepper [*Schinus terebinthifolius*]—its impact on people, animals and the environment. Econ Bot 32:353–359

Morton JF (1987) Fruits of Warm Climates. Published by the Author, Miami, Florida.

Morton JF (1994) Lantana, or red sage (*Lantana camara* L., [Verbenaceae]), notorious weed and popular garden flower; some cases of poisoning in Florida. Econ Bot 48:259–270

Moss EH (1983) Flora of Alberta: A Manual of Flowering Plants, Conifers, Ferns and Fern Allies Found Growing without Cultivation in the Province of Alberta, Canada. 2nd ed (revised by J.G. Packer). Univ Toronto Press, Toronto, Canada

Moss TK (1994) Ice plant [*Mesembryanthemum crystallinum*] eradication and native landscape restoration. Proceedings of the 46th California Weed Conference. California Weed Conference, Fremont, California; pp. 155–156

Moul, ET (1948) A dangerous weedy *Polygonum* [*perfoliatum*] in Pennsylvania. Rhodora 50:64–66

Moulton MP, Pimm SL (1986) Species introductions into Hawaii. In: Mooney HA, Drake JA (eds) Ecology of Biological Invasions of North America and Hawaii. Springer-Verlag, Berlin, Germany; pp. 213–249

Mowatt J (1981) Control of large-leaved privet (*Ligustrum lucidum*) and small-leaved privet (*Ligustrum sinense*) in urban brushland Australia. In: Wilson BJ, Swarbrick JT (eds) Proceedings of the Sixth Australian Weeds Conference, Gold Coast, Queensland, 13–18 September 1981. Queensland Weed Society for the Council of

Australian Weed Science Societies, Brisbane (?); pp. 165–168

Mudie PJ, Byrne R (1980) Pollen evidence for historic sedimentation rates in California coastal marshes. Estuarine Coastal Mar Sci 10:305–316

Mueller-Dombois D (1973) A non-adapted vegetation interferes with water removal in a tropical rain forest area in Hawaii. Trop Ecol 14:1–18

Mueller-Dombois D, Ellenberg H (1974) Aims and Methods of Vegetation Ecology. John Wiley, New York

Mueller-Dombois D, Whiteaker LD (1990) Plant associations with *Myrica faya* and two other pioneer trees on a recent volcanic surface in Hawaii Volcanoes National Park. Phytocoenologia 19:29–41

Muenscher WC (1955) Weeds. Macmillan, New York

Mühlenbach V (1979) Contributions to the synanthropic (adventive) flora of the railroads in St. Louis, Missouri, U.S.A. Ann Missouri Bot Gard 66:1–108

Mullahey JJ, Cornell J (1994) Biology of tropical soda apple (*Solanum viarum*), an introduced weed in Florida. Weed Technol 8:465–469

Mullahey JJ, Cornell JA, Colvin DL (1993a) Tropical soda apple (*Solanum viarum*) control. Weed Technol 7:723–727

Mullahy JJ, Nee M, Wunderlin RP, Delaney KR (1993b) Tropical soda apple (*Solanum viarum*): a new weed threat in subtropical regions. Weed Technol 7:783–786

Muller-Scharer H, Schroeder D (1993) The biological control of *Centaurea* spp. in North America: do insects solve the problem? Pestic Sci 37:343–353

Mulligan GA, Findlay JN (1974) The biology of Canadian weeds. 3. *Cardaria draba, C. chalepensis,* and *C. pubescens.* Canad J Pl Sci 54:149–169

Mulroy TW, Dungan ML, Rich RE, Mayerle BC (1992) Wildland weed control in sensitive natural communities: Vandenberg Air Force Base, California. Proceedings of the 44th California Weed Conference. California Weed Conference, Fremont, California; pp. 166–180

Mulvaney MJ (1991) Far from the Garden Path: An Identikit Picture of Woody Ornamental Plants Invading South-eastern Australia Bushland. Ph.D. dissertation. Australian National Univ, Canberra, Australia

Mun HT, Whitford WG (1989) Effects of nitrogen amendment on annual plants in the Chihuahuan Desert. Pl & Soil 120:225–231

Myers JH, Sabath MD (1981) Genetic and phenotypic variability, genetic variance, and the success of establishment of insect introductions for the biological control of weeds. In: Delfosse ES (ed) Proceedings of the Fifth International Symposium on Biological Control of Weeds, 22–29 July 1980, Brisbane, Australia. CSIRO, Melbourne, Australia; pp. 91–102

Myers RL (1983) Site susceptibility to invasion by the exotic tree *Melaleuca quinquenervia* in southern Florida. J Appl Ecol 20:645–658

Myers RM, Henry RD (1979) Changes in the alien flora in two west-central Illinois counties during the past 140 years. Amer Midl Naturalist 101:225–230

Nadeau LB, King JR (1991) Seed dispersal and seedling establishment of *Linaria vulgaris* Mill. Canad J Pl Sci 71:771–782

Napier RW, Gershenfeld MK (1985) Groups: Theory and Experience. 3rd ed. Houghton Mifflin, Boston, Massachusetts

Napompeth B (1992) Brief review of biological control activities in Thailand. In: Hirose Y (ed) Biological Control in South and East Asia. IOBC/SEARS, Kyushu Univ Press, Fukuoka, Japan; pp. 51–68

Nauman CE, Austin DF (1978) Spread of the exotic fern *Lygodium microphyllum* in Florida. Amer Fern J 68:65–66

Nechols JR, Andres LA, Beardsley JW, Goeden RD, Jackson CG (eds) (1995) Biological Control in the Western United States. Accomplishments and Benefits of Regional Research Project W-84, 1964–1989. Univ Calif Div Agric Nat Res Publ 3361

Nellessen JE, Ungar IA (1993) Physiological comparisons of old-field and coal-mine-spoil populations of *Andropogon virginicus* (broomsedge). Amer Midl Naturalist 130:90–105

Nesheim ON (1993) Understanding pesticide regulation—new pesticide registration. Aquatics 15(4):19–20

Netherland MD (1991) Herbicide concentration/exposure time relationships for Eurasian watermilfoil and hydrilla. In: Proceedings, 25th Annual Meeting, Aquatic Plant Control Research Program, 26–30 November 1990, Orlando, Florida. Misc Paper A-91-3. U.S. Army Engineer Waterways Experiment Station, Vicksburg, Mississippi; pp. 205–209

Netherland MD (1992) Herbicide concentration/exposure time relationships for Eurasian watermilfoil and hydrilla. In: Proceedings, 26th Annual Meeting, Aquatic Plant Control Research Program, 18–22 November 1991, Dallas, Texas. Misc Paper A-92-2. U.S. Army Engineer Waterways Experiment Station, Vicksburg, Mississippi; pp. 79–85

Netherland MD, Getsinger KD (1992) Efficacy of triclopyr on Eurasian watermilfoil: concentration and exposure time effects. J Aquatic Pl Managem 30:1–5

Netherland MD, Green WR, Getsinger KD (1991) Endothall concentration and exposure time relationships for the control of Eurasian watermilfoil and hydrilla. J Aquatic Pl Managem 29:61–67

Netherland MD, Shearer J (1995) Integrated use of herbicides and pathogens for submersed plant control. In: Proceedings, 29th Annual Meeting of the Aquatic Plant Control Research Program, 14–17 November 1994, Vicksburg, Mississippi. Misc Paper A-95-3. U.S. Army Engineer Waterways Experiment Station, Vicksburg, Mississippi; pp. 23–29

Newman RM, Holmberg KL, Biesboer DD, Penner BG (1996) Effects of a potential biological control agent, *Eurhychiopsis lecontei*, on Eurasian watermilfoil in experimental tanks. Aquatic Bot 53:131–150

Newman RM, Maher LM (1995) New records and distribution of aquatic insect herbivores of watermilfoils (Haloragaceae: *Myriophyllum* spp.) in Minnesota. Entomol News 106:6–12

Newroth PR (1987) Help protect lakes from exotic aquatic plants. LakeLine 7(4):44–49

Newroth PR (1993) Application of aquatic vegetation identification, documentation and mapping in Eurasian watermilfoil control projects. Lake Reservoir Managem 7:185–196

Newroth PR, Maxnuk MD (1993) Benefits of the British Columbia aquatic plant management program. J Aquatic Pl Managem 31:210–213

Newsome AE, Noble IR (1986) Ecological and physiological characters of invading species. In: Groves RH, Burdon JJ (eds) Ecology of Biological Invasions: An Australian Perspective. Cambridge Univ Press, New York; pp. 1–20

[NHDES] New Hampshire Department of Environmental Services (1989) Weed Watchers—An Association to Halt the Spread of Exotic Aquatic Plants. Techn Bull NHDES-WSPCD-1989-6. New Hampshire Department of Environmental Services, Concord, New Hampshire

Nichols S, Cottam G (1972) Harvesting as a control for aquatic plants. Water Resources Bull 8:1205–1210

Nichols SA (1974) Mechanical and Habitat Manipulation for Aquatic Plant Management: A Review of Techniques. Wisconsin Dep Nat Resources Techn Bull 77

Nichols SA (1984) Macrophyte community dynamics in a dredged Wisconsin lake. Water Resources Bull 20:573–576

Nichols SA (1991) The interaction between biology and the management of aquatic macrophytes. Aquatic Bot 41:225–252

Nichols SA, Engel S, McNabb T (1988) Developing a plan to manage lake vegetation. Aquatics 10(3):10–19

Niering WA, Dreyer GD, Egler FE, Anderson JP Jr (1986) Stability of a *Viburnum lentago* shrub community after 30 years. Bull Torrey Bot Club 113:23–27

Niering WA, Egler FE (1955) A shrub community of *Viburnum lentago*, stable for 25 years. Ecology 36:356–360

Niering WA, Goodwin RH (1974) Creation of relatively stable shrublands with herbicides: arresting "succession" on rights-of-way and pastureland. Ecology 55:784–795

Niering WA, Warren RS (1977) Our Dynamic Tidal Marshes: Vegetation Changes as Revealed by Peat Analysis. Connecticut Arbor Bull 12

Nilsen ET, Muller WH (1980a) A comparison of the relative naturalizing ability of two *Schinus* species (Anacardiaceae) in southern California. I. Seed germination. Bull Torrey Bot Club 107:51–56

Nilsen ET, Muller WH (1980b) A comparison of the relative naturalizing ability of two *Schinus* species (Anacardiaceae) in southern California. II. Seedling establishment. Bull Torrey Bot Club 107:232–236

Noble IR (1989) Attributes of invaders and the invading process: terrestrial and vascular plants. In: Drake JA, Mooney HA, di Castri F, Groves RH, Kruger FJ, Rejmánek M, Williamson M (eds) Biological Invasions: A Global Perspective. John Wiley, New York; pp. 301–314

Noble IR, Slatyer RO (1980) The use of vital attributes to predict successional changes in plant communities subject to recurrent disturbances. Vegetatio 43:5–21

Nowak CL, Nowak RS, Tausch RJ, Wigand PE (1994) Tree and shrub dynamics in northwestern Great Basin woodland and shrub steppe during the Late-Pleistocene and Holocene. Amer J Bot 81:265–277

Nowierski RM (1995) Dalmatian toadflax. In: Nechols JR, Andres LA, Beardsley JW, Goeden RD, Jackson CG (eds) Biological Control in the Western United States. Accomplishments and Benefits of Regional Research Project W-84, 1964–1989. Univ Calif Div Agric Nat Resources Publ 3361; pp. 312–317

[NRC] National Research Council (1992) Restora-

Nuzzo VA (1991) Experimental control of garlic mustard (*Alliaria petiolata* (Bieb.) Cavara & Grande) in northern Illinois using fire, herbicide, and cutting. Nat Areas J 11:158–167

Nuzzo VA (1993) Distribution and spread of the invasive biennial *Alliaria petiolata* (garlic mustard) in North America. In: McKnight BN (ed) Biological Pollution: The Control and Impact of Invasive Exotic Species. Indiana Academy of Science, Indianapolis, Indiana; pp. 137–145

Nuzzo VA (1994) Response of garlic mustard (*Alliaria petiolata* (Bieb.) Cavara & Grande) to summer herbicide treatment. Nat Areas J 14: 309–310

Nuzzo VA, Kennay J, Fell G (1991) Vegetation management guideline: garlic mustard, *Alliaria petiolata* (Bieb.) Cavara & Grande. Nat Areas J 11:120–121

Nyboer R, Reeves J, Ebinger J (1976) Flowering plants new to Illinois. Trans Illinois State Acad Sci 69:194–195

[NYSDEC] New York State Department of Environmental Conservation, Federation of Lake Associations (1990) Diet for a Small Lake: A New Yorker's Guide to Lake Management. New York State Department of Environmental Conservation, Albany, New York

[NYSDEC] New York State Department of Environmental Conservation, Federation of Lake Associations (n.d.) New York Citizen's Statewide Lake Assessment Program Sampling Protocol. New York State Department of Environmental Conservation, Albany, New York

Oatley TB (1984) Exploitation of a new niche by the Rameron pigeon (*Columba arquatrix*) in Natal. In: Ledger J (ed) Proceedings of the Fifth Pan-African Ornithological Congress. Southern African Ornithological Society, Johannesburg, South Africa; pp. 323–330

Oelfke J (1993) Letter to the editor. Nat Areas J 13:3

Oliver DR (1984) Description of a new species of *Crictopus* van der Wulp (Diptera: Chironomidae) associated with *Myriophyllum spicatum*. Canad Entomol 116:1287–1292

Olsen RW, Schreiner EG, Parker L (1991) Management of exotic plants in Olympic National Park. In-house report. U.S. National Park Service

Olsson IU (1991) Accuracy and precision in sediment chronology. Hydrobiologia 214:25–34

O'Neill RV, DeAngelis DL, Waide JB, Allen TFH (1986) A Hierarchical Concept of Ecosystems. Princeton Univ Press, Princeton, New Jersey

Orians GH (1986) Site characteristics favoring invasions. In: Mooney HA, Drake JA (eds) Ecology of Biological Invasions of North America and Hawaii. Springer-Verlag, New York; pp. 133–148

Osenberg CW, Schmitt RJ, Holbrook SJ, Abu-Saba KE, Flegal AR (1994) Detection of environmental impacts: natural variability, effect size, and power analysis. Ecol Applic 4:16–30

[OTA] Office of Technology Assessment, U.S. Congress (1993) Harmful Non-Indigenous Species in the United States, OTA-F-565. U.S. Government Printing Office, Washington, D.C.

Overpeck JT, Webb T III, Webb RS (1992) Mapping eastern North American vegetation change of the past 18 ka: no-analogs and the future. Geology 20:1071–1074

Ozimek T, Gulati RD, van Donk E (1990) Can macrophytes be useful in biomanipulation of lakes? The Lake Zwemlust example. Hydrobiologia 200/201:399–407

Painter DS (1988) Long-term effects of mechanical harvesting on Eurasian watermilfoil. J Aquatic Pl Managem 26:25–29

Pakeman RJ, Marrs RH (1992) The conservation value of bracken *Pteridium aquilinum* (L.) Kuhn-dominated communities in the UK, and an assessment of the ecological impact of bracken expansion or its removal. Biol Conservation 62:101–114

Palmer MW (1995) How should one count species? Nat Areas J 15:124–135

Palmer MW, Wade GL, Neal P (1995) Standards for the writing of floras. BioScience 45:339–345

Panetta FD (1989) Emergence and early establishment of *Chondrilla juncea* L. (skeleton weed) in the Western Australian wheatbelt. Austral J Ecol 14:115–122

Panetta FD (1993) A system of assessing proposed plant introductions for weed potential. Pl Protect Quart 8:10–14

Panetta FD, Dodd J (1987) The biology of Australian weeds. 16. *Chondrilla juncea* L. J Austral Inst Agric Sci 53:83–95

Panetta FD, Mitchell ND (1991) Homoclime analysis and the prediction of weediness. Weed Res 31:273–284

Parker RJ, McAvoy L, Lime DW, Thompson JL (1993) An Employee Delphi Study to Aid in Developing a Social Science Research Agenda for the Midwest Region: Identification of Emerging Issues. Univ Minnesota Cooperative Park Studies Unit Report, St. Paul, Minnesota

Parkes A, Teller JT, Flenley JR (1992) Environmental history of the Lake Vaihiria drainage basin, Tahiti, French Polynesia. J Biogeogr 19: 431–447

Parrish JAD, Bazzaz FA (1978) Pollination niche separation in a winter annual community. Oecologia 35:133–140

Parrotta JA (1992) The role of plantation forests in rehabilitating degraded tropical ecosystems. Agric Ecosyst Environm 41:115–133

Parson JJ (1972) Spread of African pasturegrasses to the American tropics. J Range Managem 25:12–17

Patterson DT (1986) Responses of soybean (*Glycine max*) and three C_4 grass weeds to CO_2 enrichment during drought. Weed Sci 34:203–210

Patterson DT, Flint EP (1980) Potential effects of global atmospheric enrichment on the growth and competitiveness of C_3 and C_4 weed and crop plants. Weed Sci 28:71–75

Patterson DT, Flint EP (1982) Interacting effects of CO_2 and nutrient concentration. Weed Sci 30:389–394

Patterson DT, Flint EP, Beyers JL (1984) Effects of CO_2 enrichment on competition between a C_4 weed and a C_3 crop. Weed Sci 32:101–105

Patterson WA III, Edwards KJ, Maguire DJ (1987) Microscopic charcoal as a fossil indicator of fire. Quatern Sci Rev 6:3–23

Pemberton RW (1985) Native weeds as candidates for biological control research. In: Delfosse ES (ed) Proceedings of the Sixth International Symposium on Biological Control of Weeds, 19–25 August 1984, Vancouver, Canada; pp. 869–877

Pemberton RW (1990) Northeast Asia as a source of biological control agents of North American weeds. In: Delfosse ES (ed) Proceedings Seventh International Symposium on the Biological Control of Weeds, 6–11 March 1988, Rome, Italy. CSIRO, Melbourne, Australia; pp. 651–657

Penfound WT, Earle TT (1948) The biology of the waterhyacinth. Ecol Monogr 18:447–472

Pennington W, Cambray RS, Fisher EM (1973) Observations on lake sediments using fallout ^{137}Cs as a tracer. Nature 242:324–326

Perkins MA (1984) An evaluation of pigmented nylon film for use in aquatic plant management. In: Lake and Reservoir Management. EPA 440/5-84-001. U.S. Environmental Protection Agency, Washington, D.C.; pp. 467–471

Perkins MA, Sytsma MD (1987) Harvesting and carbohydrate accumulation in Eurasian watermilfoil. J Aquatic Pl Managem 25:57–62

Perlin J (1991) A Forest Journey: The Role of Wood in the Development of Civilization. Harvard Univ Press, Cambridge, Massachusetts

Perrins J, Williamson M, Fitter A (1992a) A survey of differing views of weed classification: implications for regulation of introductions. Biol Conservation 60:47–56

Perrins J, Williamson M, Fitter A (1992b) Do annual weeds have predictable characters? Acta Oecol 13:517–533

Perry MC, Uhler FM (1982) Food habits of diving ducks in the Carolinas. In: Proceedings of the Annual Conference of the Southeastern Association of Fish and Wildlife Agencies 36; pp. 492–504

Peters ES, Lowance SA (1974) Fertility and management treatments to control broomsedge in pastures. Weed Sci 22:201–205

Peters RH (1991) A Critique for Ecology. Cambridge Univ Press, Cambridge, U.K.

Peters RL (1992) Conservation of biological diversity in the face of climate change. In: Peters RL, Lovejoy TE (eds) Global Warming and Biological Diversity. Yale Univ Press, New Haven, Connecticut; pp. 15–30

Peters RL, Darling JDS (1985) The greenhouse effect and nature reserves. BioScience 35:707–717

Peterson GM (1978) Pollen spectra from surface sediments of lakes and ponds in Kentucky, Illinois and Missouri. Amer Midl Naturalist 100:333–340

Peterson SA (1982) Lake restoration by sediment removal. Water Resources Bull 18:423–435

Peterson SA, Smith WL, Malueg KW (1974) Full-scale harvest of aquatic plants: nutrient removal from a eutrophic lake. J Water Pollut Control Fed 46:697–707

Petit LJ, Petit DR, Smith KG (1990) Precision, confidence, and sample size in the quantification of avian foraging behavior. In: Morrison ML, Ralph CJ, Verner J, Jehl JR Jr (eds) Avian Foraging: Theory, Methodology, and Applications. Stud Avian Biol 13:193–198

Pfeiffer EE (n.d.) Weeds and What They Tell. Bio-Dynamic Farming and Gardening Association, Kimberton, Pennsylvania

Pickett STA, Collins SL, Armesto JJ (1987) Models, mechanisms and pathways of succession. Bot Rev 53:335–371

Pickett STA, Parker VT, Fiedler PL (1992) The new paradigm in ecology: implications for conservation biology above the species level. In: Fiedler PL, Jain SK (eds) Conservation Biology: The Theory and Practice of Nature Conservation, Preservation and Management. Chapman and Hall, New York; pp. 65–88

Pickett STA, White PS (eds) (1985) The Ecology of Natural Disturbance and Patch Dynamics. Academic Press, Orlando, Florida

Pilcher JR (1993) Radiocarbon dating and the palynologist: a realistic approach to precision and accuracy. In: Chambers FM (ed) Climate Change and Human Impact on the Landscape. Chapman and Hall, London, U.K.; pp. 23–32

Pine RT, Anderson LWJ, Hung SSO (1990) Control of aquatic plants in static and flowing water by yearling triploid grass carp. J Aquatic Pl Managem 28:36–40

Pine RT, Anderson LWJ (1991) Plant preferences of triploid grass carp. J Aquatic Pl Managem 29:80–82

Piper GL (1985) Biological control of weeds in Washington: status report. In: Delfosse ES (ed) Proceedings of the Sixth International Symposium on Biological Control of Weeds, 19–25 August 1984, Vancouver, British Columbia, Canada. Agriculture Canada, Ottawa, Canada; pp. 817–826

Piper GL, Andres LA (1995a) Canada thistle. In: Nechols JR, Andres LA, Beardsley JW, Goeden RD, Jackson CG (eds) Biological Control in the Western United States. Accomplishments and Benefits of Regional Research Project W-84, 1964–1989. Univ Calif Div Agric Nat Resources Publ 3361; pp. 233–236

Piper GL, Andres LA (1995b) Rush skeletonweed. In: Nechols JR, Andres LA, Beardsley JW, Goeden RD, Jackson CG (eds) Biological Control in the Western United States. Accomplishments and Benefits of Regional Research Project W-84, 1964–1989. Univ Calif Div Agric Nat Resources Publ 3361; pp. 252–255

Piper GL, Rosenthal SS (1995) Diffuse knapweed. In: Nechols JR, Andres LA, Beardsley JW, Goeden RD, Jackson CG (eds) Biological Control in the Western United States. Accomplishments and Benefits of Regional Research Project W-84, 1964–1989. Univ Calif Div Agric Nat Resources Publ 3361; pp. 237–241

Platt KB (1959) Plant control—Some possibilities and limitations. II. Vital statistics of range management. J Range Managem 12:194–200

Porsild AE, Cody WJ (1980) Vascular Plants of Continental Northwest Territories, Canada. National Museum of Natural Sciences, National Museums of Canada, Ottawa, Canada

Prather TS, Callihan RH, Thill DC (1991) Common Crupina: Biology, Management and Eradication. Univ Idaho Agric Exp Sta Current Inform Ser 880

Prentice IC (1985) Pollen representation, source area, and basin size: toward a unified theory of pollen analysis. Quatern Res 23:76–86

Prentice IC (1988) Records of vegetation in space and time: the principles of pollen analysis. In: Huntley B, Webb T III (eds) Vegetation History. Kluwer, Dordrecht, Netherlands; pp. 17–42

Prentice IC (1992) Climate change and long-term vegetation dynamics. In: Glenn-Lewin DC, Peet RK, Veblen TT (eds) Plant Succession: Theory and Prediction. Chapman and Hall, London, U.K.; pp. 293–339

Prentice IC, Bartlein PJ, Webb T III (1991) Vegetation and climate change in eastern North America since the last glacial maximum. Ecology 72:2038–2056

Preston CD (1986) An additional criterion for assessing native status. Watsonia 16:83

Pridham AMS, Bing A (1975) Japanese-bamboo [*Polygonum cuspidatum*]. Plants Gard 31(2):56–57

Pursh F (1814) Flora America Septentrionalis. White, Cochrane, and Co., London

Putz FE, Canham CD (1992) Mechanisms of arrested succession in shrublands: root and shoot competition between shrubs and tree seedlings. Forest Ecol Managem 49:267–275

Pyšek P, Prach K, Rejmánek M, Wade M (eds) (1995) Plant Invasions. SPB Academic Publishing, Amsterdam, Netherlands

Quimby PC Jr, Kay SH (1983) Biocontrol of alligatorweed with insects. Proc Mississippi Entomol Assoc 2:27–29

Radford AE, Ahles HE, Bell CR (1968) Manual of the Vascular Plants of the Carolinas. Univ North Carolina Press, Chapel Hill, North Carolina

Radosevich SR, Holt JS (1984) Weed Ecology: Implications for Vegetation Management. John Wiley, New York

Ramakrishnan PS, Vitousek PM (1989) Ecosystem-level processes and the consequences of biological invasions. In: Drake JA, Mooney HA, di Castri F, Groves RH, Kruger FJ, Rejmánek M, Williamson M (eds) Biological Invasions: A Global Perspective. John Wiley, New York; pp. 281–300

Randall JM (1993) Exotic weeds in North American and Hawaiian natural areas: The Nature Conservancy's plan of attack. In: McKnight BN (ed) Biological Pollution: The Control and Impact of Invasive Exotic Species. Indiana Academy of Science, Indianapolis, Indiana; pp. 159–172

Randall JM (1995) Assessment of the invasive weed problem on preserves across the United States. Endangered Species Update 12(4,5):4–6

Randall JM (n.d.) Weed control for the preservation of biological diversity. Weed Technol

Randall JM, Marinelli J (eds) (1996) Invasive Plants. Brooklyn Botanic Garden, Brooklyn, New York

Random House Webster's College Dictionary (1991) Random House, New York

Rapoport EH (1993) The process of plant colonization in small settlements and large cities. In: McDonnell MJ, Pickett STA (eds) Humans as Components of Ecosystems: The Ecology of Subtle Human Effects and Populated Areas. Springer-Verlag, New York; pp. 190–207

Rasmussen GA, Smith RP, Scifres CJ (1985) Seedling growth responses of buffelgrass (*Pennisetum ciliare*) to tebuthiuron and honey mesquite (*Prosopis glandulosa*). Weed Sci 34:88–93

Raven PH (1988) The California flora. In: Barbour MG, Major J (eds) Terrestrial Vegetation of California. New expanded edition. California Native Plant Society, [s.l.]; pp. 109–137

Redman DE (1995) Distribution and habitat types for Nepal microstegium [*Microstegium vimineum* (Trin.) Camus] in Maryland and the District of Columbia. Castanea 60:270–275

Rees NE (1978) Interactions of *Rhinocyllus conicus* and thistles in the Gallatin Valley. In: Frick KE (ed) Biological Control of Thistles in the Genus *Carduus* in the United States: A Progress Report. U.S.D.A. Science and Education Administration, Agricultural Research Service, Washington, D.C.; pp. 31–38

Rees NE, Quimby PC, Piper GL, Turner CE, Coombs EM, Spencer NR, Knutson LV (1996) Biological Control of Weeds in the West. Western Society of Weed Science, Montana Department of Agriculture, U.S.D.A. Agricultural Research Service, Montana State Univ, Bozeman, Montana

Reichard S (n.d.) Predicting invasions of woody plants Introduced into North America. Conservation Biol

Reilly W, Kaufman KR (1979) The social and economic impacts of leafy spurge in Montana. In: Proceedings of the Leafy Spurge Symposium, North Dakota Cooperative Extension Service, Fargo, North Dakota

Rejmánek M (1989) Invasibility of plant communities. In: Drake JA, Mooney HA, di Castri F, Groves RH, Kruger FJ, Rejmánek M, Williamson M (eds) Biological Invasions: A Global Perspective. John Wiley, New York; pp. 369–388

Rejmánek M (1995) What makes a species invasive? In: Pyšek P, Prach K, Rejmánek M, Wade PM (eds) Plant Invasions. SPB Academic Publishing, The Hague, Netherlands

Rejmánek M (1996) Species richness and resistance to invasion. In: Orians GH, Dirzo R, Cushman JH (eds) Biodiversity and Ecosystem Processes in Tropical Forests. Springer-Verlag, New York; pp. 153–172

Rejmánek M, Randall JM (1994) Invasive alien plants in California: 1993 summary and comparison with other areas in North America. Madroño 41:161–177

Rejmánek M, Richardson DM (n.d.) What attributes make some plant species more invasive? Ecology 77:1655–1661

Rejmánek M, Thomsen CD, Peters ID (1991) Invasive vascular plants in California. In: Groves RH, di Castri F (eds) Biogeography of Mediterranean Invasions. Cambridge Univ Press, Cambridge, U.K.; pp. 81–101

Remillard MM, Welch RA (1993) GIS technologies for aquatic macrophyte studies: modeling applications. Landscape Ecol 8:163–175

Rice EL (1972) Allelopathic effects of *Andropogon virginicus* and its persistence in old fields. Amer J Bot 59:752–755

Rice PM, Bedunah DJ, Carlson CE (1992) Plant Community Diversity After Herbicide Control of Spotted Knapweed. U.S.D.A. Forest Service Res Paper INT-460. Intermountain Research Station, Ogden, Utah

Richardson DM, Cowling RM, LeMaitre DC (1990) Assessing the risk of invasive success in *Pinus* and *Banksia* in South African mountain fynbos. J Veg Sci 1:629–642

Richardson DM, Williams PA, Hobbs RJ (1994) Pine invasions in the southern hemisphere: determinants of spread and invadability. J Biogeogr 21:511–527

Ridings WH, Mitchell DJ, Schoulties CL, El-Gholl NE (1978) Biological control of milkweed vine in Florida citrus groves with pathotype of *Phytophtora citrophtora*. In: Freeman TE (ed) Proceedings of the IV International Symposium on Biological Control of Weeds, 30 August–2 September 1976, Gainesville, Florida. Institute of Food and Agricultural Sciences, Univ Florida, Gainesville, Florida; pp. 224–240

Rieger JP, Kreager DA (1989) Giant reed (*Arundo donax*): a climax community of the riparian zone. In: Abell DL (Technical Coordinator). Proceedings of the California Riparian Systems Conference: Protection, Management, and Restoration for the 1990s, 22–24 September 1988, Davis, California. Gen Techn Rep PSW-110. Pacific Southwest Forest and Range Experiment Station, Berkeley, California; pp. 222–225

Ripple SR (1990) Germination, Establishment, and Survival of Riparian Tree Species in the Lower Box of the Gila River, New Mexico. M.S. thesis. New Mexico State Univ, Las Cruces, New Mexico

Robbins WW, Crafts AS, Raynor RN (1942) Weed Control. McGraw-Hill, New York

Roberts TL, Vankat JL (1991) Floristics of a chronosequence corresponding to old field-deciduous forest succession in southwestern Ohio. II. Seed banks. Bull Torrey Bot Club 118:377–384

Robertson DJ, Robertson MC, Tague T (1994) Colonization dynamics of four exotic plants in a northern Piedmont natural area. Bull Torrey Bot Club 121:107–118

Robinson TW (1965) Introduction, Spread, and Areal Extent of Saltcedar *Tamarix* in the Western States. U.S. Geol Surv Profess Paper 491-A

Robles M, Chapin FS III (1995) Comparison of the influence of two exotic species on ecosystem processes in the Berkeley Hills. Madroño 42:349–357

Robocker MC (1974) Life History, Ecology, and Control of Dalmatian Toadflax [*Linaria dalmatica*]. Wash Agric Exp Sta Techn Bull 79

Roman CT, Niering WA, Warren RS (1984) Salt marsh vegetation change in response to tidal restriction. Environm Managem 8:141–150

Romesburg HC (1981) Wildlife science: gaining reliable knowledge. J Wildlife Managem 45:293–313

Room PM (1981) Biogeography, apparency and exploration for biological control agents in exotic ranges of weeds. In: Delfosse ES (ed) Proceedings of the Fifth International Symposium on Biological Control of Weeds, 22–29 July 1980, Brisbane, Australia. CSIRO, Melbourne, Australia; pp. 113–124

Rosenstreter R (1994) Displacement of rare plants by exotic grasses. In: Monsen SB, Kitchen SG (1994) Proceedings—Ecology and Management of Annual Rangelands. U.S.D.A. Forest Serv Gen Techn Rep INT-GTR-313. Intermountain Research Station, Ogden, Utah; pp. 170–175

Rosenthal SS, Buckingham GR (1982) Natural enemies of *Convolvulus arvensis* in western Mediterranean Europe. Hilgardia 50:1–19

Rosenthal SS, Clement SS, Hostettler N, Mimmocchi T (1988) Biology of *Tyta luctuosa* (Lepidoptera: Noctuidae) and its potential value as biological control agent for the weed *Convolvulus arvensis*. Entomophaga 33:185–192

Rosenthal SS, Piper GL (1995a) Bull thistle. In: Nechols JR, Andres LA, Beardsley JW, Goeden RD, Jackson CG (eds) Biological Control in the Western United States. Accomplishments and Benefits of Regional Research Project W-84, 1964–1989. Univ Calif Div Agric Nat Resources Publ 3361; pp. 231–241

Rosenthal SS, Piper GL (1995b) Russian knapweed. In: Nechols JR, Andres LA, Beardsley JW, Goeden RD, Jackson CG (eds) Biological Control in the Western United States. Accomplishments and Benefits of Regional Research Project W-84, 1964–1989. Univ Calif Div Agric Nat Resources Publ 3361; pp. 256–257

Rosenthal SS, Platts BE (1990) Host specificity of *Aceria* (*Eriophyes*) *malherbe*, a biological control agent for the weed *Convolvulus arvensis* (Convolvulaceae). Entomophaga 35:459–463

Rowlands PG (1989) History and treatment of the saltcedar problem in Death Valley National Monument. In: Kunzmann MR, Johnson RR, Bennett PS (eds) Proceeding Tamarisk Conference, 2–3 September 1987, Univ Arizona, Tucson. USDI Natl Park Serv, Natl Park Res Stud Unit Spec Rep 9; pp. 46–56

Roy J (1990) In search of the characteristics of plant invaders. In: di Castri F, Hansen J, Debussche M (eds) Biological Invasions in Europe and the Mediterranean Basin. Kluwer, Dordrecht, Netherlands; pp. 335–352

Roy J, Navas ML, Sonie L (1991) Invasion by annual brome grasses: a case study challenging the homocline approach to invasions. In: Groves RH, di Castri F (eds) Biogeography of Mediterranean invasions. Cambridge Univ Press, New York; pp. 207–224

Ruesink JL, Parker IM, Groom MJ, Kareiva PM (1995) Reducing the risks of nonindigenous species introductions. BioScience 45:465–477

Rundel PW, Jarrell WM (1989) Water in the environment. In: Pearcy RW, Ehleringer J, Mooney HA, Rundel PW (eds) Plant Physiological Ecology: Field Methods and Instrumentation. Chapman and Hall, London; pp. 29–56

Russell EWB, Davis RB, Anderson RS, Rhodes TE, Anderson DS (1993) Recent centuries of vegetational change in the glaciated north-eastern United States. J Ecol 81:647–664

Rutherford MC, Pressinger FM, Musil CF (1986) Standing crops, growth rates and resource use efficiency in alien plant invaded ecosystems. In: Macdonald IAW, Kruger FJ, Ferrar AA (eds) The Ecology and Management of Biological Invasions in Southern Africa. Oxford Univ Press, Cape Town, South Africa; pp. 189–199

Ryan FJ (1989) Isoenzymic variability in monoecious *Hydrilla* in the United States. J Aquatic Pl Managem 27:10–15

Ryan FJ, Holmberg DL (1994) Keeping track of *Hydrilla*. Aquatics 16(2):14–20

Ryan FJ, Thullen JS, Holmberg DL (1991) Nongenetic origin of isozymic variability in subterranean turions of monoecious and dioecious hydrilla. J Aquatic Pl Managem 29:3–6

Sala A, Devitt DE, Smith SD (1996) Water use by *Tamarix ramosissima* and associated phreatophytes in a Mojave Desert floodplain. Ecol Applic 6:888–898

Salisbury E (1961) Weeds and Aliens. Collins, London

Sanders RW, Stuessy TF, Marticorena C (1982) Recent changes in the flora of the Juan Fernandez Islands, Chile. Taxon 31:284–289

Sands DPA, Harley KLS (1981) Importance of geographic variation in agents selected for biological control of weeds. In: Delfosse ES (ed) Proceedings of the Fifth International Symposium on Biological Control of Weeds, 22–29 July 1980, Brisbane, Australia. CSIRO, Melbourne, Australia; pp. 81–89

Saner MA, Clements DR, Hall MR, Doohan DJ, Crompton CW (1995) The biology of Canadian weeds. 105. *Linaria vulgaris* Mill. Canad J Pl Sci 75:525–537

Sasek TW, Strain BR (1988) Effects of carbon dioxide enrichment on the growth and morphology of kudzu (*Pueraria lobata*). Weed Sci 36:28–36

Sasek TW, Strain BR (1989) Effects of carbon dioxide enrichment on the expansion and size of kudzu (*Pueraria lobata*) leaves. Weed Sci 37:23–28

Sasek TW, Strain BR (1990) Implications of atmospheric CO_2 enrichment and climatic change for the geographical distribution of two introduced vines in the U.S.A. Climatic Change 16:31–51

Sasek TW, Strain BR (1991) Effects of CO_2 enrichment on the growth and morphology of a native and an introduced honeysuckle vine. Amer J Bot 78:69–75

Sauer JD (1988) Plant Migration: The Dynamics of Geographic Patterning in Seed Plant Species. Univ California Press, Berkeley, California

Savino JF, Marschall EA, Stein RA (1992) Bluegill growth as modified by plant density: an exploration of underlying mechanisms. Oecologia 89:152–160

Savino JF, Stein RA (1989) Behavior of fish predators and their prey: habitat choice between open water and dense vegetation. Environm Biol Fishes 24:287–293

Sawyers C (1989) Native plants under siege. Garden 13(2):12–13, 15, 32

Schardt J (1993) Management results from 1992. Aquatics 15(1):12–14

Schierenbeck KA (1995) The threat to the California flora from invasive species: problems and possible solutions. Madroño 42:168–174

Schierenbeck KA, Mack RN, Sharitz RR (1994) Effects of herbivory on growth and biomass allocation in native and introduced species of *Lonicera*. Ecology 75:1661–1672

Schiffman PM (1994) Promotion of exotic weed establishment by endangered giant kangaroo rats (*Dipodomys ingens*) in a California grassland. Biodivers Conservation 3:524–537

Schimming WK, Messersmith CG (1988) Freezing resistance of overwintering buds of four perennial weeds. Weed Sci 36:568–573

Schlesinger WH (1991) Biogeochemistry: An Analysis of Global Change. Academic Press, San Diego, California

Schlesinger WH (1994) The vulnerability of biotic diversity. In: Socolow R, Andrews C, Berkhout F, Thomas V (eds) Industrial Ecology and Global Change. Cambridge Univ Press, Cambridge, U.K.; pp. 245–260

Schmidt W (1989) Plant dispersal by motor car. Vegetatio 80:147–152

Schmitz DC, Brown TC (1994) An Assessment of Invasive Non-Indigenous Species in Florida's Public Lands. Florida Department of Environmental Protection, Tallahassee, Florida

Schmitz DC, Schardt JD, Leslie AJ, Dray FA, Osborne JA, Nelson BV (1993) The ecological impact and management history of three invasive alien aquatic plant species in Florida. In: McKnight BN (ed) Biological Pollution: The Control and Impact of Invasive Exotic Species. Indiana Academy of Science, Indianapolis, Indiana; pp. 173–194

Schneider SH (1993) Scenarios of global warming. In: Kareiva PM, Kingsolver JG, Huey RB (eds) Biotic Interactions and Global Change. Sinauer Associates, Sunderland, Massachusetts; pp. 9–23

Schofield EK (1989) Effects of introduced plants and animals on island vegetation: examples from the Galápagos Archipelago. Conservation Biol 3:227–238

Schramm HL Jr, Jirka KJ (1989) Effects of aquatic macrophytes on benthic macroinvertebrates in two Florida lakes. J Freshwater Ecol 5:1–12

Schroeder D (1983) Biological control of weeds. In: Fletcher WW (ed) Recent Advances in Weed Research. Commonwealth Agricultural Bureaux, Farnham Royal, Slough, U.K.; pp. 41–78

Schroeder RL, Keller ME (1990) Setting objectives—a prerequisite of ecosystem management. In: Mitchell RS, Sheviak CJ, Leopold DJ (eds) Ecosystem Management: Rare Species and Significant Habitats. New York State Mus Bull 471:1–4

Schwartz MW (1994) Issues of scale and values create potentially conflicting goals for conserving biodiversity. Nat Areas J 14:213–216

Schwartz MW, Hermann SM (1993) The continuing population decline of Torreya taxifolia Arn. Bull Torrey Bot Club 120:275–286

Schwartz MW, Hermann SM, Vogel C (1995) The catastrophic loss of Torreya taxifolia: assessing environmental induction of disease hypotheses. Ecol Applic 5:501–516

Schwartz MW, Randall JM (1995) Valuing natural area and controlling non-indigenous plants. Nat Areas J 15:98–100

Schwegman JE (1988) Exotic invaders. Illinois Outdoor Highlights 16(6):6–11

Scifres CJ (1974) Salient features of huisache [Acacia farnesiana] seed germination. Southwest Naturalist 18:383–392

Scott JK, Panetta FD (1993) Predicting the Australian weed status of southern African plants. J Biogeogr 20:87–93

Scott R, Marrs RH (1984) Impact of Japanese knotweed and methods of control. In: Aspects of Applied Biology. Association of Applied Biologists, Wellesbourne, Warwick, U.K.; pp. 291–296

Segerström U, Bradshaw R, Hörnberg G, Bohlin E (1994) Disturbance history of a swamp forest refuge in northern Sweden. Biol Conservation 68:189–196

Seki H, Takahashi M, Ichimura S-E (1979) Impact of nutrient enrichment in a waterchestnut ecosystem at Takahama-Iri Bay of Lake Kasumigaura, Japan. Water Air Soil Pollut 12:383–391

Self DW (1986) Exotic plant inventory, rating and management planning for Point Reyes National Seashore. In: Thomas LK (ed) Proceedings of the Conference on Science in the National Parks, 13–18 July 1986. George Wright Society and U.S. National Park Service, [s.l.]; pp. 85–95

Shafroth PB, Auble AG, Scott ML (1995a) Germination and establishment of the native plains cottonwood (Populus deltoides Marshall subsp. monilifera) and the exotic Russian-olive (Elaeagnus angustifolia L.). Conservation Biol 9:1169–1175

Shafroth PB, Friedman JM, Ischinger LS (1995b) Effects of salinity on establishment of Populus fremontii (cottonwood) and Tamarix ramosissima (saltcedar) in southwestern United States. Great Basin Naturalist 55:58–65

Shearer JF (1993) Biocontrol of hydrilla and milfoil using plant pathogens. In: Proceedings, 27th Annual Meeting, Aquatic Plant Control Research Program, 16–19 November 1992, Bellevue, Washington. Misc Paper A-93-2. U.S. Army Engineer Waterways Experiment Station, Vicksburg, Mississippi; pp. 79–81

Shearer JF (1994) A historical perspective of biocontrol of the submersed macrophytes Myriophyllum spicatum and Hydrilla verticillata using plant pathogens. In: Proceedings, 28th Annual Meeting, Aquatic Plant Control Research Program, 15–18 November 1993, Baltimore, Maryland. Misc Paper A-94-2. U.S. Army Engineer Waterways Experiment Station, Vicksburg, Mississippi; pp. 211–213

Shearer JF (1995) The use of pathogens for the management of hydrilla and Eurasian watermilfoil. In: Proceedings, 29th Annual Meeting, Aquatic Plant Control Research Program, 14–17 November 1994, Vicksburg, Mississippi. Misc Paper A-95-3, U.S. Army Engineer Waterways Experiment Station, Vicksburg, Mississippi; pp. 124–129

Sheley RL, Larson LL, Johnson DE (1993) Germination and root dynamics of range weeds and forage species. Weed Technol 7:234–237

Silver WL, Brown S, Lugo A (1996) Biodiversity and biogeochemical cycles. In: Orians GH, Dirzo R, Cushman H (eds) Biodiversity and Ecosystem Processes in Tropical Forests. Springer-Verlag, New York; pp. 49–67

Simberloff D (1981) Community effects of introduced species. In: Nitecki MM (ed) Biotic Crises in Ecological and Evolutionary Time. Academic Press, New York

Simberloff D, Cox J (1987) Consequences and costs of conservation corridors. Conservation Biol 1:63–71

Simmonds FJ, Bennett FD (1966) Biological control of Opuntia spp. by Cactoblastis cactorum in the Leeward Islands (West Indies). Entomophaga 11:183–189

Sinclair ARE (1991) Science and the practice of wildlife management. J Wildlife Managem 55:767–773

Singer DK, Jackson ST, Madsen BJ, Wilcox DA (n.d.) Differentiating climatic and successional influences on long-term development of a marsh. Ecology

Singer FJ, Swank WT, Clebsch EEC (1984) Effects of wild pig rooting in a deciduous forest. J Wildlife Managem 48:464–473

Sinha S, Sharma A (1984) *Lantana camara* L.—a review. Feddes Repert 95:621–633

Siver PA, Coleman AM, Benson GA, Simpson JT (1986) The effects of winter drawdown on macrophytes in Candlewood Lake, Connecticut. Lake Reservoir Managem 2:69–73

Skinner LC, Rendall WJ, Fuge EL (1994) Minnesota's Purple Loosestrife Program: History, Findings, and Management Recommendations. Minnesota Dep Nat Res Spec Publ 145

Smart RM, Doyle RD, Madsen JD, Dick GO (1996) Establishing Native Submersed Aquatic Plant Communities in Southern Reservoirs. Techn Rep A-96-2. U.S. Army Engineer Waterways Experiment Station, Vicksburg, Mississippi

Smart RM, Doyle R (1995) Ecological Theory and the Management of Submersed Plant Communities. Inform Exch Bull A-95-3, U.S. Army Engineer Waterways Experiment Station, Vicksburg, Mississippi

Smith CS, Barko JW (1990) Ecology of Eurasian watermilfoil. J Aquatic Pl Managem 28:55–64

Smith CW (1985) Impact of alien plants on Hawaii's native biota. In: Stone CP, Scott JM (eds) Hawaii's Terrestrial Ecosystems: Preservation and Management. Cooperative National Park Resources Studies Unit, Univ Hawaii, Honolulu, Hawaii; pp. 180–250

Smith TM, Shugart HH, Bonan GB, Smith JB (1992) Modeling the potential response of vegetation to global climate change. Advances Ecol Res 22:93–116

Sokal RR, Rohlf FJ (1981) Biometry. W.H. Freeman, New York

Soni P, Vasistha HB, Kumar O (1989) Biological diversity in surface mined areas after reclamation. Indian Forester 115:475–482

Sorensen AE (1986) Seed dispersal by adhesion. Ann Rev Ecol Syst 17:443–463

Sorsa KK, Nordheim EV, Andrews JH (1988) Integrated control of Eurasian water milfoil, *Myriophyllum*, by a fungal pathogen and a herbicide. J Aquatic Pl Managem 26:12–17

Soulé ME (1990) The onslaught of alien species, and other challenges in the coming decades. Conservation Biol 4:233–239

[SDSWPC] South Dakota State Weed and Pest Commission and South Dakota Department of Agriculture (1993) Purple Loosestrife Management Plan for South Dakota, Pierre, South Dakota (unpublished)

Spear RW, Davis MB, Shane LCK (1994) Late Quaternary history of low- and mid-elevation vegetation in the White Mountains of New Hampshire. Ecol Monogr 64:85–109

Spencer DF, Whitehand LC (1993) Experimental design and analysis in field studies of aquatic vegetation. Lake Reservoir Managem 7:165–174

Sprecher SL, Netherland MD (1995) Methods for Monitoring Herbicide-Induced Stress in Submersed Aquatic Plants: A Review. Misc Paper A-95-1. U.S. Army Engineer Waterways Experiment Station, Vicksburg, Mississippi

Sprecher SL, Stewart AB, Brazil JM (1993) Peroxidase changes as indicators of herbicide-induced stress in aquatic plants. J Aquatic Pl Managem 31:45–50

Stanley JG, Miley WW, Sutton DL (1978) Reproductive requirements and likelihood for naturalization of escaped grass carp in the United States. Trans Amer Fish Soc 107:119–128

Steadman DW (1995) Prehistoric extinctions of Pacific island birds: biodiversity meets zooarchaeology. Science 267:1123–1131

Stebbins GL (1965) Colonizing species of the native California flora. In: Baker HG, Stebbins GL (eds) The Genetics of Colonizing Species. Academic Press, New York; pp. 173–195

Stebbins GL (1985) Polyploidy, hybridization, and the invasion of new habitats. Ann Missouri Bot Gard 72:824–832

Stewart RM (1994) HERBICIDE simulation model for evaluating fate processes effects. In: Proceedings, 28th Annual Meeting, Aquatic Plant Control Research Program, 15–18 November 1993, Baltimore Maryland. Misc Paper A-94-2. U.S. Army Engineer Waterways Experiment Station, Vicksburg, Mississippi; pp. 69–76

Stewart RM (1995) Mechanical Harvester Simulations. Joint Agency Guntersville Project Report Summary. U.S. Army Engineer Waterways Experiment Station, Vicksburg, Mississippi (unpublished)

Stewart RM, Boyd WA (1992) User's Manual for INSECT (Version 1.0): A Simulation of Waterhyacinth Plant Growth and *Neochetina* Weevil Development and Interaction. Instruct Rep A-92-1. U.S. Army Engineer Waterways Experiment Station, Vicksburg, Mississippi

Stewart RM, Boyd WA (1994) Simulation model evaluation of sources of variability in grass carp stocking requirements. In: Proceedings of the Grass Carp Symposium, 7–9 March 1994,

Gainesville, Florida. U.S. Army Engineer Waterways Experiment Station, Vicksburg, Mississippi; pp. 85–92

Stewart-Oaten A, Murdoch WM, Parker KR (1986) Environmental impact assessment: "pseudoreplication" in time? Ecology 67:929–940

Steyermark JA (1963) Flora of Missouri. Iowa State Univ Press, Ames, Iowa

Stickney PF (1972) *Crupina vulgaris* (Compositae: Cynareae), new to Idaho and North America. Madroño 21:402

Stiles EW (1982) Expansions of mockingbird and multiflora rose in the northeastern United States and Canada. Amer Birds 36:358–364

Stone KM, Matthews ED (1977) Soil survey of Allegany County, Maryland. U.S.D.A. Soil Conservation Service, Washington, D.C.

Story JM (1995) Spotted knapweed. In: Nechols JR, Andres LA, Beardsley JW, Goeden RD, Jackson CG (eds) Biological Control in the Western United States. Accomplishments and Benefits of Regional Research Project W-84, 1964–1989. Univ Calif Div Agric Nat Resources Publ 3361; pp. 258–263

Strausbaugh PD, Core EL (1982) Flora of West Virginia. 2nd ed. Seneca Books, Grantsville, West Virginia

Stritch LR (1990) Landscape scale restoration of barrens-woodland within the oak-hickory forest mosaic. Restoration Managem Notes 8:73–77

Stubbendieck J, Butterfield CH, Flessner TR (1992) An assessment of exotic plant species at Pipestone National Monument and Wilson's Creek National Battlefield. Final Report. U.S. National Park Service, Omaha, Nebraska

Stuckey RL (1980) Distributional history of *Lythrum salicaria* (purple loosestrife) in North America. Bartonia 47:3–20

Stuiver M, Reimer PJ (1986) A computer programme for radiocarbon age calibration. Radiocarbon 28:1022–1030

Stuiver M, Reimer PJ (1993) Extended ^{14}C data base and revised CALIB 3.0 ^{14}C age calibration program. Radiocarbon 35:215–230

Stumpf JA, Stubbendieck J, Butterfield CH (1995) An assessment of exotic plants at Scotts Bluff National Monument and Effigy Mound National Monument. Final report. U.S. National Park Service, Omaha, Nebraska

Sudbrock A (1993) Tamarisk control. I. Fighting back. An overview of the invasion, and a low-impact way of fighting it. Restoration Managem Notes 11(1):31–34

Sugita S (1994) Pollen representation of vegetation in Quaternary sediments: I. Theory and methods in patchy vegetation. J Ecol 82:881–897

Sugita S, MacDonald GM, Larsen CPS (n.d.) Reconstruction of fire disturbance and forest succession from fossil pollen in lake sediments: potential and limitations. In: Clark JS (ed) Sediment Records of Biomass Burning and Global Change. Springer-Verlag, Berlin

Sukopp H, Trautmann W (1981) Causes of the decline of threatened plants in the Federal Republic of Germany. In: Synge H (ed) The Biological Aspects of Rare Plant Conservation. John Wiley and Sons, New York; pp. 113–116

Supkoff DM, Joley DB, Marois JJ (1988) Effect of introduced biological control organisms on the density of *Chondrilla juncea* in California. J Appl Ecol 25:1089–1095

Sutton DL, Van Diver VV (1986) Grass Carp: A Fish for Biological Management of Hydrilla and Other Aquatic Weeds in Florida. Florida Agric Exp Sta Bull 867

Sutton MA, Pitcairn CER, Fowler D (1993) The exchange of ammonia between the atmosphere and plant communities. Advances Ecol Res 24:301–393

Suzuki JI (1994) Growth dynamics and shoot height and foliage structure of a rhizomatous perennial herb, *Polygonum cuspidatum*. Ann Bot (Oxford) 73:629–638

Swain AM (1973) A history of fire and vegetation in northeastern Minnesota as recorded in lake sediments. Quatern Res 3:383–396

Swarbrick JT, Finlayson CM, Cauldwell AJ (1981) The biology of Australian weeds 7. *Hydrilla verticillata* (L.f.) Royle. J Austral Inst Agric Sci 1981:183–190

Swink F, Wilhelm G (1994) Plants of the Chicago Region. 4th ed. The Morton Arboretum, Lisle, Illinois

Tanner EVJ, Kapos V, Healey JR (1991) Hurricane effects on forest ecosystems in the Caribbean. Biotropica 23:513–521

Tarver DP (1980) Water fluctuation and the aquatic flora of Lake Miccosukee. J Aquatic Pl Managem 18:19–23

Tayutivutikul J, Kusigemati K (1992a) Biological studies of insects feeding on the kudzu plant, *Pueraria lobata* (Leguminosae) I. List of feeding species. Mem Fac Agric Kagoshima Univ 28:89–124

Tayutivutikul J, Kusigemati K (1992b) Biological studies of insects feeding on the kudzu plant, *Pueraria lobata* (Leguminosae) II. Seasonal abundance, habitat and development. Mem Fac Agric Kagoshima Univ 28:37–89

Tayutivutikul J, Yano K (1989) Biology of insects associated with the kudzu plant, *Pueraria lobata* (Leguminosae) 1. *Chauliops fallax* (Hemiptera, Lygaeidae). Jap J Entomol 57:831–842

Tayutivutikul J, Yano K (1990) Biology of insects associated with the kudzu plant, *Pueraria lobata* (Leguminosae) 2. *Megacopta punctissimum* (Hemiptera, Plataspidae). Jap J Entomol 58:533–539

Temple SA (1990) The nasty necessity: eradicating exotics. Conservation Biol 5:113–115

Templeton GE (1982) Status of weed control with plant pathogens. In: Charudattan R, Walker HL (eds) Biological Control of Weeds with Plant Pathogens. John Wiley, New York; pp. 29–44

Templeton GE, TeBeest DO, Smith RJ Jr (1979) Biological weed control with mycoherbicides. Annual Rev Phytopathol 17:301–310

Theriot EA, Cofrancesco AF Jr, Shearer JF (1993) Pathogen biocontrol research for aquatic plant management. In: Proceedings, 27th Annual Meeting, Aquatic Plant Control Research Program, 16–19 November 1992, Bellevue, Washington. Misc Paper A-93-2. U.S. Army Engineer Waterways Experiment Station, Vicksburg, Mississippi; pp. 82–84

Thill DC, Zamora DL, Kambitsch DL (1986) The germination and viability of excreted common crupina (*Crupina vulgaris*) achenes. Weed Sci 34:237–241

Thomas KJ (1981) The role of aquatic weeds in changing the pattern of ecosystems in Kerala. Environm Conservation 8:63–66

Thomas LK (1980) The Impact of Three Exotic Plant Species on a Potomac Island. U.S.D.I. Natl Park Serv Sci Monogr Ser 13

Thomas WL Jr (ed) (1956) Man's Role in Changing the Face of the Earth. Univ Chicago Press, Chicago, Illinois

Thompson DQ, Stuckey RL, Thompson EB (1987) Spread, Impact, and Control of Purple Loosestrife (*Lythrum salicaria*) in North American Wetlands. Fish and Wildife Research; 2. U.S. Fish and Wildlife Service, Washington, D.C.

Thompson K, Grime JP (1979) Seasonal variation in the seed banks of herbaceous species in ten contrasting habitats. J Ecol 67:893–921

Thompson RS (1988) Western North America. Vegetation dynamics in the western United States: modes of response to climatic fluctuations. In: Huntley B, Webb T III (eds) Vegetation History. Kluwer, Dordrecht, Netherlands; pp. 415–458

Thorp RW, Wenner AM, Barthell JF (1994) Flowers visited by honey bees and native bees on Santa Cruz Island. In: Halvorson WL, Meander GL (eds) The Fourth California Island Symposium: Update on the Status of Resources. Santa Barbara Museum of Natural History, Santa Barbara, California; pp. 351–364

Tilman D (1988) Plant Strategies and the Dynamics and Structure of Plant Communities. Monogr Populat Biol 26

Titus JE (1993) Submersed macrophyte vegetation and distribution within lakes: line transect sampling. Lake Reservoir Managem 7:155–164

Tobiessen P, Swart J, Benjamin S (1992) Dredging to control curly-leaf pondweed: a decade later. J Aquatic Pl Managem 30:71–72

Topp GC, Davis JL (1985) Measurement of soil water content using time-domain reflectometry (TDR): a field evaluation. J Soil Sci Soc Amer 49:19–24

Traveset A (1991) Pre-dispersal seed predation in Central American *Acacia farnesiana*: factors affecting the abundance of co-occurring bruchid beetles. Oecologia 87:570–576

Truelson RL (1984) Use of Bottom Barriers to Control Nuisance Aquatic Plants. Water Management Branch, British Columbia Ministry of Environment, Victoria, British Columbia, Canada

Trujillo EE (1985) Biological control of hamakua pa-makani with *Cercosporella* sp. in Hawaii. In: Delfosse ES (ed) Proceedings of the VI International Symposium on Biological Control of Weeds, 19–25 August 1984, Vancouver, Canada. Agriculture Canada, Ottawa, Canada; pp. 661–671

Tunison JT (n.d.[a]) Studies on the ecology and management of Faya Tree, Hawaii Volcanoes National Park, 1986–1995: Conclusions. Technical Report. Cooperative National Park Resources Studies Unit, Univ Hawaii, Manoa, Hawaii

Tunison JT (n.d.[b]) Studies on the ecology and management of Faya Tree, Hawaii Volcanoes National Park, 1986–1995: Introduction. Technical Report. Cooperative National Park Resources Studies Unit, Univ Hawaii, Manoa, Hawaii

Tunison JT (1993) Element Stewardship Abstract: *Psidium cattleianum*. Computer printout. The Nature Conservancy, Honolulu, Hawaii

Tunison JT, Castro LF, Loh RL (n.d.) Faya Tree Dieback: Distribution, Demography, and Associated Ecological Factors. Technical Report. Cooperative National Park Resources Studies Unit, Univ Hawaii, Manoa, Hawaii

Tunison JT, Smith CW, Stone CP (1992) Alien plant management in Hawaii: conclusions. In: Stone CP, Smith CW, Tunison JT (eds) Alien

Plant Invasions in Native Ecosystems of Hawaii: Management and Research. Cooperative National Park Resources Studies Unit, Univ Hawaii, Honolulu, Hawaii; pp. 821–833

Turner BL II (1994) Local faces, global flows: the role of land use and land cover change in global environmental change. Land Degrad Rehab 5: 71–78

Turner CE, Herr JC (1996) Impact of *Rhinocyllus conicus* on a non-target, rare, native thistle (*Cirsium fontinale*) in California. In: Moran VC, and Hoffmann JH (eds) Proceedings of the Ninth International Symposium on Biological Control of Weeds, 19–26 January 1996, Stellenbosch, South Africa. University of Cape Town, Cape Town, South Africa; p. 103

Turner CE, Johnson JB, McCaffrey JP (1995) Yellow starthistle. In: Nechols JR, Andres LA, Beardsley JW, Goeden RD, Jackson CG (eds) Biological Control in the Western United States. Accomplishments and Benefits of Regional Research Project W-84, 1964–1989. Univ Calif Div Agric Nat Resources Publ 3361; pp. 270–275

Turner CE, McEvoy PB (1995) Tansy ragwort. In: Nechols JR, Andres LA, Beardsley JW, Goeden RD, Jackson CG (eds) Biological Control in the Western United States. Accomplishments and Benefits of Regional Research Project W-84, 1964–1989. Univ Calif Div Agric Nat Resources Publ 3361; pp. 264–269

Turner CE, Pemberton RW, Rosenthal SS (1987) Host utilization of native *Cirsium* thistles (Asteraceae) by the introduced weevil *Rhinocyllus conicus* (Coleoptera: Curculionidae) in California. Environm Entomol 16:111–115

Turner MG, Gardner RH, O'Neill RV (1995) Ecological dynamics at broad scales. BioScience Suppl 1995:S29–S35

Turner NC (1981) Techniques and experimental approaches for the measurement of plant water status. Pl & Soil 58:339–366

Turner RM (1974) Quantitative and Historical Evidence of Vegetation Changes along the Upper Gila River, Arizona. U.S. Geol Surv Profess Paper 655-H

Tyser RW (1992) Vegetation associated with two alien plant species in a fescue grassland in Glacier National Park, Montana. Great Basin Naturalist 52:189–193

Tyser RW, Key CH (1988) Spotted knapweed in natural area fescue grasslands: an ecological assessment. Northwest Science 62:151–160

Tyser RW, Worley CA (1992) Alien flora in grasslands adjacent to road and trail corridors in Glacier National Park, Montana (U.S.A.). Conservation Biol 6:253–262

Underwood AJ (1994) On beyond BACI: sampling designs that might reliably detect environmental disturbances. Ecol Applic 4:3–15

[UNEP] United Nations Environment Programme (1995) Global Biodiversity Assessment. Cambridge Univ Press, Nairobi, Kenya

Upadhyaya MK, Turkington R, McIlvride D (1986) The biology of Canadian weeds. 75. *Bromus tectorum* L. Canad J Pl Sci 66:689–709

[USDCES] United States Department of Commerce, Economics, and Statistics. (1994) 1992 Census of agriculture. Vol. 1, Part 51. Table 44. U.S. Government Printing Office, Washington, D.C.

[USDI, BLM] United States Department of Interior, Bureau of Land Management, Oregon State Office (1994) Noxious Weed Strategy for Oregon/Washington. August 1994

[USDI, BLM] United States Department of Interior. Bureau of Land Management, Bruneau Resource Area (1995) Snake River Birds of Prey National Conservation Area Draft Management Plan

[USDI, FWS] United States Department of the Interior. Fish and Wildlife Service, Division of Refuges (1995) Environmental Assessment of the proposed release of three exotic insects to control purple loosestrife *Lythrum salicaria*. Washington, D.C.

[USDI, NPS] United States Deptartment of the Interior, National Park Service (1991) Natural Resources Management Guidelines, NPS-77. Natural Resources Publication Office, Denver, Colorado

Usher MB (1986) Invasibility and wildlife conservation: invasive species on nature reserves. Philos Trans R Soc Lond B Biol Sci 314:695–710

Usher MB (1988) Biological invasions of nature reserves: a search for generalisations. Biol Conservation 44:119–135

Usher MB (1991) Biological invasion into tropical nature reserves. In: Ramakrishnan (ed) Ecology of Biological Invasion in the Tropics. International Scientific Publications, New Delhi, India

Vail D (1994) Symposium introduction: management of semiarid rangelands—impacts of annual weeds on resource values. In: Monsen SB, Kitchen SG (eds) Proceedings—Ecology and Management of Annual Rangelands. U.S.D.A. Forest Serv Gen Techn Rep INT-GTR-313. Intermountain Research Station, Ogden, Utah; pp. 3–4

Van TK (1988) Integrated control of waterhyacinth with *Neochetina* and paclobutrazol. J Aquatic Pl Managem 26:59–61

Van TK, Vandiver VV Jr (1992) Response of monoecious and dioecious hydrilla to bensulfuron methyl. J Aquatic Pl Managem 30:41–44

Van Devender TR, Martin PS, Thompson RS, Cole KL, Jull AJT, Long A, Toolin LJ, Donahue DJ (1985) Fossil packrat middens and the tandem accelerator mass spectrometer. Nature 317:610–613

Van Dyke JM, Leslie AJ Jr, Nall LE (1984) The effects of the grass carp on the aquatic macrophytes of four Florida lakes. J Aquatic Pl Managem 22:87–95

van Hylckama TEA (1974) Water Use by Saltcedar as Measured by the Water Budget Method. U.S. Geol Surv Profess Paper 491-E

VanGundy AB (1984) Managing Group Creativity: A Modular Approach to Problem-solving. American Management Association, New York

Vankat JL, Snyder GW (1991) Floristics of a chronosequence corresponding to old field-deciduous forest succession in southwestern Ohio. I. Undisturbed vegetation. Bull Torrey Bot Club 118:365–376

Van Zant KL, Webb T III, Peterson GM, Baker RG (1979) Increased *Cannabis/Humulus* pollen, an indicator of European agriculture in Iowa. Palynology 3:227–233

Vermeij GJ (1989) Invasion and extinction: the last three million years of North Sea pelecypod history. Conservation Biol 3:274–281

Versfeld DB, van Wilgen BW (1986) Impact of woody aliens on ecosystem properties. In: Macdonald IAW, Kruger FJ, Ferrar AA (eds) The Ecology and Management of Biological Invasions in Southern Africa. Oxford Univ Press, Cape Town, South Africa; pp. 239–246

Virginia RA, Jarrell WM (1983) Soil properties in a mesquite-dominated Sonoran Desert ecosystem. J Soil Sci Soc Amer 47:138–144

Vitousek PM (1986) Biological invasions and ecosystem properties: can species make a difference? In: Mooney HA, Drake J (eds) Ecology of Biological Invasions of North America and Hawaii. Springer-Verlag, New York; pp. 163–176

Vitousek PM (1990) Biological invasions and ecosystem processes: towards an integration of population biology and ecosystem studies. Oikos 57:7–13

Vitousek PM (1992) Effects of alien plants on native ecosystems. In: Stone CP, Smith CW, Tunison JT (eds) Alien Plant Invasions in Native Ecosystems of Hawaii: Management and Research. Univ Hawaii Press, Honolulu, Hawaii; pp. 29–41

Vitousek PM (1994) Beyond global warming: ecology and global change. Ecology 75:1861–1876

Vitousek PM, Walker LR (1989) Biological invasion by *Myrica faya* in Hawaii: plant demography, nitrogen fixation, ecosystem effects. Ecol Monogr 59:247–265

Vitousek PM, Walker LR, Whiteaker LD, Matson PA (1993) Nutrient limitation to plant growth in primary succession in Hawaii Volcanoes National Park. Biogeochemistry 23:197–215

Vitousek PM, Walker LR, Whiteaker LD, Mueller-Dombois D, Matson PA (1987) Biological invasion by *Myrica faya* alters ecosystem development in Hawaii. Science 238:802–804

Vivrette NJ, Muller CH (1977) Mechanism of invasion and dominance of coastal grassland by *Mesembryanthemum crystallinum*. Ecol Monogr 47:301–318

Vogt GB, Quimby PC, Kay SH (1992) Effects of Weather on the Biological Control of Alligatorweed in the Lower Mississippi Valley Region, 1973–83. U.S.D.A. Agric Res Serv Techn Bull 1766

Wace N (1977) Assessment of dispersal of plant species—the car borne flora in Canberra. Proc Ecol Soc Austral 10:167–186

Wagner FH (1989) Grazers, past and present. In: Huenneke LF, Mooney H (eds) Grassland Structure and Function: California Annual Grassland. Kluwer, Dordrecht, Netherlands; pp. 151–162

Wagner FH, Kay CE (1993) "Natural" or "healthy" ecosystems: are U.S. national parks providing them? In: McDonnell MJ, Pickett STA (eds) Humans as Components of Ecosystems: The Ecology of Subtle Human Effects and Populated Areas. Springer-Verlag, New York; pp. 257–270

Wagner WH (1993) Problems with biotic invasions: a biologist's viewpoint. In: McKnight BN (ed) Biological Pollution: The Control and Impact of Invasive Exotic Species. Indiana Academy of Sciences, Indianapolis, Indiana; pp. 1–8

Wagner WL, Herbst DR, Sohmer SH (1990) Manual of the Flowering Plants of Hawai'i. Univ Hawaii Press, Honolulu, Hawaii

Walker LR (1993) Nitrogen fixers and species replacements in primary succession. In: Miles J, Walton DH (eds) Primary Succession on Land. Blackwell, Oxford, U.K.; pp. 249–272

Walker LR (1994) Effects of fern thickets on woodland development on landslides in Puerto Rico. J Veg Sci 5:525–532

Walker LR, Chapin III FS (1987) Interactions

among processes controlling successional change. Oikos 50:131–135
Walker LR, Vitousek PM (1991) An invader alters germination and growth of a native dominant tree in Hawaii. Ecology 72:1449–1455
Wapshere AJ (1974) A strategy for evaluating the safety of organisms for biological weed control. Ann Appl Biol 77:201–211
Wapshere AJ (1975) A protocol for programmes for biological control of weeds. Pestic Articles News Summ 21:295–303
Wapshere AJ (1981) Recent thoughts on exploration and discovery for biological control of weeds. In: Delfosse ES (ed) Proceedings of the Fifth International Symposium on Biological Control of Weeds, 22–29 July 1980, Brisbane, Australia. CSIRO, Melbourne, Australia; pp. 75–79
Waring RH, Cleary BD (1967) Plant moisture stress: evaluation by pressure bomb. Science 155:1253–1254
Watson AK (1980) The biology of Canadian weeds. 43. *Acroptilon* (*Centaurea*) *repens* (L.) DC. Canad J Pl Sci 60:993–1004
Watson AK (1985). Introduction. The leafy spurge problem. *In* Leafy Spurge. Weed Science Society of America, Champaign, Illinois
Watson AK, Renney AJ (1974) The biology of Canadian weeds. 6. *Centaurea diffusa* and *Centaurea maculosa*. Canad J Pl Sci 54:687–701
Watts WA (1973) Rates of change and stability in vegetation in the perspective of long periods of time. In: Briks HJB, West RG (eds) Quaternary Plant Ecology. Blackwell, Oxford, U.K.; pp. 195–206
Watts WA, Bradbury JP (1982) Paleoecological studies at Lake Patzcuaro on the west-central Mexican Plateau and at Chalco in the Basin of Mexico. Quatern Res 17:56–60
Watts WA, Winter TC (1966) Plant macrofossils from Kirchner Marsh, Minnesota—a paleoecological study. Geol Soc Amer Bull 77:1339–1360
Webb DA (1985) What are the criteria for presuming native status? Watsonia 15:231–236
Webb MA, Elder HS, Howells RG (1994) Grass carp reproduction in the lower Trinity River, Texas. In: Proceedings of the Grass Carp Symposium, 7–9 March 1994, Gainesville, Florida. U.S. Army Engineer Waterways Experiment Station, Vicksburg, Mississippi; pp. 29–32
Webb SL, Kaunzinger CK (1993) Biological invasion of the Drew University (New Jersey) Forest Preserve by Norway maple (*Acer platanoides* L.). Bull Torrey Bot Club 120:343–349

Webb T III (1973) A comparison of modern and presettlement pollen from southern Michigan (U.S.A.). Rev Palaeobot Palynol 16:137–156
Webb T III (1982) Temporal resolution in Holocene pollen data. Third N Amer Paleontol Conven Proc 2:569–572
Webb T III (1986) Is vegetation in equilibrium with climate? How to interpret late-Quaternary pollen data. Vegetatio 67:75–91
Webb T III (1988) Eastern North America. In: Huntley B, Webb T III (eds) Vegetation History. Kluwer, Boston, Massachusetts; pp. 385–414
Webb T III (1992) Past changes in vegetation and climate: lessons for the future. In: Peters RL, Lovejoy TE (eds) Global Warming and Biological Diversity. Yale Univ Press, New Haven, Connecticut; pp. 59–75
Webb T III (1993) Constructing the past from late-Quaternary pollen data: temporal resolution and a zoom lens space-time perspective. In: Kidwell SM, Behrensmeyer AK (eds) Taphonomic Approaches to Time Resolution in Fossil Assemblages. Short Courses Paleontol 6. Paleontol Soc, Knoxville, Tennessee; pp. 79–101
Webb T III, Bartlein PJ (1992) Global changes during the last 3 million years: climatic controls and biotic responses. Annual Rev Ecol Syst 23:141–173
Webb T III, Bartlein PJ, Harrison SP, Anderson KH (1993) Vegetation, lake levels, and climate in eastern North America for the past 18,000 years. In: Wright HE Jr, Kutzbach JE, Webb T III, Ruddiman WF, Street-Perrott FA, Bartlein PJ (eds) Global Climates since the Last Glacial Maximum. Univ Minnesota Press, Minneapolis, Minnesota; pp. 415–467
Webb T III, Cushing EJ, Wright HE Jr (1983) Holocene changes in the vegetation of the Midwest. In: Wright HE Jr (ed) Late Quaternary Environments of the United States. Volume 2. The Holocene. Univ Minneosta Press, Minneapolis, Minnesota; pp. 142–165
Webb T III, McAndrews JH (1976) Corresponding patterns of contemporary pollen and vegetation in central North America. Geol Soc Amer Mem 145:267–299
Weiss PW, Noble IR (1984) Status of coastal dune communities invaded by *Chrysanthemoides monilifera*. Austral J Ecol 9:93–98
Wenner AM, Thorp RW (1994) Removal of feral honey bee (*Apis mellifera*) colonies from Santa Cruz Island. In: Halvorson WL, Meander GL

(eds) The Fourth California Island Symposium: Update on the Status of Resources. Santa Barbara Museum of Natural History, Santa Barbara, California; pp. 513–522

Werner PA, Rioux R (1977) The biology of Canadian weeds. 24. *Agropyron repens* (L.) Beauv. Canad J Pl Sci 57:905–919

Westbrooks RG (1993) Biological pollution: a historical perspective of nonindigenous invasive plants of grazing lands in the United States. In: Proceedings of the Grazing Lands Forum. An Explosion in Slow Motion: Noxious Weeds and Invasive Alien Plants on Grazing Lands, Eighth Forum, 2 December 1993. U.S.D.A. Economic Research Service, Washington, D.C.; pp. 17–22

Westbrooks RG (n.d.) Federal regulatory efforts to minimize the introduction and impacts of exotic pest plants in the United States. Proceedings of the California Exotic Pest Plant Council

Wester LL (1992) Origin and distribution of adentive alien flowering plants in Hawaii. In: Stone CP, Smith CW, Tunison JT (eds) Alien Plant Invasions in Native Ecosystems of Hawaii: Management and Research. Univ Hawaii Press, Honolulu, Hawaii; pp. 99–154

Wester LL, Wood HB (1977) Koster's curse (*Clidemia hirta*), a weed pest in Hawaiian forests. Environm Conservation 4:35–41

Westerdahl HE (1987) Herbicide concentration/exposure time relationships. In: Proceedings, 21st Annual Meeting, Aquatic Plant Control Research Program, 17–21 November 1986, Mobile, Alabama. Misc Paper A-87-2. U.S. Army Engineer Waterways Experiment Station, Vicksburg, Mississippi; pp. 169–172

Westerdahl HE, Getsinger KD (1988a) Aquatic Plant Identification and Herbicide Use Guide. Volume I. Aquatic Herbicides and Application Equipment. Techn Rep A-88-9. U.S. Army Engineer Waterways Experiment Station, Vicksburg, Mississippi

Westerdahl HE, Getsinger KD (1988b) Aquatic Plant Identification and Herbicide Use Guide. Volume II. Aquatic Plants and Susceptibility to Herbicides. Techn Rep A-88-9. U.S. Army Engineer Waterways Experiment Station, Vicksburg, Mississippi

Westman WE (1990a) Managing for biodiversity. BioScience 40:26–33

Westman WE (1990b) Park management of exotic plant species: problems and issues. Conservation Biol 4:251–260

Westman WE, Malanson GP (1992) Effects of climate change on Mediterranean-type ecosystems in California and Baja California. In: Peters RL, Lovejoy TE (eds) Global Warming and Biological Diversity. Yale Univ Press, New Haven, Connecticut; pp. 258–276

Whelan CJ, Dilger ML (1992) Invasive exotic shrubs: a paradox for natural areas managers? Nat Areas J 12:109–110

White AS (1983) The effects of thirteen years of annual prescribed burning on a *Quercus ellipsoidalis* community in Minnesota. Ecology 64:1081–1085

White AS (1986) Prescribed Burning for Oak Savanna Restoration in Central Minnesota. Research Paper NC-266. U.S.D.A. Forest Service, North Central Forest Experiment Station, St. Paul, Minnesota

White DJ, Haber E, Keddy C (1993) Invasive Plants of Natural Habitats in Canada. Canadian Museum of Nature and Canadian Wildlife Service, Environment Canada, Ottawa, Canada

White DW, Stiles EW (1991) Bird dispersal of fruits of species introduced into eastern North America. Canad J Bot 70:1689–1696

Whiteaker LD, Gardner DE (1985) The distribution of *Myrica faya* Ait. in the state of Hawaii. Techn Rep 55. Cooperative National Park Resources Studies Unit, Univ Hawaii, Manoa, Hawaii

Whiteaker LD, Gardner DE (1987) The phenology and stand structure of *Myrica faya* Ait. in Hawaii. Techn Rep 62. Cooperative National Park Resources Studies Unit, Univ Hawaii, Manoa, Hawaii

Whitehead DR, Sheehan MC (1985) Holocene vegetational changes in the Tombigbee River Valley, eastern Mississippi. Amer Midl Naturalist 113:122–137

Whitney GG, Adams SD (1980) Man as a maker of new plant communities. J Appl Ecol 17:431–448

Wilcox DA (1995) Wetland and aquatic macrophytes as indicators of anthropogenic hydrologic disturbance. Nat Areas J 15:240–248

Wilcut JW, Dute RR, Roland R, Truelove B (1988) Factors limiting the distribution of cogongrass, *Imperata cylindrica*, and torpedograss, *Panicum repens*. Weed Sci 36:577–582

Wilcut JW, Truelove B, Davis DE (1988) Temperature factors limiting the spread of cogongrass (*Imperata cylindrica*) and torpedograss (*Panicum repens*), Weed Sci 36:49–55

Wilde GR, Riechers RK, Johnson J (1992) Angler attitudes toward control of freshwater vegetation. J Aquatic Pl Managem 30:77–79

References

Wile I (1975) Lake restoration through mechanical harvesting of aquatic vegetation. Verh Int Vereinigung Theor Angew Limnol 19:660–671

Wiley MJ, Gorden RW, Waite SW, Powless T (1984) The relationship between aquatic macrophytes and sport fish production in Illinois ponds: a simple model. Amer J Fish Managem 4:111–119

Williams CE (1993a) Age structure and importance of naturalized *Paulownia tomentosa* in a central Virginia streamside forest. Castanea 58:243–249

Williams CE (1993b) Alien invasion. Pennsylvania Wildlife 14(5):16–19

Williams CE (1993c) The exotic empress tree, *Paulownia tomentosa*: an invasive pest of forests? Nat Areas J 13:221–222

Williams CE (1996) Alien plant invasions and forest ecosystem integrity: a review. In: Majumdar SK, Miller EW, Brenner FJ (eds) Forests—A Global Perspective. Pennsylvania Academy of Science, Easton, Pennsylvania; pp. 169–185

Williams CE, Ralley JJ, Taylor DH (1992) Consumption of invasive Amur honeysuckle *Lonicera maackii* (Rupr.) Maxim., by small mammals. Nat Areas J 12:86–89

Williams DG, Mack RN, Black RA (1995) Ecophysiology of introduced *Pennisetum setaceum* on Hawaii: the role of phenotypic plasticity. Ecology 76:1569–1580

Williams M (1989) Americans and Their Forests: A Historical Geography. Cambridge Univ Press, New York

Williams M (1993) An exceptionally powerful biotic factor. In: McDonnell MJ, Pickett STA (eds) Humans as Components of Ecosystems: The Ecology of Subtle Human Effects and Populated Areas. Springer-Verlag, New York; pp. 24–39

Williams T (1994) Invasion of the aliens. Audubon 96(5):24–26, 28, 30, 32

Willson MF, Rice BL, Westoby M (1990) Seed dispersal spectra: a comparison of temperate plant communities. J Veg Sci 1:547–562

Wilson JB (1994) Who makes the assembly rules? J Veg Sci 5:275–278

Wilson SD (1989) The suppression of native prairie by alien species introduced for revegetation. Landscape Urban Planning 17:113–119

Wiltshire PEJ, Edwards KE (1993) Mesolithic, early Neolithic, and later prehistoric impacts on vegetation at a riverine site in Derbyshire, England. In: Chambers FM (ed) Climate Change and Human Impact on the Landscape. Chapman and Hall, London, U.K.; pp. 157–168

Wong SL, Clark B (1979) The determination of desirable and nuisance plant levels in streams. Hydrobiologia 63:223–230

Woods KD (1993) Effects of invasion by *Lonicera tatarica* L. on herbs and tree seedlings in four New England forests. Amer Midl Naturalist 130:62–74

Woods KD, Davis MB (1989) Paleoecology of range limits: beech in northern Michigan and Wisconsin. Ecology 70:681–696

Woodward FI, Bazzaz FA (1988) The responses of stomatal density to CO_2 partial pressure. J Bot 39:1771–1781

Wunderlin RP (1982) Guide to the Vascular Plants of Central Florida. Univ Presses of Florida, Tampa, Florida

Wurdack JJ (1980) Melastomataceae. Volume 13. In: Harding G, Sparre B (eds) Flora of Ecuador. Department of Systematic Botany, Univ Göteborg, Göteborg, and Section for Botany, Riksmuseum, Stockholm, Sweden

Yang P, Foote D, Jones V (n.d.) Preliminary investigations of the the effect of the two-spotted leafhopper, *Sophonia rufofascia* (Kuoh and Kuoh) on faya tree. Technical Report. Cooperative National Park Resources Studies Unit, Univ Hawaii, Manoa, Hawaii

Yost SE, Antenen S, Hartvigsen G (1991) The vegetation of the Wave Hill natural area, Bronx, New York. Bull Torrey Bot Club 118:312–325

Young JA (1992) Ecology and management of medusahead (*Taeniatherum caput-medusae* ssp. *asperum* [Simk.] Melderis). Great Basin Naturalist 52:245–252

Young JA (1994) History and use of semiarid plant communities—changes in vegetation. In: Monsen SB, Kitchen SG (1994) Proceedings—Ecology and Management of Annual Rangelands. Gen U.S.D.A Forest Serv Gen Techn Rep INT-GTR-313. Intermountain Research Station, Ogden, Utah; pp. 5–8

Young JA, Evans RA (1970) Invasion of medusahead [*Taeniatherum caput-medusae*] into the Great Basin. Weed Sci 18:89–97

Young JA, Evans RA (1978) Population dynamics after wildfires in sagebrush grasslands. J Range Managem 31:283–289

Young JA, Evans RA (1989) Seed production and germination dynamics in California annual grasslands. In: Huenneke LF, Mooney HA (eds) Grassland Structure and Function. Kluwer, Boston, Massachusetts; pp. 39–46

Zamora DL, Thill DC, Eplee RE (1989) An eradication plan for plant invasions. Weed Technol 3:2–12

Zangerl AR, Bazzaz FA (1984) The response of plants to elevated CO_2. II. Competitive interactions among annual plants under varying light and nutrients. Oecologia 62:412–417

Zar JH (1984) Biostatistical Analysis. 2nd ed. Prentice-Hall, Englewood Cliffs, New Jersey

Zedler PH, Gautier CR, McMaster GS (1983) Vegetation change in response to exteme events: the effect of a short fire interval between fires in a California chaparral and coastal shrub. Ecology 64:809–818

Zimmerman UD, Ebinger JE, Diekroeger KC (1993) Alien and native woody species invasion of abandoned crop land and reestablished tall grass prairie in east-cental Illinois. Trans Illinois Acad Sci 86:111–118

Zocchi R (1971) Contributo alla conoscenza dell'entomofauna delle tamerici in Italia. Redia 52:31–129

Zwölfer H, Harris P (1971) Host specificity determination of insects for biological control of weeds. Annual Rev Entomol 16:159–178

Index

A
Abutilon theophrasti, 23, 191
Acacia, 45, 50, 71
  confusa, 256
  cyclops, 90
  farnesiana, 256
  longifolia, 59
  mearnsii, 256
  melanoxylon, 256
Acer, 39, 46
  negundo, 120
  platanoides, 59
  rubrum, 51
  saccharinum, 120
  saccharum, 43, 49, 51, 59, 64
Achillea millefolium, 45
Acroptilon, 185, 186
  repens, 186
Aeschynomene virginica, 176
Ageratina riparia, 183
Ageratum, 46
Agropyron
  cristatum, 238
  repens, 33, 257
Agrostis capillaris, 60
Ailanthus altissima, 23, 117, 191, 215, 224, 235, 257
Albizia
  julibrissin, 191
  lebbek, 33
Allenrolfea, 76
Alliaria officinalis. See *Alliaria petiolata*
Alliaria petiolata, 23, 26, 117, 118, 122, 123, 124, 125, 127, 257
  community response, 122
  competitive interaction, 126
  germination, 126
  life history, 118
  removal experiment, 118
Alnus, 39, 47, 53
Alternanthera philoxeroides, 152, 179
Amaranthus, 13, 191
Ambrosia, 41, 42, 43, 46, 50, 51
  trifida, 13
Ammophila arenaria, 61, 257
Ampelopsis brevipedunculata, 257
Andropogon, 232
  virginicus, 61, 75, 257
Anthropic effects
  cultivation, 13, 38, 45
  deforestation, 42
  degree of, 27
  disturbance, 15, 34, 38, 41, 42, 53, 62, 66, 137
  nitrogen, 98, 143
  novel habitat, 13, 29, 49, 63
  water level, 50
Applied ecology, 105
Aquatic ecosystems
  benthic barrier, 162
  biological control, 151
  diversity, 150
  drawdown, 161
  herbicide, 151, 154, 166
  mechanical management, 159
  nutrients, 160, 163
  oxygen exchange, 146
  physical management, 161
  quarantine, 163

seed bank, 154
  weed watcher, 164
  wildlife habitat, 146
Arctostaphylos, 234
Aristotelia chilensis, 60
Artemisia, 50
Arundo donax, 178, 244, 247, 257
Astragalus applegatei, 21
Atriplex, 76
Avena, 92
  fatua, 87, 90
Azolla caroliniana, 145

B
Baccharis, 245
  neglecta, 188
  pilularis, 64
  salicifolia, 177
BACI. *See* Before-after control-impact
Before-after control-impact (BACI), 110
Betula, 39, 43, 46, 51, 53
  alleghaniensis, 64
  lutea. *See* Betula, alleghaniensis
Bidens cernua, 50, 52
Biological control
  augmentation, 176
  classical, 175
  community response, 179
  conflicts, 127, 176
  defined, 174
  domestic field studies, 181
  early history, 173
  evaluation, 182
  facilities, 174
  future research, 191
  Hawaiian rangeland, 183
  in lakes, 151
  miconia, 251, 253
  nature reserves, 189, 205
  nontarget effects, 179
  overseas testing, 180
  public perceptions, 192
  release and establishment, 182
  research protocol, 179
  southwestern rangeland, 188
  western rangeland, 184
Boerhavia, 44
Brasenia schreberi, 50, 52
Brassica rapa, 89
Bromus, 92, 221
  inermis, 258
  madritensis, 90
  tectorum, 62, 71, 75, 229, 258
Broussonetia, 46

C
Callirhoe involucrata, 13
Calluna vulgaris, 142
Cannabis, 41, 50, 52, 53
  sativa, 190
Capsella bursa-pastoris, 45
Cardaria draba, 104, 258
Carduus, 179, 185
  acanthoides, 185
  nutans, 179, 185
  pycnocephalus, 185
Carpinus, 43
Carpobrotus edulis, 73
Carya, 43, 46
Castanea, 41, 46
  dentata, 49, 51
Casuarina, 224, 233, 234
  equisetifolia, 61, 258
Ceanothus, 234
Celastrus orbiculatus, 237, 258
Centaurea, 23, 24, 183, 185, 186, 234
  diffusa, 186
  maculosa, 62, 63, 186, 234, 258
  repens, 258
  solstitialis, 230, 258
Cephalanthus occidentalis, 50, 52
Chenopodium, 13
  album, 19
Chondrilla juncea, 182, 187, 259
Chromolaena odorata, 60
Chrysanthemoides monilifera, 59
Cirsium, 24, 179, 185
  arvense, 23, 186, 236
  cymosum, 179
  fontinale, 179
  vulgare, 19, 186
Cleome, 44
Clidemia hirta, 91, 183, 259
Climate change, 14, 34
  air pollution, 97, 143
  competitive interaction, 96
  and disturbance, 102
  increase in CO_2, 96
  species migration, 15, 46, 101
  ultraviolet radiation, 97
Coffea arabica, 33
Community susceptibility to invasion, 2, 66
Conant, Patrick, 249
Conicosia pugioniformis, 259
Conium maculatum, 185
Conservation goals, 16, 21, 23, 29, 49
Control program, 237
  Everglades, 239
  Melastone Action Committee, 251

Convolvulus arvensis, 188
Corispermum, 44, 54
Cornus, 39
  florida, 127
Cortaderia jubata, 259
Crupina vulgaris, 195, 259
Cultural landscape, 34
Cynara cardunculus, 33, 62
Cynodon dactylon, 172, 232, 259
Cyperus, 50
  diandrus, 52
  odoratus, 52
  rivularis, 52
  rotundus, 183
Cytisus scoparius, 63, 185, 234, 259

D

Deforestation, 45, 49
DeLoach, Jack C., 172
Dennstaedtia, 51
Dioscorea
  batatas, 259
  bulbifera, 259
Dispersal
  animal-mediated, 89, 216, 224, 231
  human-mediated, 9, 14, 66, 88, 98, 100, 144, 216, 226, 253
  and succession, 135
Distichlis spicata, 84
Disturbance, 19
  altered, 61, 99
  animal-mediated, 91
  anthropic, 137
  characteristics of, 79
  and climate change, 102
  and invader persistence, 139
  and invasion, 15, 66, 70, 139
Diversity
  genetic, 16
  species, 16, 146, 238
Dulichium arundinaceum, 52

E

Echium, 221, 224
Ecological systems
  anthropic, 28, 141
  impacted natural, 28
  intact natural, 27
Ecological value threshold, 30
Egeria densa, 145
Ehrharta
  calycina, 259
  erecta, 260
Eichhornia crassipes, 60, 145, 148, 149, 260

Elaeagnus
  angustifolia, 23, 173, 183, 260
  umbellata, 260
Eleocharis, 50
  calva, 52
  geniculata, 52
  olivacea, 52
  palustris, 52
  parvula, 178, 234
  smallii, 52
Elodea densa. See Egeria densa
Emex
  australis, 183
  spinosa, 183
Endangered species, 13, 16
Environmental impact statement (EIS), 201
Epipactis helleborine, 58
Equisetum, 53
  fluviatile, 50, 52
Eragrostis, 115, 172
  lehmanniana, 63, 136
Eriogonum, 50
Erodium cicutarium, 45, 50, 54, 90, 92, 234
Eschscholzia californica, 92
Eucalyptus, 45, 50, 59, 73, 89
Euonymus, 92, 93
  fortunei, 260
Eupatorium adenophorum, 63, 183
Euphorbia, 44
  cyparissias, 187
  esula, 22, 183, 236, 260
  esula/virgata, 178
Everglades, 239
Exotic Pest Plant Council, 228
  role, 239
Experimental design
  control, 113, 120
  manipulative, 72, 112, 128
  mensurative, 112
  principles, 106, 108, 112
  replication, 113

F

Fagus, 39, 43, 46, 51
  grandifolia, 10, 47, 59, 72
Federal Noxious Weed Act, 217, 231, 241
Festuca ovina, 142
Ficus, 46, 100
  microcarpa, 260
Flora
  county records, 12
  Neolithic, 9
  neosysanthropic, 14
  paleosynanthropic, 14

pre-Columbian, 9
Floras, standards, 8
Foeniculum vulgare, 89
Frandsen, Paul R., 244
Fraxinus, 39, 43, 46, 61
    pennsylvanica, 120
Froelichia gracilis, 13

G

Genista monspessulana, 234, 260
Geographic information system (GIS), 168
Glyceria, 52
Grazing, effects, 32, 45, 49, 62, 90, 92, 115, 231
Gutierrezia, 177
Gymnosteris nudicaulis, 234

H

Habitat, pre-European, 9, 17, 141
Halogeton glomeratus, 187
Hedera helix, 60, 237, 260
Hedychium gardnerianum, 261
Helianthus
    annuus, 13
    salicifolius, 13
Hepatica acutiloba, 125
Herbicide
    aquatic, 155
    compared to biocontrol, 175
    concentration/exposure time relationship, 156
    nontarget effects, 170
    trial, 239, 247, 252
Hiebert, Ronald D., 195
Hieracium, 60
Hordeum, 53
Horticulture industry, 216
Huenneke, Laura Foster, 95
Humulus, 41
Hydrilla verticilata, 92, 145, 147, 149, 183, 261
Hydrocotyle ranunculoides, 13
Hypericum
    concinnum, 184
    perforatum, 142, 172, 173
Hypothetico-deductive method, 105

I

Ilex aquifolium, 59
Impatiens, 119, 127
Imperata brasiliensis, 234
Integrated pest management (IPM), 198
Introduction
    human-mediated, 9, 144, 216
    and integration, 87
    new, 100, 190
    and niche shift, 92

policy strategies, 217, 241
preventing, 134, 165, 190, 226, 253
rate of, 16
species, 11
Invasion
    areal extent, 234
    and community change, 63, 136, 177
    community effects, 66, 126, 134, 150, 177, 245
    database, 115
    and disturbance, 58, 66, 70, 91, 118, 126, 136, 232, 245
    economic cost, 230, 236
    and fire, 232, 236, 246
    and herbivory, 92
    incipient, 24
    monitoring, 108, 170, 225
    number of species, 229
    and nutrient status, 97
    and oxygen exchange, 146
    predicted, 24, 218
    and primary production, 71
    rate of spread, 231, 240
    and soil water, 74, 178
    temporary, 137
    wildlife habitat, 146, 178, 232
Invasion of
    Everglades, 239
    forest, 32, 59, 64, 82, 90, 92, 117, 139, 229, 235
    grassland, 60, 62, 64, 75, 90, 136, 229, 234, 235
    human-altered habitat, 13
    islands, 60, 67, 82, 249
    lakes, 60, 92, 145, 150
    nature reserves, 1, 21, 22, 31, 92, 189, 196, 237
    New World, 12
    prairie, 21, 142
    riparian ecosystems, 177, 245
    shrubland, 60
    wetlands, 75, 83, 231, 232, 235
Invasive species
    and aquatic habitat, 146
    competitive interaction, 58, 59, 64, 72, 83, 96, 126, 137, 174, 234
    disturbance regimen, 79
    ecological value, 29, 141
    economic impact, 19, 22, 82
    functional roles, 27, 65
    identification, 167, 225
    indigenous, 47, 64
    life history, 19, 23, 65, 70, 118, 146, 151, 196, 206, 221, 250
    pool, 2

pre-adaptation, 89
prediction, 218
wildlife habitat, 23, 32
IPM, 198
Ipomoea aquatica, 191
Iris pseudacorus, 61
Isatis tinctoria, 261
Iva
 ciliata, 13
 xanthifolia, 7, 13

J
Jackson, Stephen T., 37
Juglans, 39, 43
 nigra, 47
Juniperus, 50, 53, 75, 177
 osteosperma, 47
 virginiana, 22, 47

K
Kallstroemia, 44
Kochia scoparia, 236, 261

L
Landscape change, fragmentation, 98
Lantana camara, 62, 173, 261
Larrea tridentata, 188
Lasthenia californica, 92
Leonorus cardiaca, 127
Lepidium latifolium, 261
Leptospermum scoparium, 177
Lespedeza cuneata, 261
Ligustrum, 237
 sinense, 191, 262
 vulgare, 262
Linaria
 dalmatica, 262
 genistifolia subsp. dalmatica, 187
 vulgaris, 187, 262
Liriodendron tulipifera, 237
Lolium, 61
Lonicera, 32
 japonica, 23, 26, 60, 65, 92, 95, 96, 117, 127, 191, 234, 262
 maackii, 90, 117, 133, 142, 235, 262
 sempervirens, 92, 96
 tatarica, 59, 65, 262
Loope, Lloyd L., 249
Luken, James O., 133
Lupinus sulphureus kincaidii, 234
Lycopodium annotinum, 53
Lygodium
 japonicum, 191, 234
 microphyllum, 191, 234, 262

Lythrum salicaria, 12, 23, 26, 61, 93, 145, 178, 189, 228, 231, 262

M
Madsen, John D., 145
Management
 adaptive, 24
 assessment, 166, 182
 conflicts, 30, 32, 91, 93, 142, 144, 176, 193, 242
 cost/benefit analysis, 168, 202
 database, 115, 169
 experimental design, 106
 fire, 22, 99
 flood, 99
 goals, 16, 23, 24, 27, 29, 49, 73, 74, 78, 80, 81, 103, 134, 139, 143, 206
 hand removal, 127
 herbicide, 127, 142, 154, 166, 175, 252
 historical, 140
 institutional, 163
 integrated plan, 158, 164
 and invasion, 142
 mechanical, 159
 monitoring, 25, 106, 127, 166, 168, 170, 225
 physical, 161
 plan, 164, 208, 241, 254
 principles, 143
 prioritization, 167, 196, 203
 regional scale, 207
 regulation, 166
 seed bank response, 141, 154
 shrub removal, 32
Management decisions
 cost/benefit analysis, 202
 environmental assessment, 200
 expert opinion technique, 202
 nature reserves, 204
 plant ranking system, 203, 210
 process, 198, 200
 types, 198
Marrubium vulgare, 89, 90
McCarthy, Brian C., 117
Medeiros, Arthur C., 249
Medicago polymorpha, 90
Melaleuca, 59
 quinquenervia, 145, 178, 189, 232, 239, 263
Melastome Action Committee, 251
Melia azedarach, 191
Melilotus, 23
Melinis minutiflora, 263
Menispermum canadense, 127
Mesembryanthemum crystallinum, 62, 232, 263

Miconia, 46
  calvescens, 249, 263
Microstegium vimineum, 263
Migration of species, 47
Mimosa
  pigra, 59, 263
  pilleta, 239
Mollugo verticillata, 44, 54
Monitoring, 25, 106, 108, 225, 238
Morrenia odorata, 176
Morrison, Michael L., 104
Morus, 46, 50
Myrica faya, 62, 71, 82, 91, 228, 232, 263
Myriophyllum spicatum, 60, 145, 146, 149, 167, 263

N
Najas flexilis, 52
Nassella, 92
National Park Service Organic Act, 206
Native
  criteria, 8
  defined, 8
Native Americans, 9, 42
Naturalization, 89
Neyraudia reynaudiana, 234, 263
Niche shift, 92, 99
Nicotiana glauca, 264
Nitrogen pollution, 98, 143
Nuphar, 52
Nutrient cycling
  analytical methods, 74, 76
  nitrogen fixation, 62, 73, 80, 98, 232
  nutrient removal, 160
  salt accumulation, 62, 73, 76, 232
  sequestration, 72, 73, 160, 163

O
Oenothera, 44
Onopordum, 185
Opuntia
  ficus-indica, 183
  inermis, 173
  littoralis, 177
  oricola, 177
  tricantha, 177
  vulgaris, 173
Ostrya, 43, 46

P
Paederia foetida, 264
Paleoecological record
  human disturbance, 42
  indigenous species, 64
  post-European, 45
  precision of, 38
  pre-European, 42
  spatial scale, 41
  species migration, 10, 47, 58, 64, 101
  temporal scale, 39
Panicum
  maximum, 264
  repens, 264
Passiflora, 92
  mollissima, 60, 184, 237, 264
Paulownia tomentosa, 117, 191
Peganum harmala, 183
Pennisetum
  clandestinum, 264
  setaceum, 264
Perilla frutescens, 264
Phaseolus, 42
Phleum pratense, 63
Phragmites
  australis, 22, 23
  communis. See Phragmites, australis
Picea, 51, 53
  mariana, 10
Pinus, 39, 46, 50, 51, 53, 64, 72, 219, 225
  pinaster, 59
  radiata, 59
  strobus, 43
Pistia stratiotes, 60, 182
Plantago, 39, 43, 45, 46
  lanceolata, 45, 54
  major, 45, 54
Plant removal
  efficacy, 127, 142
  experimental design, 106
  from lakes, 159
Plant status, 8
Platanthera praeclara, 178
Platanus occidentalis, 120
Poa
  annua, 45
  pratensis, 172
  secunda, 92
Podocarpus, 45
Polygonum, 39, 191
  aviculare, 45
  cuspidatum, 264
  lapathifolium, 52
  perfoliatum, 264
Polyploidy, 223
Populus, 33, 38, 43, 76, 177, 245
Portulaca, 43
  oleracea, 37, 43, 44, 54, 191
Potamogeton, 146, 149

Index

crispus, 161
Predictive characteristics
  flowering period, 224
  geographic origin, 221
  growth rate, 223
  historical performance, 222
  latitudinal range, 221
  plant height, 222
  polyploidy, 223
  reproductive system, 223
  seed germination, 224
  seed production, 224
  taxonomic relationship, 221
Predictive methods, models, 218
Presettlement condition, 21, 49, 53, 140
Proserpinaca palustris, 50, 52
Prosopis, 71, 76
  glandulosa, 177
  pubescens, 177
  velutina, 178
Protea, 72
Prunus serotina, 237
Psidium, 46
  cattleianum, 60, 92, 265
  guajava, 265
Pteridium, 43
  aquilinum, 142, 177
Pueraria
  lobata. See Pueraria, montana
  montana, 26, 58, 60, 96, 172, 189, 265

Q
Quercus, 22, 39, 43, 46, 47, 51, 64
  petraea, 59
  rubra, 49

R
Randall, John M., 18
Ranunculus
  ficaria, 265
  flabellaris, 50, 52
  pennsylvanicus, 14
Raphanus sativus, 89
Rate of spread, 231
Regeneration foci, 32, 33
Regulation
  biocontrol, 182
  Federal Noxious Weed Act, 217, 231
  management, 23, 166
  new, 226
  quarantine, 163, 190
Reichard, Sarah E., 215
Restoration
  Everglades, 239

  forest, 31, 33, 85
  indigenous plant, 154
  lake, 154
  Santa Ana River, 246
  system function, 81
Rhamnus
  cathartica, 32, 92, 265
  frangula, 265
Rhododendron ponticum, 59
Robinia pseudoacacia, 23, 33
Rosa
  bracteata, 191
  multiflora, 26, 93, 191, 216, 265
  rubiginosa, 31
Rottboellia
  cochinchinensis, 191
  exaltata. See Rottboellia, cochinchinensis
Rubus argutus, 183
Rumex, 39, 43, 45, 46, 53, 191
  acetosa, 53
  acetosella, 45, 53, 54
  crispus, 33

S
Salix, 61, 76, 177, 245
Salsola, 45, 50, 54, 187
Salvia aethiopis, 182, 187
Salvinia molesta, 60
Santa Ana River, 246
Sapium sebiferum, 23, 191, 266
Scaevola
  plumieri, 234
  sericea, 266
Schiffman, Paula M., 87
Schinus terebinthifolius, 23, 190, 234, 266
Schwartz, Mark W., 7
Scientific method
  induction, 105
  retroduction, 105
Scirpus
  acutus, 52
  californicus, 45
  longii, 178, 234
  validus, 52
Secale cereale, 53
Seed bank, 141, 154, 224
Senecio
  jacobaea, 182
  mikanioides, 234, 266
Shepherdia, 39
Shrub stratum, 32, 59, 65
Silybum marianum, 179, 185
Smith, Stanley D., 69
Soil disturbance, 91, 100

Soil water
  measurement, 76
  utilization, 74, 83, 178
Solanum
  mauritianum, 90
  viarum, 231, 266
Sorghum
  bicolor, 236
  halepense, 23, 172, 236
Sparganium, 50, 52
Species, endangered, 16, 21, 234
Species diversity, 62, 66, 102, 150, 238
Species range, 10, 12
  climate change, 14, 34
  expansion, 12, 13, 14, 42, 88
  human effects, 42, 144
Species richness, estimating, 12
Sphaeralcea, 44, 50
Sporobolus airoides, 84
Sporormiella, 41, 50
Statistical methods, 108, 110, 120
  classification and regression trees, 220
  discriminant analysis, 220
  multiple logistic regression, 220
  ordination, 123
  power, 110
Stellaria media, 45
Succession
  altered, 63, 80, 83
  climax, 135
  facilitation, 31, 33, 135
  general models, 134
  inhibition, 60, 135
  invader persistence, 137
  nonequilibrium, 66, 135, 139
  stability, 135
  tolerance, 135

T
Taeniatherum caput-medusae, 266
Tamarix, 23, 61, 62, 69, 75, 82, 83, 173, 183, 232, 266
  aphylla, 178
Taxodium, 234
Team Arundo, 246
Tibouchina herbacea, 251
Tidestromia, 44

Tilia, 43
  americana, 64
Tillandsia, 234
Torreya taxifolia, 10
Toxicodendron, 177
  radicans, 191
Tradescantia fluminensis, 60
Trapa natans, 145
Tribulus
  cistoides, 183
  terrestris, 182, 183
Trifolium, 62
Triticum, 53
Tsuga, 43, 46, 51
  canadensis, 12, 49, 64, 72
Typha, 50, 233

U
Ulex europaeus, 63, 183, 237, 267
Ulmus, 43, 46
  americana, 237
  rubra, 120
Urban/rural gradient, 30, 141

V
Vines, 60
Vitis vulpina, 127

W
Walker, Lawrence R., 69
Weed
  environmental, 20
  defined, 19
  natural area, 19, 23
  noxious, 23
Williams, Charles E., 26
Woods, Kerry D., 56

X
Xanthium
  occidentale, 97
  spinosum, 90

Z
Zea, 43, 46
Zizania aquatica, 52